Green Infrastructure

Rohana Hassan · Nor Hayati Abdul Hamid ·
Ahmad Kamil Arshad · Anizahyati Alisibramulisi ·
Muhd Norhasri Muhd Sidek ·
Norshariza Mohamad Bhkari · Ekarizan Shaffie
Editors

Green Infrastructure

Materials and Applications

Springer

Editors
Rohana Hassan
IIESM
Universiti Teknologi MARA
Shah Alam, Selangor, Malaysia

Nor Hayati Abdul Hamid
IIESM
Universiti Teknologi MARA
Shah Alam, Selangor, Malaysia

Ahmad Kamil Arshad
IIESM
Universiti Teknologi MARA
Shah Alam, Selangor, Malaysia

Anizahyati Alisibramulisi
IIESM
Universiti Teknologi MARA
Shah Alam, Selangor, Malaysia

Muhd Norhasri Muhd Sidek
IIESM
Universiti Teknologi MARA
Shah Alam, Selangor, Malaysia

Norshariza Mohamad Bhkari
IIESM
Universiti Teknologi MARA
Shah Alam, Selangor, Malaysia

Ekarizan Shaffie
IIESM
Universiti Teknologi MARA
Shah Alam, Selangor, Malaysia

ISBN 978-981-16-6382-6 ISBN 978-981-16-6383-3 (eBook)
https://doi.org/10.1007/978-981-16-6383-3

This Springer imprint is published by the registered company Springer Nature Singapore Pte Ltd.
The registered company address is: 152 Beach Road, #21-01/04 Gateway East, Singapore 189721, Singapore

Preface

This book highlights green infrastructure materials and their applications along with their limitations in contributing towards green industries. It starts by discussing modifications in concrete design by adding recycled polyethylene terephthalate bottles as partial replacement of aggregates, mortar, ordinary Portland cement, alternative binder for paving blocks and others. This mixed concrete can be used as sustainable wall panels and concrete blocks. These types of sustainable wall panels can be used to construct green buildings and to save the environment through reduction of waste plastic bottles at landfills.

The initial chapters discuss the repair and retrofitting of concrete structure by using palm oil fly ash (POFA) which is the byproduct of palm oil industries as alkaline and physical activated POFA cement paste. A review of concrete performance on tin slag polymer concrete as green structural material for sustainable future and its study on the effect of waste paper sludge ash (WPSA) in addition to fresh and hardened ultra-high-performance concrete are explained and elaborated.

Chapters "Flexural Performance of Strengthened Glued Laminated (GLULAM) Timber Beam Using Glass Fibre-Reinforced Polymer (GFRP)"–"Analysis of the Flexural Strength of Reinforced Beam with Bamboo by Empirical Modeling Using Statistical Model" present the physical and mechanical properties of timber as one of the green material that contributes to green infrastructure construction and sustainability. The experimental results on flexural performance of strengthened glued laminated (glulam) timber beam using glass fibre reinforced polymer (GFRP); overview on bending and rolling shear properties of cross laminated timber (CLT), together with delamination test for Mengkulang timber species using method A and C are presented. Further analysis on flexural strength of reinforced beam with bamboo by empirical modelling using statistical model is also presented in the book.

The two following chapters in this book discuss road pavement materials and their testing and evaluation. It includes innovative new and waste materials which can be recycled to prolong the life span of road pavements. In order for the pavement to withstand the ever-increasing loadings from vehicles in terms of intensity and axle applications, pavement materials must have adequate strength to withstand these loads and must also be durable to withstand the effects of environment, namely moisture and temperature. Chapter "Application of Pavement Evaluation for Road Maintenance and Rehabilitation" describes the process of pavement evaluation and the application of its results for use in the maintenance and rehabilitation of roads. Meanwhile, chapter "Performance Evaluation of Stone Mastic Asphalt Containing Steel Fibre as Additive" presents the findings of a study on addition of steel fibre in stone mastic asphalt (SMA) application. SMA is a gap-graded asphalt mixture that consists of a large proportion of coarse aggregate, a high percentage of asphalt binder and a substantial amount of filler which provides a durable surface course.

Chapters "Post-construction Complexity Factors Impacting Infrastructure Project Performance in Malaysia"–"Governance Practices in Poverty Alleviation Projects: Case Study from Stewardship-Driven Perspective and Sustainability Context" present the post-construction complexity factors that impact the infrastructure project performance in Malaysia, followed by the SWOT analysis of green technology application for the development of low carbon cities and a sharing of a case study from stewardship-driven perspective and sustainability context through governance practices in poverty alleviation projects.

Discussion on interception loss of tree canopy as green infrastructure and an evaluation of parameters for sustainability assessment of green infrastructure in the urban water system are presented in chapters "Interception Loss of Tree Canopy as Green Infrastructure" and "Evaluation of Parameters for Sustainability Assessment of Green Infrastructure in the Urban Water System", respectively. This contemporary book also covers a green community-based social enterprise (CBSE) for B40 in Sabah, through a sharing of freshwater lobster production-green Aquaponics Perennial System (flp-gaps) in chapter "Freshwater Lobster Production-Green Aquaponics Perennial System (FLP-GAPS): A Green Community-Based Social Enterprise (CBSE) for B40 in Sabah". Last but not least, chapter "An Edible Cutleries Using Green Materials: Sorghum Flour" presents the use of sorghum flour as one of the green materials in a form of edible cutleries performance.

This comprehensive book entitled "Green Infrastructure Materials and Applications" is suitable for engineering students to develop competency in theory, practical and communication skills. It presents a framework for engineers to add on different materials into practice. The chapters in the book also present the link between materials and the application processes. The various engineering materials presented in this book will inspire and stimulate discussions in academia and industry to develop and promote green infrastructure materials and their applications for the future.

Shah Alam, Malaysia

Rohana Hassan
Nor Hayati Abdul Hamid
Ahmad Kamil Arshad
Anizahyati Alisibramulisi
Muhd Norhasri Muhd Sidek
Norshariza Mohamad Bhkari
Ekarizan Shaffie

Summary

This book explains the meaning of green infrastructure and limits its attention to the contribution of materials and applications since green infrastructure is varied and diverse. The development of green infrastructure is gathering momentum internationally both in theory and policy, and there is development agreement that green infrastructure provides an attractive opportunity for the delivery of significant environmental, social and economic benefits. This chapter of the book covers four perspectives on green infrastructure. First, the exploration of contested green infrastructure materials that include concrete, wood and pavement. Second, the discussion of opposition to the ambiguity of managing construction green infrastructure is based on a broader debate on the application and process of construction impact. Third, an examination of the relevant concepts and factors that contribute to the application towards green infrastructure in nature. Finally, contributions related to how the study of bread cutters used using sorghum flour will also provide implications for defining green infrastructure explicitly given its ever-changing and disputed nature will be critically examined. Therefore, this contribution is by no means certain, except in an effort to provide a separate and holistic perspective on the concept of "green infrastructure" of engineering.

Contents

Recycled Plastic Bottles as Sustainable Materials

N. H. Hamid and M. M. H. Shamsudin

Abstract Recycle Polyethylene Terephthalate (PET) plastic bottles can be used as partial replacement of aggregates to form plastic bricks, silica plastic blocks and PET bricks. By replacing from 0.5 to 5% of PET recycled plastic as partial replacement of fine and coarse aggregate into concrete mixture can be moulded as silica plastic blocks, colour plastic bricks, RePlast bricks and burnt brick to construct a single-storey house and temporary shelters. Furthermore, by adding 0.5–1.5% of recycled PET fibre in concrete with water cement ratio of 0.65 can increase the compressive strength of concrete. The advantages of using recycled bottles as green construction materials can reduce landfill pollution, conserve natural resources, improve the environment and save sea life from dying. Furthermore, the bottle houses are bioclimatic in design which means it is cold outside and warm inside or vice versa. It can be concluded that by substituting recycled plastic bottles into concrete mixture and transforming into green recycled bricks, they can be used to construct the houses in rural or slump areas where the cost of construction materials are very expensive and limited.

Keywords Recycle plastic bottle · Recycled polyethylene terephthalate · Burnt brick · Green recycled bricks · Green infrastructure material

1 Introduction

In Malaysia, about 40% of the income households are categorized under the B40 group where their incomes are less than RM4, 850 per month. Most of them cannot afford to buy terrace houses ranging between RM490,000 and RM750,000 in Klang

N. H. Hamid (✉)
Institute for Infrastructure Engineering and Sustainable Management (IIESM), Universiti Teknologi MARA, 40450 Shah Alam, Selangor, Malaysia
e-mail: norha454@uitm.edu.my

M. M. H. Shamsudin
School of Civil Engineering, Universiti Teknologi MARA, 40450 Shah Alam, Selangor, Malaysia

R. Hassan et al. (eds.), *Green Infrastructure*,
https://doi.org/10.1007/978-981-16-6383-3_1

Valley which is the highest population area in Malaysia. Thus, the Malaysian government needs to build low-cost houses for the B40 group ranging from RM 72,000 to RM 150,000. Waste minimization is common in the construction project site in Malaysia where 73% of the waste materials such as plastic waste is used and recycled. The net benefit of reusing and recycling of plastic waste is estimated at 2.5% of the total project budget (Beguma et al. 2006). The contractors and builders need to search for the lowest cost of sustainable green materials for the construction of these types of houses. Significantly, by 2018, Malaysia produces more than 0.94 million tons of mismanaged plastic waste per year (Chen et al. 2021). At the same time, Malaysia also became a top importer of plastic waste where a total of 105 thousand tons of plastic was imported in 2017 with an increase of 68% from 2016 (Periathamby et al. 2009). Thus, one of the alternative ways is to use recycled plastic bottles for the construction of wall panels, brick, joist, ceiling and columns of the houses. The first bottle house was constructed in 1902 by William F. Peck in Tonopah, Nevada using 10,000 alcohol bottles (Jalaluddin 2017). Andreas Froese is a German social entrepreneur and environmentalist who has developed ECOTEC technology which transforms the reuse of plastic bottles to recycle bricks to erect dwelling houses (Patel et al. 2016). He also constructed the first house in Africa using plastic bottles bound together with string in the Yelwa Village, Nigeria (Hemraj et al. 2018). PET bottles are filled with sand, soil or even rubble to give some firmness and they are placed one by one, creating a house based on a simple and functional design of a building structure. Due to poor soil fertility as a result of the presence of plastic bottles and polythene bags, Butakoola Village Association for Development (BUVAD) in Uganda constructed a bottle house in 2010. Then, Mr. Arthur Huang from Taiwan had built a boat-shaped exhibition hall known as Eco-Ark using 1.8 million recycled plastic bottles. After that, Samarpan Foundation constructed a primary school in New Delhi, India using hundreds of used PET bottles instead of conventional bricks in May 2011 (Jalaluddin 2017).

2 Physical and Chemical Properties of Pet Bottles

Polyethylene terephthalate (PET) is created by two combinations of monomers which are purified terephthalic and modified ethylene glycol (Faraca et al. 2019). PET is the most common thermoplastic polymer resin of the polyester family and is used in fibres for clothing, containers for liquids and foods and thermoforming for manufacturing. It consists of polymerized units of the monomer ethylene terephthalate, with repeating ($C_{10}H_8O_4$) units (Faisal et al. 2016). It is normally used for cereal box liners, microwave food trays, plastic vessels and for implantation in medicinal and plastic bottles. Plastic is heat resistant, chemically stable, resistant against acid, base, some solvents, oils and fats. The tests of physical properties of PET bottles washing include hygroscopic, transient and total moisture, as well as specific and bulk density. Table 1 shows physical properties of PET bottles after burning in a furnace and tested in the laboratory (Department of the Environment and Energy

Table 1 Physical properties of PET bottle ashes (Beata et al. 2019)

Quantity	Unit	Value
Hygroscopic moisture	%	43.7 ± 1.5
Transient moisture	%	6.0 ± 0.76
Total moisture	%	45.1 ± 4.03
Analytical moisture	%	12.1 ± 1.5
Bulk density	kg/m^3	602.3 ± 63.2
Specific density	kg/dm^3	1.27 ± 0.14

Table 2 The composition of inorganic substance in PET bottle ash (Beata et al. 2019)

Component	SiO_2	Al_2O_3	Fe_2O_3	CaO	Na_2O	MgO	K_2O	P_2O_5	TiO_2	SO_3	MnO	LOI
Weight (%)	10.32	4.49	3.13	16.0	0.67	1.32	0.3	0.69	0.07	0.27	0.11	61.2

2018). Meanwhile, Table 2 shows the chemical composition of inorganic substances in PET bottle ash (Beata et al. 2019).

According to Van Krevelen (1990), PET is a polymer that has tensile modulus elasticity and flexural modulus elasticity of 2.9 MPa and 2.4 MPa, respectively. Correspondingly, the characteristics of PET are high chemical resistance and have melting points approximately up to 260 °C. However, the issue concerned is the basic properties of PET such as wetting tension and alkali resistance to ensure the performance of PET itself is suitable to be used in concrete or mortar. Wetting tension value is used to evaluate the adhesion or bonding between PET and concrete or mortar matrix. Likewise, the durability of PET is measured through the alkali resistance to perform well in reacting with cement that contains a high concentration of alkali.

Ochi et al. (2007) have created an understanding of determining properties of PET and comparing it with Polypropylene (PP) and polyvinyl alcohol (PVA). It was found that the wetting tension of PET is 40 mN/m which is slightly higher than PP 35 (mN/m) but this value is lower than PVA (45 mN/m). Furthermore, PET was immersed in an alkali solution (sodium hydroxide) and the tensile strength was tabulated in Table 3. The experimental result revealed that the characteristic of PET

Table 3 Result for alkali resistance of PET by measure tensile strength (Ochi et al. 2007)

Fibre	Diameter (mm)	Length (mm)	Tensile strength (MPa)		Strength ratio[a] %
			Before exposure	After exposure for 12 h	
PET	0.75	30	352	348	99
PP	0.21	30	170	147	86
PVA	0.71	30	360	202	56

[a]Strength after exposure/strength before exposure

is sufficient for alkali resistance which experiences only minor deterioration through tensile strength after immersion as compared to PP and PVA.

3 Plastic Waste as Potential Construction Material

According to Azhdarpour et al. (2016), PET recycled plastic bottles as a building material has gained popularity in the recent decades. Polyethylene Terephthalate (PET) is categorized as a form of plastic that is mostly being thrown away right after a single use (Saxena et al. 2018). In Malaysia, the usage of plastic in various industries is increasing gradually day by day. This phenomenon happens due to urbanization at the same time resulting in an enormous amount of plastic wastes and causes significant environmental problems like pollution, flooding, climate change and diseases. In 2020, the estimated value of annual food plastic packaging consumption in Malaysia amounted to 148,000 metric tons and the annual per capita plastic packaging consumption in Malaysia was equal to 16.78 kg (Müller 2021). According to the study done by Wahab et al. (2007), the rate of plastic consumption in Malaysia was 4.6 tons/day for PET. However, these figures are expected to increase in the future.

Most of the plastic bottles which are considered as solid wastes are dumped at landfills as shown in Fig. 1. The massive proportion of the PET is harmful because it is not being composed readily by nature due to the slow biodegradation phenomenon

Fig. 1 Plastic bottles as solid waste are dumped at landfill

(Ghernouti et al. 2015). These plastic wastes will take more than 100 years to biodegrade inside the soils. Developed countries like Australia generated around 67 million tons of waste in 2016–2017 and only 37 million tons (55%) were recycled (Department of the Environment and Energy 2018). Contradictorily, it was reported that one of the biggest cities, Beijing, China produces 25,000 tons of rubbish in a day (Guo et al. 2020). Various methods of disposing of plastic wastes were introduced such as burning, landfill and dumping wastes into the oceans. Despite recycled plastic bottles, the majority are sent out to landfill which have negative impacts on the environment, social and economy. Furthermore, harmful gases that are being produced from the process of incineration to dispose of PET waste can also disrupt human health (Ghernouti et al. 2015). Many studies have shown that more than 15% of a building's energy consumption and carbon emissions are from virgin materials (Sandanayake et al. 2018, 2020; Huang et al. 2019). Furthermore, excessive usage of virgin materials in building materials results in depleting natural resources and leads to additional environmental burden (Meyer 2009; Hossain and Poon 2018). Therefore, by replacing virgin materials with waste products addresses key issues, excessive waste generation and save virgin material usage.

The use of plastic as one of the construction materials is an environmentally friendly method of disposal. Mixing plastic in construction materials can also absorb desirable properties in the finished products making favourable economic sense. For example, the usage of PET as construction material which can reduce the need for fine aggregate, improve corrosion resistance and makes the concrete lighter.

The construction industry area that produced high consumption capacity is likely to be the most suitable industry to reuse PET waste (Saxena et al. 2018). A lot of research work had been conducted by putting recycled plastic bottles in concrete and mortar as replacement of aggregates and fibre reinforcement (Saikia and De Brito 2012; Usman et al. 2018; Merli et al. 2019). Currently, the impacts of plastic and rubber aggregates on the physical, mechanical, durability and functional properties of concrete are studied (Li et al. 2020) and the state-of-the-art on the fresh and hardened properties of self-compacting concrete containing PAs and PFs (Faraj et al. 2020). According to research conducted by Mahdi et al. (2013) and Jo and Park (2006), recycled PET was used as a binder in concrete by depolymerizing the PET using the glycolysis process to produce unsaturated resin to replace cement. Figure 2 shows the recycled PET use as replacement of aggregates, meanwhile Fig. 3 shows the PET waste was transformed fibre by previous researchers.

There are also applications of recycled plastic waste especially on PET in concrete structures' components such as beams, slabs, precast concrete panels and concrete footpaths. Foti and Paparella (2014) investigated the dynamic behaviour of concrete reinforced with waste PET plastic by cutting drinking bottles and arranged in a grid structure to reinforced concrete slabs subjected to shock and impact loading. Bending tests were performed on concrete beams with PET bars placed at the bottom of the beams under two points load (Foti 2016) with advantages against corrosion. Adding a circular ring of plastic bottle with 5 mm thickness or 10 mm thickness to the reinforced concrete beams did not lower the deflection behaviour as compared with control reinforced concrete beam specimens. During the cracking stage, concrete

Fig. 2 Recycled PET use as replacement of aggregate (Islam et al. 2016)

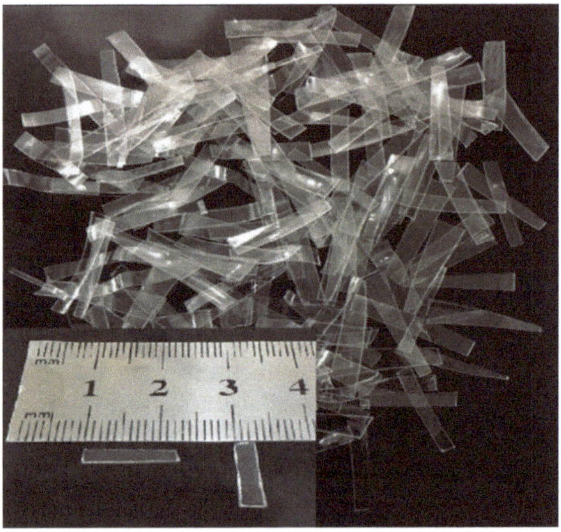

Fig. 3 Recycled PET use as fibre (Alfahdawi et al. 2019)

beams containing ring plastic bottles with 10 mm showed that the strength of the first crack improved by 32.3% as compared to normal concrete beams (Khalid et al. 2018). Cracking due to drying shrinkage was delayed in the PET fibre-reinforced concrete beams as compared to such cracking in non-reinforced specimens without fibre reinforcement which indicates crack controlling and bridging characteristics of the recycled PET fibres (Kim et al. 2010). Short fibres from recycled PET bottles proved to be an excellent alternative to restrain plastic shrinkage in concrete and building cements materials (Pelisser et al. 2010). Dai et al. (2011) investigated the behaviour of concrete confined by fibre-reinforced polymer (FRP) jackets formed with polyethylene naphthalate (PEN) and PET fibres derived from post-consumer bottles. Besides that, PET is said to be suitable to be utilized in concrete and mortar because the characteristics of PET itself which have high tensile strength, high alkali resistance and high value of wetting tension are believed to further result in good bonding with concrete matrix (Ochi et al. 2007).

4 Performance of Recycled PET Fibre in Concrete

The use of recycled PET as concrete-reinforcing fibre has been steadily developed for a decade. In order to overcome the shortcoming of the concrete in the construction field especially on tensile strength and resistance of crack, recycled PET were used (Borg et al. 2016; Kim et al. 2010). However, the level of performance of fibre as concrete reinforcing depends on various factors such as type, amount and quality of the fibre. Besides that, the shape and dimension of the fibre also have a significant effect on the bonding between the materials in the concrete mix as well (Borg et al. 2016). The shape, amount and dimension of recycled PET fibre lead to various findings on the concrete performance. Table 4 shows the characteristics of the PET fibre and its findings from previous researchers (Ochi et al., 2007; Kim et al., 2010; Fraternali et al., 2011; Foti 2011; Irwan et al., 2013; Shamsudin et al., 2021).

The straight shape of plastic fibres typically does not provide enough ability to resist crack due to poor bonding contact within concrete materials. The smooth form and sides of waste plastic is a significant problem due to its adhesion with the concrete mix (Saxena et al. 2018). Therefore, transforming the shape of fibre to several patterns such as crimped, embossed and twisted can improve the bond strength between the recycled PET fibres and concrete matrix (Kim et al. 2010). Figure 4 until Fig. 7 show the various shapes of fibre which had been used by previous researchers (Borg et al.,

Table 4 Summary of previous research on shape of PET fibre

Researcher	Percentage of PET fibre (%)	The shape of PET fibre	Water cement ratio	Compressive strength of concrete
Ochi et al. (2007)	0.50, 1.00, 1.50	Monofilament	0.55, 0.60, 0.65	Increase
Kim et al. (2010)	0.50, 0.75, 1.00	Strip	0.41	Decrease
Fraternali et al. (2011)	1.00	Straight and crimp	0.53	Increase
Foti (2011)	0.26	Strip and circular	0.70	Decrease
Irwan et al. (2013)	0.50, 1.00, 1.50	Irregular	0.65	Increase
Shamsudin et al. (2021)	0.5, 1.0, 1.5	Straight	0.53	Increase

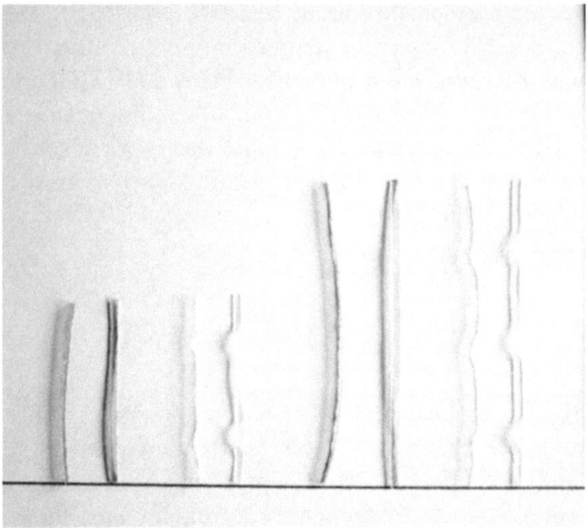

Fig. 4 Embossed and straight fibre shape with length 30 and 50 mm (Borg et al. 2016)

2016; Kim et al., 2010; Foti, 2011; Fraternali et al., 2011). Embossed fibres show the highest bond strength as compared with crimped and straight ones as shown in Fig. 4. The circular fibre shape had been tested by Foti (2011) as shown in Fig. 6. Fraternali et al. (2011) found in his study that the crimp shape of PET fibre can improve the compressive strength and ductility of the concrete as compared to concrete with straight fibre shape as shown in Fig. 7. In addition, the bonding strength of fibre can be enhanced through the coating process. Kim et al. (2010) treated the PET fibre by coating with maleic anhydride grafted polypropylene to produce a rough surface as shown in Fig. 5.

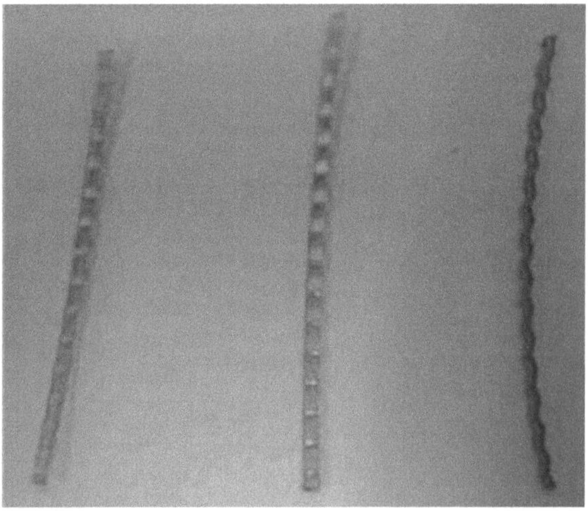

Fig. 5 Crimped fibre shape (Kim et al. 2010)

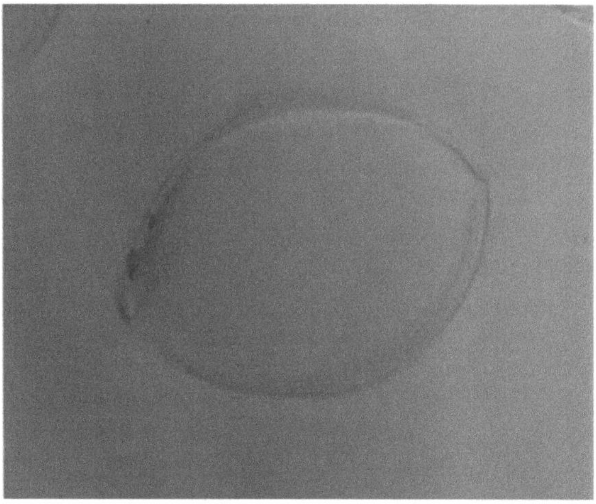

Fig. 6 Circular fibre shape (Foti 2011)

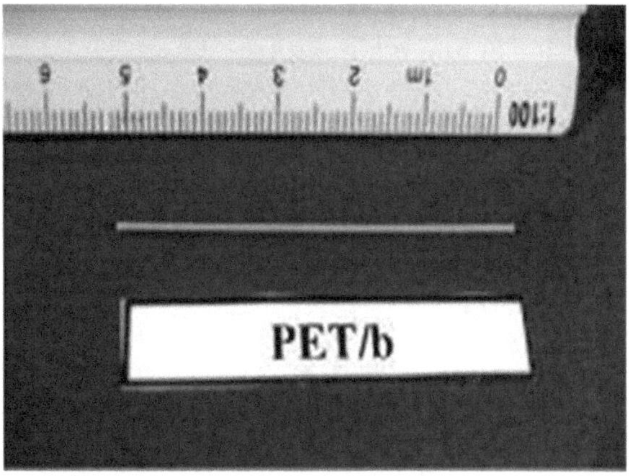

Fig. 7 Straight fibre shape (Fraternali et al. 2011)

Based on experimental work conducted by Irwan et al. (2013) using 0.5% of recycled PET fibre in concrete, it shows better performance as compared to plain concrete in 28 days. However, its performance was dropped by utilizing more than 1.0% to 1.5% volume fraction of PET in the concrete. This study showed that a vast amount of PET can reduce the strength of the concrete. Therefore, it is concluded that the fibre content will affect the strength of the concrete causing difficulty during the mixing process which affects poor compaction, poor workability, fibres did not distribute uniformly and lead to an increase of void volume (Khalid et al. 2018).

5 Recycled Plastic Bottle as Plastic Bricks

Nowadays, recycled material is widely used in the construction industry to promote green construction in order to overcome the pressing issue on the environment as well as enhancing performance of structure. Some of the advantages of using recycled plastic bottles are lower amount of waste, lower greenhouse gas emissions, lower pollution rates, saves on energy, uses less resources, saves money in waste management and maintains sustainability of resources. Plastic shows various benefits through its qualities such as do not corrode, lighter, chemically inert, good in plastic shrinkage cracking resistance and easy to mix with concrete and mortar matrix (Foti 2011). Furthermore, it can reduce the emission of CO_2 to the atmosphere, save energy in producing bricks in the factory and reduce air pollution after burning bottles at damping sites. Furthermore, it also can contribute to conserving natural resources, protecting ecosystems and wildlife, reducing demand for raw materials and can stop disastrous climate change. On top of that, recycled plastic bottles can be processed to become plastic bricks as replacement for conventional bricks. They are thinner, lighter, have superb heat insulating properties and insulating against noise as compared with conventional bricks. Each brick helps rid the world of discarded plastic and is cheaper and more fuel efficient to manufacture than conventional bricks. It is also less energy intensive than recycling the plastic into other forms. The shape and size of bricks from recycled plastic bottles will be discussed in the following topics.

There are a lot of studies that have been conducted on using recycled plastic bottles as plastic bricks. The plastic bottles with 500 mL were used to create voids at equal distance between them in the masonry units and analysed the compressive strength. The experimental results showed 57% difference in the strength by using plastic bottles as compared with local concrete blocks and it can be used as plastic bricks for the construction of houses (Sina and Amani 2016). Rhino Machines is the India-based company that has launched the Silica Plastic Blocks which is a sustainable building brick that is made from recycling foundry dust/sand waste (80%) and mixed plastic waste (20%) (Juliana 2021). Figure 8 shows the hollow Silica Plastic Blocks which were used for the construction of houses in India.

A young Kenyan material engineer, Nzambi Matee decided to invent recycled plastic bits mixed with sand into sustainable bricks and building materials through the company known as Gjenge Makers Ltd. The company gets the plastic waste from packaging factories or other plastic recyclers and mixes with sand as the binder. The extruder machine does the mixing of plastic waste with sand at very high temperatures and then compresses it. This company converts high density polyethylene, low density polyethylene and polypropylene plastics into solid bricks with different colours and sizes. This plastic brick is five times stronger than normal brick due to plastic being fibrous in nature and has excellent properties. This company produces 1000 to 1500 plastic bricks per day. Figure 9 shows some examples of plastic bricks with different shapes and colours which are produced using recycled plastic bits

Fig. 8 Silica plastic blocks produced by Rhino Machines (Jo and Park 2006)

Fig. 9 Different shapes and colours of plastic bricks produced by Gjenge Makers Ltd. (Juliana 2021)

and sand. These plastic bricks are used for the construction of houses, walls, pavement and driveways (Juliana 2021). The PET bricks were developed and patented by the National Council of Scientific and Technological Research (CONICET) using recycled PET bottles, cement and various additives. These bricks have the same properties as common ceramic bricks, but are lighter and have better insulating and sound-proofing properties (Bricks Made from Recycled PET Bottles 2018). Figure 10 shows the PET bricks made from recycled PET bottles patented by CONICET which can be used for the construction of houses (Bricks Made from Recycled PET Bottles 2018). RePlast bricks were developed and created by Peter Lewis under the company called ByFusion in New Zealand. The RePlast brick has the same size as a standard concrete block and is made from different kinds of scrap plastics that do not end up being a waste again. The RePlast system does not require the plastic to be washed or sorted and the permanent nature of construction helps to prevent the plastic from simply ending up in the oceans again (Neha 2016). Figure 11 shows RePlast bricks are produced from different kinds of plastic waste by ByFusion Company.

Fig. 10 PET bricks were made from recycled plastic PET bottles in Cordoba, Argentina (Bricks Made from Recycled PET Bottles 2018)

Fig. 11 RePlast bricks are produced from different kinds of plastic waste by ByFusion Company which is located in New Zealand (Neha 2016)

6 Construction of Houses Using Plastic Bricks

The plastic bricks as described above are used to construct single-storey houses especially in third world countries because the construction materials such as reinforcement bars, cements, concrete binders and others are very expensive and difficult to buy. Figure 12 shows a single-storey house and temporary shelters were constructed in Colombia using plastic bricks which were made from recycled plastic bottles. Plastic bricks are produced using an extrusion process where the plastic is melted and emptied into a final mould by creating a 3 kg brick similar in size as standard brick. This house is thermo-caustic and earthquake-resistant, satisfying the current seismic code of practice of Colombia. The total cost of this house is 20 million Colombian pesos (RM28, 291) per unit and was constructed in 5 days only. The temporary shelters were built using plastic bricks in 28 days at Guapi, Southwest of Colombia for 42 families displaced by armed conflict (Nicolás 2017). Figure 13 shows two single-storey houses built using recycled PET plastic bottles filled with sand. The PET plastic bottles were filled with sand, sealed and then paste them with a mixture made of earth, clay, sawdust and a little cement to provide additional strength

Fig. 12 A single-storey house and temporary shelters were built in Guapi, Colombia (Nicolás 2017)

Fig. 13 Single-storey houses were constructed using recycled plastic bottles refilled and compacted with sand and capped (Keiren 2020)

and durability. This house is an eco-friendly, sustainable and ecological environment which was constructed using 8000 PET bottles with composting toilets and a solar water heating system. The name of the entrepreneur who constructed these types of houses is Andreas Froese and the founder of "ECO-TEC" who built more than 50 eco-friendly projects in Honduras, Columbia and Bolivia.

7 The Manufacturing Process of Plastic Brick

The manufacturing process of the plastic brick blocks involves a modular platform which is portable and removable from one place to another place which is designed to operate using gas or electricity. The plastic trashes which are collected from factories and households are transported to the site or factory by lorry. The shredder is used to reduce the size of plastic bottles and trashes into smaller pieces before compacting them into plastic brick. The plastic bottles and wastes did not require sorting or washing. It is just enough to compress these small pieces of plastic into the plastic brick or block using a water boiler and compactor. The super-heated water and compression machine is used to fuse plastic pieces into plastic blocks. This process is non-toxic and did not use any adhesives during the construction process. Rebar steel rods, metal plate and tape are used during construction of non-load bearing walls, retaining walls and road pavements.

8 The Performance of Plastic Brick

Burnt brick is a vital building material that is widely used throughout the world. Compressed earth and burnt brick technology can be traced to the ancient Egyptian and Babylonian empires. Materials including straw, broken ceramic tiles, broken blocks and waste concrete have all been utilized to either strengthen, or act as a filler in clay bricks. Guettalal et al. (2016) came out with a study on the effects of cork granules in compressed earth bricks on the mechanical properties of bricks. This investigation concluded that with the increasing content of cork granules, the compressive and tensile strengths of cement-stabilized compressed earth blocks decrease. However, the compressive and tensile strength are still seen as acceptable when the cork's minimum mass content was used.

The previous studies on the utilization of waste materials in both compressed and burnt bricks are being reviewed by Zhang (2013). The materials that were investigated with bricks were Class F fly ash, Hematite tailings, Municipal solid waste incineration sludge, Granite sawing waste, Gold mill tailing, Paper production residue, Rice husk ash, Cigarette butt, Waste tea, River sediment, Saw dust, Waste marble powder, Foundry sand by products waste, Kraft pulp production residue, Sugarcane bagasse waste and Petroleum effluent treatment plant sludge. From his review, it can be deduced that one of the useful methods of recycling waste materials is to use it in building materials such as bricks. Despite the fact that the commercial production of bricks from waste products is still limited, the inclusion of these materials in bricks will be beneficial if there is standardization and commercialization being set according to their usage.

Akinyele et al. (2020) studied the possibility of using various proportions of PET as an additional material in burnt bricks. Figure 14 shows PET burnt brick before firing shrinkage test and Table 5 indicates the result of the utilization of various PET percentages in burnt brick including of 0%, 5% and 10% of PET in terms of firing shrinkage, water absorption, dry density and compressive strength. It was observed through the findings that PET material melted during firing due to its low melting point of 250°C. Samples containing more than 10% PET content have been investigated and collapsed during the firing process. In contrast, samples containing less than 10% PET content did not collapse but are deformed in shape. Furthermore, the compressive strength of the bricks containing PET was also low but samples comprising more than 5% of PET content have better performance in terms

Fig. 14 Burnt brick before firing (Akinyele et al. 2020)

of structural efficiency when compared with control samples. This finding deduced that bricks that consist of 5% PET or less may perform well and further detailed research is needed. The main reason causing the poor performance of PET bricks under loading is revealed through microstructure properties. From the findings and observations, it indicates that the use of PET content of less than 5% can be utilized as a substitute in burnt bricks, aside from well-monitoring of the temperature during the firing process.

A separate research was carried out by (Akinyele and Toriola 2018), with the utilization of crushed plastic (PET) as a replacement of fine aggregate in Sandcrete bricks at 0, 5, 10, 15 and 20% content. All of the brick samples were subjected to mechanical and water absorption tests. Both compressive and flexural strength tests revealed that brick samples containing 5% PET outperform all other brick samples, including the control mix. In addition, it has lower density and water absorption in comparison to the control sample. The research work concluded that shredded PET can be used in Sandcrete bricks if it is less than 5% replacement.

9 Future and Challenges of Using Pet Plastic Bottle as Plastic Bricks in Malaysian Construction Industry

Many researchers found out that the recycled plastic aggregate and fibre as partial replacement can improve the concrete properties. However, there are several challenges regarding using PET plastic bottles as plastic bricks in Malaysian construction industries. One of the biggest challenges is separating and grading the recycle bottles from household waste which are collected by garbage truck. Furthermore, the mentality and discipline of Malaysian citizens to segregate the recycled plastic bottle from other waste materials is still low as compared with developed countries such as Japan, United Kingdom and others. Another challenge is the construction industries in Malaysia are reluctant to recycle plastic bricks manufactured locally

Table 5 Experiment result (Akinyele et al. 2020)

Proportion of PET (%)	Firing Shrinkage (%)	Water absorption (%)	Dry density (kg/m^3)	Compressive strength (N/mm^2)	Modulus of rupture (N/mm^2)	Structural efficiency (× 10^3 m)	Bending stress (N/mm^2)	Shear stress (N/mm^2)
0	2.11	10.29	1674	5.15	13.20	3.08	20.42	14.12
5	2.18	9.43	1404	2.30	11.96	1.64	18.64	12.86
10	2.28	6.57	1330	0.85	8.53	0.64	13.28	9.22

because the clients have many choices of bricks in the market for them to choose for their houses. Due to very low demand for plastic bricks, the cost of setting up the factory, producing plastic bricks and transportation are very expensive. Thus, it can be concluded that the construction industries in Malaysia are not ready to produce and use plastic bricks for the construction of houses, apartments and commercial buildings.

10 Conclusions

Based on the following explanation and discussion on recycled PET plastic bottles as construction materials, the following conclusions and recommendations are as listed below:

a. PET bottle itself consists of good physical and mechanical properties, in which it is non-corrosive, lighter; has high tensile strength, high alkali resistance, high value of wetting tension; and is chemically inert, good in plastic shrinkage cracking resistance and easy to mix with concrete and mortar matrix.
b. The utilization of plastic bottles as one of the construction materials can help to prevent catastrophic impacts, particularly on environmental issues such as climate change, contaminated soil caused by leachate, human health and extinction of aquatic life. Furthermore, this method has the potential to reduce a significant amount of plastic waste generated each year in Malaysia.
c. The construction cost of houses can be reduced in comparison to conventional houses, making it affordable particularly for B40 income households as the cost of material being utilized is much lower due to the use of recycled bottles.
d. Performance of 5% recycled PET bottles in brick due to fire shrinkage, compressive strength and flexural strength outperforms other proportions and controls. This result demonstrates that PET bottles can be used as an additional material in brick.

Acknowledgements Special thanks goes to Ministry of Higher Education of Malaysia (MOHE), Putrajaya, Malaysia who funded this research work under Fundamental Research Grants with file No: 600-RMI/FRGS 5/3 (0091/2016) and RMI (Research Management Institute) for managing this research work in the most efficient ways. Gratitude and appreciation to the laboratory staff members for their invaluable assistance during the course of this experimental research work.

References

Akinyele J, Igba UT, Adigun BG (2020) Effect of waste PET on the structural properties of burnt bricks. Sci Afr 7:2468–2276

Alfahdawi IH, Osman SA, Hamid R, AI AL-Hadithi I (2019) Influence of PET wastes on the environment and high strength concrete properties exposed to high temperatures. Constr Build Mater 225:358–370
Azhdarpour AM, Nikoudel MR, Taheri M (2016) The effect of using polyethylene terephthalate particles on physical and strength-related properties of concrete: a laboratory evaluation. Constr Build Mater 109:55–62
Beata J, Kielbasa P, Korenko M, Drozdz T (2019) Physical and chemical properties of waste PET bottles washing as a component of solid fuels. Energ J, MDPI:1–17
Beguma RA, Siwar C, Pereira JJ, Jaafar AH (2006) A benefit–cost analysis on the economic feasibility of construction waste minimisation: the case of Malaysia. Resour, Conserv Recycl J 48(2006):86–98
Borg RP, Baldacchino O, Ferrara L (2016) Early age performance and mechanical characteristics of recycled pet fibre reinforced concrete. Constr Build Mater 108:29–47
Bricks Made from Recycled PET Bottles (2018) Material district. https://materialdistrict.com/article/bricks-recycled-pet-bottles/. Accessed 15 Apr 2021
Chen HL, Nath T, Chong S, Foo V, Gibbins C, Lechner AM (2021) The plastic waste problem in Malaysia: management, recycling and disposal of local and global plastic waste. SN Appl Sci Springer Nat J 3:437
Dai JG, Bai YL, Teng J (2011) Behavior and modeling of concrete confined with FRP composites of large deformability. J Compos Constr 15:963–973
Department of the Environment and Energy (2018) Australian Government. Analysis of Australia's municipal recycling infrastructure capacity October 2018 commonwealth of Australia 2018. https://www.environment.gov.au/system/files/resources/f0196d2e-9040-4547-8cb6-8b433923b53d/files/waste-stocktake-report.pdf. Accessed 11 Aug 2020
Faisal SK, Irwan JM, Othman N, & Ibrahim MW (2016) Flexural toughness of ring-shaped waste bottle fibre concrete. In: MATEC web of conferences, vol 47, p 01002. EDP Sciences
Faraca G, Edjabou A, Astrup T (2019) Plastic waste from recycling centres: characterisation and evaluation of plastic recyclability. Waste Manage 95:388–398
Faraj RH, Ali HFH, Sherwani AFH, Hassan BR, Karim H (2020) Use of recycled plastic in self-compacting concrete: A comprehensive review on fresh and mechanical properties. J. Build. Eng. 30, 101283
Foti D (2011) Preliminary analysis of concrete reinforced with waste bottles PET fibres. Constr Build Mater 25(2011):1906–1915
Foti D (2016) Innovative techniques for concrete reinforcement with polymers. Constr Build Mater 112:202–209
Foti D, Paparella F (2014) Impact behavior of structural elements in concrete reinforced with PET grids. Mech Res Commun 57:57–66
Fraternali F, Ciancia V, Chechile R, Rizzano G, Feo L, Incarnato L (2011) Experimental study of the thermo-mechanical properties of recycled PET fibre-reinforced concrete. Compos Struct 93(9):2368–2374
Ghernouti Y, Rabehi B, Bouziani T, Ghezraoui H, Makhloufi A (2015) Fresh and hardened properties of self-compacting concrete containing plastic bag waste fibres (WFSCC). Constr Build Mater 82:89–100
S Guettalal, M Bachar and L Azzouz (2016) Properties of the compressed-stabilized earth brick containing cork granules. J Earth Sci Clim Change 7:5. https://doi.org/10.4172/2157-7617
Guo Z, Kong X, Zhang J, Sun Y, Tu A, Jiang T (2020) Life-cycle assessment of concrete building blocks incorporating recycled concrete aggregates—a case study in China. In: Advances in construction and demolition waste recycling. Amsterdam, The Netherlands, Elsevier, pp 515–535
Hemraj N, Trilok M, Satveer M (2018) Study on utilization of plastic bottle waste—a review. Int J Adv Res Sci Eng 7
Hossain MU, Poon CS (2018) Comparative LCA of wood waste management strategies generated from building construction activities. J Clean Prod 177:387–397

Huang B, Chen Y, McDowall W, Türkeli S, Bleischwitz R, Geng Y (2019) Embodied GHG emissions of building materials in Shanghai. J Clean Prod 210:777–785

Irwan JM, Asyraf RM, Othman N, Koh HB, Annas MMK, Faisal SK (2013) The mechanical properties of PET fibre reinforced concrete from recycled bottle wastes. Adv Mater Res Trans Tech Publ, 347–351

Islam MJ, Meherier MS, Rakinul Islam AKM (2016) Effects of waste PET as coarse aggregate on the fresh and harden properties of concrete. Constr Build Mat 125:946–951

Jalaluddin M (2017) Use of plastic waste in civil constructions and innovative decorative material (eco-friendly). MOJ Civ Eng 3(05):01–10

Jo BW, Park SK (2006) Kim CH (2006) Mechanical properties of polyester polymer concrete using recycled polyethylene terephthalate. ACI Struct J 103(2):219–225

Juliana N (2021) Kenyan woman's startup recycles plastic into bricks that are stronger than concrete. Designboom, https://www.designboom.com/technology/gjenge-makers-recycled-plastic-bricks-kenya-02-08-2021/. Accessed 19 Apr 2021

Keiren (2020) Plastic bottle homes and greenhouses, Insteading, https://insteading.com/blog/plastic-bottle-homes/. Accessed 15 Apr 2021

Khalid F, Irwan J, Ibrahim MW, Othman N, Shahidan S (2018) Performance of plastic wastes in fibre-reinforced concrete beams. Constr Build Mater 183:451–464

Kim SB, Yi NH, Kim HY, Kim JHJ, Song YC (2010) Material and structural performance evaluation of recycled PET fibre reinforced concrete. Cem Concr Compos 32:232–240

Li X, Ling TC, Mo KH (2020) Functions and impacts of plastic/rubber wastes as eco-friendly aggregate in concrete–a review. Constr Build Mater 240:117869

Mahdi F, Abbas H, Khan AA (2013) Flexural, shear and bond strength of polymer concrete utilizing recycled resin obtained from post-consumer PET bottles. Constr Build Mater 44:798–811

Merli R, Preziosi M, Acampora A, Lucchetti MC, Petrucci E (2019) Recycled fibres in reinforced concrete: a systematic literature review. J Clean Prod 248:119207

Meyer C (2009) The greening of the concrete industry. Cem Concr Compos 31:601–605

Müller J (2021) Estimated annual household plastic packaging consumption Malaysia 2020 by type. Discover Statista Magazines

Neha S (2016) New building material from waste plastic: replast the future of plastics, techietonic. https://www.techietonics.com/futuretech-tonics/new-building-material-from-waste-plastic-replast-the-future-of-plastics.html. Accessed 15 Apr 2021

Nicolás Valencia (2017) This house was built in 5 days using recycled plastic bricks. ArchDaily. https://www.archdaily.com/869926/this-house-was-built-in-5-days-using-recycled-plastic-bricks. Accessed 15 Apr 2021

Ochi T, Okubo S, Fukui K (2007) Development of recycled PET fibre and its application as concrete—reinforcing fibre, cement & concrete composites, 29, 448–455

Patel PA, Shah A, Patel H (2016) Waste plastic bottles offering innovative building materials with sustainable application. Int J Innov Emerg Res Eng 3(3):38–45

Pelisser F, Neto ABdSS, La Rovere HL, de Andrade Pinto RC (2010) Effect of the addition of synthetic fibres to concrete thin slabs on plastic shrinkage cracking. Constr Build Mater 24:2171–2176

Periathamby A, Hamid FS, Khidzir K (2009) Evolution of solid waste management in Malaysia: impacts and implications of the solid waste bill. J Mater Cycles Waste Manage 11(2):96–103

Saikia N, De Brito J (2012) Use of plastic waste as aggregate in cement mortar and concrete preparation: a review. Constr Build Mater 34:385–401

Sandanayake M, Lokuge W, Zhang G, Setunge S, Thushar Q (2018) Greenhouse gas emissions during timber and concrete building construction—a scenario based comparative case study. Sustain Cities Soc 38:91–97

Sandanayake M, Gunasekara C, Law D, Zhang G, Setunge S, Wanijuru D (2020) Sustainable criterion selection framework for green building materials—an optimisation-based study of fly-ash Geopolymer concrete. Sustain Mater Technol 25:e00178

Saxena R, Siddique S, Gupta T, Sharma RK, Chaudhary S (2018) Impact resistance and energy absorption capacity of concrete containing plastic waste. Constr Build Mater 176:415–421

Shamsudin MMH, Hamid NH, Fauzi MAM (2021) Compressive and flexural strength of concrete containing recycled polyethylene terephthalate (PET). Key Eng Mater 879:13–21

Sina S, Amani A (2016) Use of recycled plastic water bottles in concrete blocks. Procedia engineering, creative construction conference, CCM2016, 25–28 June 2016, 164, pp 214–221

Usman A, Sutanto MH, Napiah M (2018) Effect of recycled plastic in mortar and concrete and the application of gamma irradiation-a review. In: Proceedings of the E3S web of conferences, Polanica-Zdrój, Poland, 16–18 April 2018; p 05027

Van Krevelen DW (1990) Properties of polymer. Elsevier The Netherlands, Asterdam

Wahab DA, Abidin A, Azhari CH (2007) Recycling trends in the plastics manufacturing and recycling companies in Malaysia. J Appl Sci 7:1030–1035

Zhang L (2013) Production of bricks from waste materials—a review. Constr Build Mater 47:643–655

Performance of Alkaline-Activated Cement Paste Toward Repairing and Rehabilitation on Concrete Structure

N. I. Halim, A. Newman, M. S. Muhd Norhasri, H. M. Saman, and N. H. Suliman

Abstract The incorporation of waste products as an additive or replacement in concrete is tremendously growing in the world of construction. The use of Palm Oil Fuel Ash (POFA) as a partial replacement in cement paste can contribute to improving the conventional repair and rehabilitation work as it enhances some particular properties of the cement paste. Three different mixes were prepared; plain cement paste, POFA cement paste with and without alkaline activator. POFA is used to partially replace cement with the inclusion of sodium silicate. The test conducted in this study includes the quality of concrete tests such as the ultrasonic pulse velocity test and mechanical properties tests such as the compression test. This study shows the presence of POFA tends to improve the compressive strength of cement paste. However, with the presence of an alkaline activator, the cement paste tends to have adequate compressive strength but improvement in setting time and indicates good quality and continuity of the filling material with the cube. On the other hand, POFA cement with an alkaline activator can be used as repair and rehabilitation material due to improvement in setting time, which still provides adequate compressive strength.

Keywords POFA · Cement replacement · Alkaline activator · Repair · Patching

---

N. I. Halim · A. Newman · M. S. Muhd Norhasri (✉) · H. M. Saman · N. H. Suliman
School of Civil Engineering, College of Engineering, Universiti Teknologi MARA Malaysia, 40450 Shah Alam, Selangor, Malaysia
e-mail: muhdnorhasri@uitm.edu.my

H. M. Saman
e-mail: hamidah929@uitm.edu.my

N. H. Suliman
e-mail: nurul_huda@uitm.edu.my

M. S. Muhd Norhasri
Institute for Infrastructure Engineering and Sustainable Management, UniversitiTeknologi MARA Malaysia, 40450 Shah Alam, Selangor, Malaysia

R. Hassan et al. (eds.), *Green Infrastructure*,
https://doi.org/10.1007/978-981-16-6383-3_2

1 Introduction

The waste disposal crisis from the industrial and agriculture industry is arising due to the formation of decomposed waste materials. One of the solutions to overcome this problem is by recycling waste into useful products. Maximizing recycling and reusing is one of the aims of sustainable waste management. Research into the innovative uses of these waste materials is continually advancing over the years. The waste material such as fly ash, rice husk ash, and palm oil fuel ash has successfully used in concrete for years (Sata et al. 2004; Meyer 2009; Satish et al. 2013; Hossain et al. 2016; Khankhaje et al. 2016; Aprianti 2017). This waste is involved as pozzolanic materials and can be used in concrete as cement replacement material. These materials also can be referred to as mineral admixtures or called pozzolans. It does not have any cementitious compounds within themselves; however, it will lead to form cementitious particles when combining with Portland cement. The contribution on waste as partial or full replacement of Portland cement can be one of the methods to resolve the landfill crisis, decreasing cost in building materials, produce a good solution to the environmental issues and waste management crisis, energy-saving, and also protect the environment from pollution (Aprianti 2017).

Numerous research stated that the use of POFA as a pozzolanic material enhanced the mechanical properties and durability of the concrete (Altwair and Kabir 2010; Safiuddin et al. 2011). Palm Oil Fuel Ash (POFA) is a product of the oil palm husk and palm kernel shell burnt in palm oil mill boiler (Nguong and Awal 2010). Zarina et al. (2013) defined POFA as a waste from the power plant that produces electricity using palm fiber, shell, and unfilled fruit bunch as the fuel that burnt at temperature 800–1000 °C. An early study on the use of POFA in concrete indicates that POFA in its original form has low pozzolanic properties, and it suggested POFA should not comprise more than 10% of cement by weight (Tay 1990). The utilization of POFA in concrete mix design reduced the disposal effects of POFA on the environment. Besides, the role of POFA as a Supplementary Cementitious Material (SCM) is reducing the amount of cement used in concrete mix and indirectly producing more economical concrete. Furthermore, it decreases the concrete permeability but increases the strength. POFA is a good SCM since the pozzolanic reactions producing more calcium silicate hydrate (C–S–H) due to silica oxide content in POFA reacts with calcium hydroxide $Ca(OH)_2$ from the hydration process (Sooraj 2013).

In recent years, there are various studies on POFA as source material for the development of cementless concrete, or called geopolymer concrete (Zarina et al. 2013; Ranjbar et al. 2014; Alengaram and Alamgirkabir 2015). POFA was discovered to have a high percentage of silica, which is one of the main elements in producing a geopolymer (Mohd Ariffin et al. 2011; Ranjbar et al. 2014; Salih et al. 2014). Geopolymer production needs alumina silicate-based materials that are rich in silicon (Si) and aluminum (Al), where it is activated by an alkaline solution. Commonly used alkaline activators in preparing geopolymers are sodium hydroxide (NaOH), potassium hydroxide (KOH), sodium silicate (Na_2SiO_3), or potassium silicate (K_2SiO_3)

(Abdulkareem and Ramli 2015). Alkaline liquid plays an important role in the polymerization process (Joshi and Kadu 2012). The polymerization process takes place, in which Silicon (Si) and Aluminium (Al) present in the source material reacts with the alkaline liquid to produce binders (Ridzuan et al. 2014).

In this study, POFA utilizes as a partial cement replacement, and sodium silicate use to enhance the properties of POFA cement. Sodium silicate previously is found to have many usages in cementitious materials, and one of them as an alkali-activator in alkali-activated cement. This study can reduce the excessive waste product that has been produced from the palm oil industry by utilizing the waste product as SCMs as cement replacement; hence the environmental issue can be counteracted. The utilization of Palm Oil Fuel Ash (POFA) in concrete promotes environmental-friendly practices and develops new concrete innovation in the construction industry. Also, the usage of POFA as SCMs can be more economical as it can reduce cement costing as well as concrete manufacturing. In concrete, sodium silicate is applied in the form of silicate mineral paint roles as setting accelerator and also improves in waterproofing and develops durability (Aïtcin 2008). The contribution of POFA and sodium silicate as a partial replacement in alkaline activator cement-based POFA improves the conventional repair and maintenance work as it enhances some particular properties of the cement itself, which are suitable for lighter external applications such as patching.

Therefore, this paper presents the effect of alkaline activators on POFA cement through core concrete cubes in improving the efficiency and effectiveness of the conventional repair and maintenance work. The core concrete cube is used to replicate the coring situation of the structure on site. The alkaline-activated POFA-based cement pastes were filled into the holes, and the qualitative assessment and performance of the concrete are evaluated.

2 Experimental Details

2.1 Sample Preparation

In order to figure out the effectiveness of the alkaline-activated POFA-based cement paste in the construction industry on repairing work, application testing was conducted on the filled cored concrete cube. Two tests were conducted, which are the ultrasonic pulse velocity (UPV) test and compressive test. A concrete cube of grade 30 is cast with a hole of size 15 mm x 15 mm at the center, as shown in Fig. 1. The design mix for the concrete cube is referred to in Table 1.

There is no specific standard for the hole size but the size 15 mm × 15 mm holed concrete cube is used in this study to simulate a core concrete structure on site. The cored concrete cube was cured for 7 days to maintain an adequate temperature of concrete during an early age, which may affect the hydration of cement and the gaining of the strength of the concrete. After 7 days of curing, the concrete cube was

Fig. 1 Cored concrete cube

Table 1 Concrete mix design (control OPC)

Mix designation	Cement (kg/m^3)	Water (kg/m^3)	Fine aggregate (kg/m^3)	Coarse aggregate (kg/m^3)
OPC	380	205	675	115

Table 2 Three different mixes applied on cored (hollow) concrete cubes

Mix designations	Description
OPC	Ordinary Portland cement
5P	Cement paste contained 5% POFA as cement replacement without the addition of sodium silicate
5P5.2S	Cement paste contained 5% POFA as cement replacement with the addition of sodium silicate

taken out from the curing tank and let dry for 2 h under the sun. Soon after the cube is dried, the hole is applied with the three different mixes of cement paste, as shown in Table 2. Once the paste is applied to the concrete hole, it is left to harden for 1 day before tested for the UPV test and compressive test.

2.2 *Application Testing on Concrete Cube Filled with Alkaline-Activated Plain and Paste Contained 5% POFA with and Without 5.2 mL Sodium Silicate Solution (Tambah Teori and Formula)*

In application testing, two tests are conducted, which are the ultrasonic pulse velocity (UPV) test and compressive test on the filled cored concrete cube.

Table 3 Classification of the quality of concrete (Aydin and Doven 2006)

Longitudinal pulse velocity (km/s)	Quality of concrete
Greater than 4.57	Excellent
3.655–4.57	Good
3.050–3.655	Doubtful
2.13–3.05	Poor
Less than 2.13	Very poor

2.2.1 Ultrasonic Pulse Velocity (UPV) Test

The quality of the concrete also can be evaluated using a non-destructive test by measuring the velocity of an ultrasonic pulse velocity that passing through the concrete. For this study, the ultrasonic pulse velocity involves measuring the 50 kHz ultrasonic pulse frequency travel time from an emitter on one side to the receiver on the other side of the concrete surface. The pulse velocity equipment will show the data of time for the pulse to transfer from the emitter to the receiver. The pulse velocity (km/s) is obtained by the distance between the two transducers (km) divided by the time travel (s) obtained from the ultrasonic pulse velocity equipment as shown in Eq. 1. The transmitter transducer and receiver transducer were calibrated before it contacts the concrete surface, and grease was applied to obtain an accurate reading. Table 3 shows the Classification of the Quality of Concrete based on the Pulse Velocity. BS EN 12,504-4, 2004, was used as a reference for this ultrasonic pulse velocity test.

$$\text{Pulse velocity} = (L/T) \quad (1)$$

L = the distance between the two transducers (meter).

T = travel time (seconds).

2.3 *Compression Test*

Three identical cube samples are prepared for each different mix, as shown in Table 4. First, the filled core concrete cube is tested on the UPV test. Then, the compressive strength with a pace rate of 3.0 N/s is executed on the filled core concrete cube

Table 4 The number of samples for cored concrete cube compressive strength

Compression strength Test (Day 1)	
OPC	3
5P	3
5P5.2S	3

to observe its mechanical properties. The procedure of curing concrete is repeated for the other two cored concrete cube samples, and the reading of these three cube samples is calculated to find the mean value of compressive strength. BS EN 12,390-3, 2002, was used as a reference for this compression test. The compressive strength can be calculated as in Eq. (2).

$$\text{Compressive Strength, MPa}\left(\text{N/mm}^2\right) = (F/A) \tag{2}$$

F = maximum load at failure (N).

A = cross-sectional area of the specimen (mm^2).

3 Result and Analysis

3.1 UPV Test on Cement-Filled Cored Concrete

Figure 2 shows pulse velocity in m/s tested on cored cube filled up with plain OPC that made of 5% POFA without the addition of sodium silicate (5P) and with the addition of Sodium silicate (5P5.2S).

From Fig. 2, cube filled with 5P5.2S paste gives the highest pulse velocity recorded 4 km/s, followed by those filled with plain OPC and 5P paste recorded 3.95 km/s and 3.89 km/s, respectively. The cube filled with 5P5.2S paste shows the highest velocity, which indicates good quality and continuity of the filling material with the cube. A decrease in the pulse velocity corresponds to an increase in porosity and oxygen permeability, which confirms that a region of low compaction, voids, or damaged material is present in the composites and leads to a reduction in the calculated pulse velocity (Panzera et al. 2012). 5P paste gives the lowest pulse velocity indicating the cube filled with 5P paste possessed defect. The slower pozzolanic reaction of

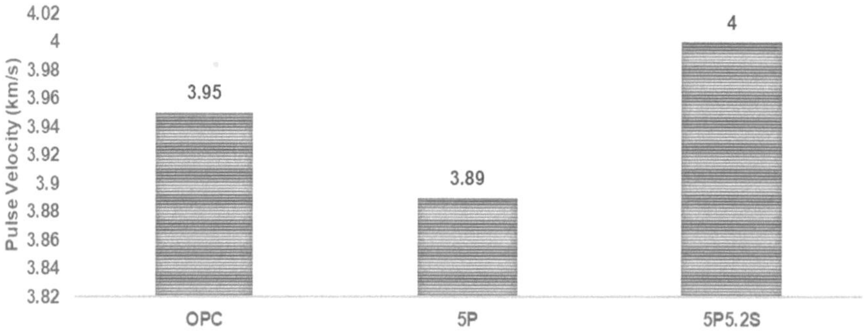

Fig. 2 Pulse velocity of concrete cube filled with alkaline-activated plain and paste contained 5% POFA with and without 5.2 mL sodium silicate solution

POFA and an inadequate amount of water needed during the hydration process gives possibly results in the poorer microstructure of concrete with more voids (Islam et al. 2015). By referring to Classification of the Quality of Concrete based on the Pulse Velocity, 5P5.2S, 5P, and OPC filled cube are within the range 3.655–4.570 km/s, which imply the good quality of concrete. However, the pulse velocity of concrete cube filled with 5P5.2S demonstrated a 1.3% increment compared with those filled with OPC and 2.9% increment compared with 5P filling. The increment of pulse velocity for 5P5.2S paste due to lesser porosity exists in the concrete filling compared to plain OPC and 5P paste.

3.2 Compressive Strength of Concrete Cube Filled with Cement Paste

Based on Fig. 3, the concrete cube filled with 5P paste indicated the highest compressive strength with the result attained is 34.07 MPa, followed by plain OPC and 5P5.2S paste with the result recorded 33.37 MPa and 32.74 MPa, respectively, at age of 1.

It showed that concrete cube filled with 5P paste marked a 2.1% increment, while concrete cube that filled with 5P5.2S paste exhibited a 1.9% decrement in compressive strength compared with that OPC. With the utilization of POFA, the compressive strength tends to increase. The explanation behind this result is due to adhesion mechanisms between CSH and POFA as fillers affect a great significance when employed as a cement replacement. However, the addition of silicate in the cement paste tends to slightly decrease its compressive strength but accelerated in hardening, which is suitable for repairing purposes such as patching as it hardens faster compare to 5P.

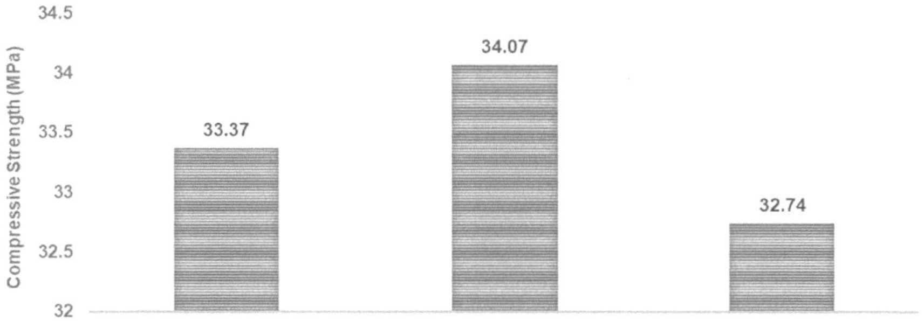

Fig. 3 Compressive strength 1 day of concrete cube filled with alkaline-activated plain and paste contained 5% POFA with and without 5.2 mL sodium silicate solution

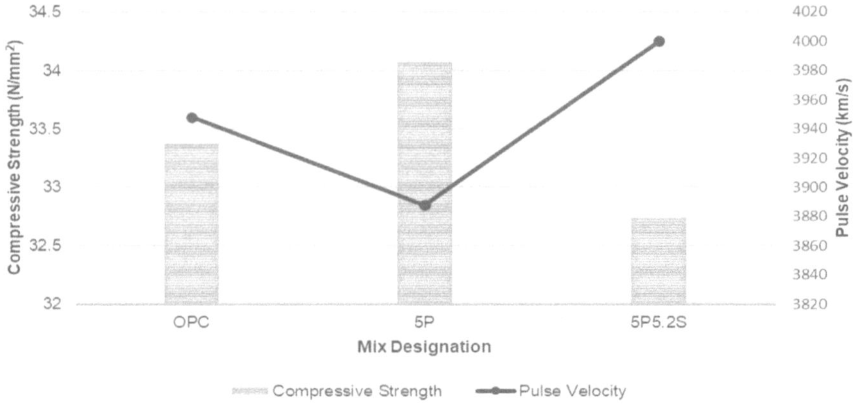

Fig. 4 Relationship between pulse velocity with compressive strength of concrete cube filled with alkaline-activated plain and paste contained 5% POFA with and without 5.2 mL sodium silicate solution.

3.3 Relationship Between Pulse Velocity with Compressive Strength of Concrete Cube Filled with Alkaline-Activated Plain and Paste Contained 5% POFA with and Without 5.2 mL Sodium Silicate Solution

Figure 4 shows the relationship between pulse velocity with compressive strength of concrete cube filled with alkaline-activated plain and paste contained 5% POFA with and without 5.2 mL sodium silicate solution.

From Fig. 4, it is concluded that the strength of the cube does not come from defects of filling as determine from the UPV test, but comes from the strength properties of the filling material itself, which is the cement paste itself. As observed from the above figure, the compressive test of 5P paste tends to have the highest 1-day compressive strength, which can be seen to affect the concrete cube strength even though it has the highest defects among all of the samples. These can be due to entrapped air pores between the ash particles (Ikpong 1993). Whereas, 5P5.2S paste has the lowest 1-day compressive strength even though it has the highest pulse velocity, which indicates that the fillings have fewer defects among the other samples.

4 Conclusion

This study highlights the analysis of concrete cubes filled with alkaline-activated plain and pastes contained 5% POFA with and without 5.2 ml sodium silicate solution through experimental investigation. Based on the test conducted, the findings are as follows:

Cube filled with 5P5.2S paste tends to show the fastest ultrasonic pulse velocity, which indicates good quality and continuity o f the filling material with the cube. On the other hand, a cube filled with 5P paste shows the highest compressive strength. These show that 5P cement paste can improve the overall strength of the cube compared to the other cement paste.

As a consequence, a cored cube filled up with 5P paste tends to show its effectiveness in repairing work by the quality of filling and the highest strength of the concrete cube. However, a cored cube filled up with 5P5.2S has the capability to act as repair cement. Faster setting time and moderate strength make it an ideal cement paste to be used as patching in construction work on patching and repairing work.

5 Recommendation

Below shows the recommendation that could potentially improve this research for future work, for example as follows:

1. Other pozzolans, such as RHA, fly ash, sawdust ash, or a combination of them, should be used to improve the strength of alkaline-activated cement paste in the future. The utilization of waste by-products could lead to saving the environment and sustainable construction.
2. In the future, it is also recommended to extend the study by using the deteriorated concrete samples rather than the new mixture to reflect the repair and rehabilitation of the existing concrete structures in the future.

Acknowledgements The author would express the appreciation post to the Research Management Institute (RMI) for the financial support granted under (600-RMC/GIP 5/3 (032/2021)) and (600-IRMI/FRGS 5/3 (176/2019)). Also, the author is indebted to the lab technicians, Mohd Faiz bin Shukor and Zairinizam bin Ahmad at the Material and Heavy Structural Lab of the Faculty of Civil Engineering, Universiti Teknologi MARA, Shah Alam, Selangor Darul Ehsan that involve throughout this research for sharing their knowledge and guidance with the laboratory work. Besides, special acknowledgment to the Institute for Infrastructure Engineering and Sustainable Management (IIESM) and Faculty of Civil Engineering, Universiti Teknologi Mara (UiTM), for providing the facilities to accomplish the laboratory work throughout this research.

References

Abdulkareem OA, Ramli M (2015) Optimization of alkaline activator mixing and curing conditions for a fly ash-based geopolymer paste system. Mod Appl Sci 9(12):61. https://doi.org/10.5539/mas.v9n12p61

Aïtcin P-C (2008) Binder for durable and sustainable concrete. Taylor & Francis

Alengaram UJ, Alamgirkabir SM (2015) Engineering properties of oil palm shell and palm oil clinker based geopolymer concrete. 38:34–48

Altwair NM, Kabir S (2010) Palm oil fuel ash (POFA): an environmental-friendly supplemental cementitious material for concrete production. In: International RILEM conference on material science, III (January 2010), pp 113–126

Aprianti SE (2017) A huge number of artificial waste material can be supplementary cementitious material (SCM) for concrete production—a review part II. J Clean Prod 142:4178–4194. https://doi.org/10.1016/j.jclepro.2015.12.115

Aydin E, Doven AG (2006) Influence of water content on ultrasonic pulse-echo measurements through high volume fly ash cement paste—physicomechanical characterization. Res Nondestr Eval 17(4):177–189. https://doi.org/10.1080/09349840600788004

Hossain MM, Karim MR, Hasan M, Hossain MK, Zain MFM (2016) Durability of mortar and concrete made up of pozzolans as a partial replacement of cement: a review. In: Construction and building materials, vol 116. https://doi.org/10.1016/j.conbuildmat.2016.04.147

Ikpong AA (1993) The relationship between the strength and non-destructive parameters of rice husk ash concrete. Cem Concr Res 23(1):387–398

Islam MMU, Mo KH, Alengaram UJ, Jumaat MZ (2015) Mechanical and fresh properties of sustainable oil palm shell lightweight concrete incorporating palm oil fuel ash. J Clean Prod 115:307–314. https://doi.org/10.1016/j.jclepro.2015.12.051

Joshi SV, Kadu MS (2012) Role of alkaline activator in development of eco-friendly fly ash based geo polymer concrete. Int J Environ Sci Dev 3(5):417–421. https://doi.org/10.7763/IJESD.2012.V3.258

Khankhaje E, Hussin MW, Mirza J, Rafieizonooz M, Salim MR, Siong HC, Warid MNM (2016) On blended cement and geopolymer concretes containing palm oil fuel ash. Mater Des 89:385–398. https://doi.org/10.1016/j.matdes.2015.09.140

Meyer C (2009) The greening of the concrete industry. Cement Concr Compos 31(8):601–605. https://doi.org/10.1016/j.cemconcomp.2008.12.010

Mohd Ariffin, M. A., Hussin, M. W., & Rafique Bhutta, M. A. (2011) Mix Design and Compressive Strength of Geopolymer Concrete Containing Blended Ash from Agro-Industrial Wastes. Adv Mater Res 339(November 2015):452–457. https://doi.org/10.4028/www.scientific.net/AMR.339.452

Nguong SK, Awal ASMA (2010) A short-term investigation on high volume palm oil fuel ash (POFA) concrete. In: 35th conference on our world in concrete and structures, pp 1–9

Panzera TH, Christoforo AL, de Paiva Cota F, Ribeiro Borges PH, Bowen CR (2012) Ultrasonic pulse velocity evaluation of cementitious materials. Adv Compos Mater - Anal Nat Man-Made Mater, May 2014. https://doi.org/10.5772/17167

Ranjbar N, Mehrali M, Behnia A, Alengaram UJ, Jumaat MZ (2014) Compressive strength and microstructural analysis of fly ash/palm oil fuel ash based geopolymer mortar under elevated temperatures. Constr Build Mater 65:114–121. https://doi.org/10.1016/j.conbuildmat.2014.04.064

Ridzuan ARM, Abdullah MMAB, Arshad MF, Mohd Tahir MF, Khairulniza AA (2014) The effect of NaOH concentration and curing condition to the strength and shrinkage performance of recycled geopolymer concrete. Mater Sci Forum 803:194–200. https://doi.org/10.4028/www.scientific.net/msf.803.194

Safiuddin M, Salam MA, Jumaat MZ (2011) Utilization of palm oil fuel ash: a review. J Civ Eng Manag 17(2):234–247. https://doi.org/10.3846/13923730.2011.574450

Salih MA, Abang Ali AA, Farzadnia N (2014) Characterization of mechanical and microstructural properties of palm oil fuel ash geopolymer cement paste. Constr Build Mater 65:592–603. https://doi.org/10.1016/j.conbuildmat.2014.05.031

Sata V, Jaturapitakkul C, Kiattikomol K (2004) Utilization of Palm Oil Fuel Ash in High-Strength Concrete. J Mater Civ Eng 16(6):623–628. https://doi.org/10.1061/(ASCE)0899-1561(2004)16:6(623)

Satish Chandra C, Takhelmayum G, Deepa T (2013) Study on flow and compressive strength properties of mortars using fly ash and lime. Int J Eng Sci Innov Technol (IJESIT) 2(6):309–315

Sooraj VM (2013) Effect of palm oil fuel ash (POFA) on strength properties of concrete. Int J Sci Res Publ 3(6):2250–3153. www.ijsrp.org
Tay J (1990) Ash from oil-palm waste as concrete material. 2(2):94–105
Zarina Y, Mustafa Al Bakri AM, Kamarudin H, Nizar IK, Rafiza AR (2013) Review on the various ash from palm oil waste as geopolymer material. Rev Adv Mater Sci 34(1):37–43

Physical Properties of Activated Pofa Cement Paste for Repair and Retrofitting Purposes

A. Newman, N. I. Halim, M. S. Muhd Norhasri, H. M. Saman, and A. Alisibramulisi

Abstract All these years, commonly used repairing and retrofitting materials are cement paste, grout, patching, and epoxy. However, due to the expensive and unreliable application, an alternative material is needed. This research emphases the use of waste material such as palm oil fuel ash (POFA) and alkaline activator (sodium hydroxide) in cement paste as repair and retrofitting material. Hence, it focuses on the physical properties of activated POFA cement paste by means of Ultrasonic Pulse Velocity and compressive strength test. In this research, three types of paste are prepared: OPC, POFA paste, and Activated POFA paste. The POFA used is sieve to 212 μm particle size, and the alkaline activator used is sodium hydroxide. These pastes are used to fill concrete cubes with a hole on the center, representing a core structure on site. The paste is let to harden for a day before being tested for quality using UPV and compressive strength using compression test. POFA cement tends to improve the structure strength by up to 2% but has defect in filling as it shows 1.42% less ultrasonic pulse velocity reading compare to OPC. Whereas, activated POFA cement shows 0.5% increase in structure strength with 0.84% less defects compare to OPC.

Keywords Cement paste · Palm oil fuel ash · Alkaline activator · Sodium hydroxide

---

A. Newman · N. I. Halim · M. S. M. Norhasri (✉) · H. M. Saman · A. Alisibramulisi
School of Civil Engineering, College of Engineering, Universiti Teknologi MARA Malaysia, 40450 Shah Alam, Selangor, Malaysia

H. M. Saman
e-mail: hamidah929@uitm.edu.my

M. S. M. Norhasri · A. Alisibramulisi
Institute for Infrastructure Engineering and Sustainable Management, Universiti Teknologi MARA Malaysia, 40450 Shah Alam, Selangor, Malaysia

R. Hassan et al. (eds.), *Green Infrastructure*,
https://doi.org/10.1007/978-981-16-6383-3_3

1 Introduction

As our economical, technological, and ecological advancement grew b y days, the usage of pozzolanic material as cement replacement also increases and is expected to continue in years to come (Abdul Awal and Warid Hussin 2011). One of the most abundant pozzolanic material is Palm Oil Fuel Ash (POFA). In numerous researches, POFA has been used as cement replacement due to its pozzolanic properties. The usage of POFA in building construction material has not only improved the performance of the concrete but also contributed in decreasing the usage of land disposal and reducing the carbon footprint due to cement production. Cement production industry has contributed up to 5% of the world's carbon dioxide emission (Carrasco 2017). Tons of carbon dioxide are release into the atmosphere from the de-carbonation process of limestone and the combustion of fossil fuel during the cement manufacturing process (Joshi and Kadu 2012).

Several researches have shown that POFA can benefit the performance of concrete. A research by Wi et al. (2018) has shown that high content of silicon dioxide or calcium oxide in POFA can produce extra hydration process that could improve the concrete strength and durability. Hamada et al. (2018) also concluded that a denser and durable cement mortar is produced from high Calcium-Silicate-Hydroxide gel formed from high content of silica in POFA. Another research by Hamada et al. (2018) have shown that due to the fine particle of POFA, cement replaced by POFA in a high amount has the ability to increase the concrete workability as well as the setting time. According to Imran et al. (2020), POFA can also improve finishing surfaces and workability due to the ability of POFA to absorb water from the mixture and retained it in the system. Mujah (2016) also verifies that when POFA is incorporated, the compressive strength of cement grout increased up to 20%.

On the other hand, cement performance can also be improved by using alkaline activator. Huseien et al. (2018) stated that alkali-activated binder production uses a moderate amount of energy where they are considered environmentally friendly. A few researches have shown the benefit of alkaline activator to cement.

Abdulkareem and Ramli (2015) concluded that alkaline activator has the ability to dissolve the acidic oxides of POFA which normal water is not able to do. Huang and Lin (2010) research also states that higher amount of Sodium Hydroxide also reduces the cement setting time. Other than that, a research by Ridzuan et al. (2014) showed that geopolymer concrete compressive strength is improved due the higher concentration of Sodium Hydroxide which produced a good bonding of aggregate and paste.

For this research, the idea of incorporating POFA and alkaline activator in cement paste is explore. The activator partially replaces the water in the design mix. Whereas POFA partially replace the cement. During the hydration process of cement, small amount of calcium hydroxide is produced when less amount of cement is used. The small amount of calcium hydroxide is insufficient for hydration with large amount of reactive silica. So, in order to activate the reaction of excessive reactive silica, sodium hydroxide is needed (Bahadure and Naik 2013). With this, the usage of alkaline activator might be able to enhance the efficiency of POFA replacement in cement.

Over the years, researchers have been focusing on material for building construction. However, the lack of attention to research on building repairing and retrofitting material has been the interest of this research. Therefore, this research focuses on the use of POFA and alkaline activator in building repairing and retrofitting material instead of construction material. This research hopes to discover new materials that can improve existing and conventional repairing and retrofitting work. The goal of this research is to evaluate the quality and strength of POFA cement paste reacted with alkaline activator. This research also hopes to contribute to the reduction in cement manufacture which resulted in increasing of carbon dioxide release, which could be relate to SDG 12 responsible consumption and production and SDG 3 good health and well-being. This research also wishes to reduce POFA waste which consume large amount of land for disposal as well as improving concrete repair works material which is directly related to SDG 15 life on land and SDG 9 industry, innovation and infrastructure.

2 Experimental Details

2.1 *Material Preparation*

The material used in this study is cement powder, mineral admixture, water, and alkaline activator. Ordinary Portland Cement with grade 42.5 is the cement powder used. It is provided by the Faculty of Civil Engineering Concrete Laboratory in UiTM Shah Alam. For mineral admixtures, POFA is used. It is obtained from the United Palm Oil SDN. BHD. factory at Sungai Kecil, Nibong Tebal, Pulau Pinang. The raw POFA obtained is sieved to 212 mm size before used in the design mix. On the other hand, the water used is in this study is normal tap water obtained from the concrete laboratory. For alkaline activator, sodium hydroxide is used. The sodium hydroxide is supply by the Institute for Infrastructure Engineering and Sustainable Management (IIESM) lab in pellet form. This pellet is mixed with distilled water to produce Sodium Hydroxide solution with molarity of 2. The 2 mol sodium hydroxide is used as a partial replacement to water in the design mix. Therefore, a proportion of 1 kg/1 L of water equal to 200 mL alkaline activator is used in the design mix. The design mix for this study is shown in Table 1.

2.2 *Sample Preparation*

A number of 9 concrete cubes of 100 mm with grade 30 are prepared for this research 100 mm. The concrete cubes are casted with a hole of size 15mmx15mm at the center,

Table 1 Design mix

Sample	Cement (kg/m^3)	POFA (kg/m^3)	Water (kg/m^3)	Activator (kg/m^3)
OPC	380	0	205	0
5P	360	20	205	0
5P5.2H	360	20	163	42

which is to be filled with cement paste shown in Table 1. This concrete cube is used to represent a core concrete structure on site. The cube is then cured for seven days. After seven days, the cube is left to dry under the sun. As soon as the cube dries, it is filled with different cement paste shown in Table 1. After that, it is left to harden for 24 h before testing.

2.3 *Sample Testing*

In this research, the effectiveness of the paste in concrete repairing is the main focus. Therefore, to access this effectiveness, two tests are conducted: ultrasonic pulse velocity test and compressive strength test.

2.3.1 Ultrasonic Pulse Velocity

Ultrasonic Pulse Velocity test is a non-destructive test used to distinguish the quality of the concrete by identifying any internal flaws, density, cracks, voids, or even poor patches inside the concrete itself.

UPV test consists of measuring the 50 kHz ultrasonic pulse travel time from an emitter on one side of the concrete surface to the receiver on the other side. The transmitter transducer and receiver transducer need to be in contact with the concrete surface and sometimes both of the contacted concrete surface, and the transducers need to be applied with grease so that an accurate reading can be obtained. In this research, a direct transmission is used since the direct transmission can provide a better reading. This test is in accordance to ASTM C597. The quality of concrete identified by referring to Table 2 shows the Classification of the Quality of Concrete

Table 2 Classification of the quality of concrete on the basis of the pulse velocity (Aydin 2016)

Longitudinal pulse velocity (m/s)	Quality of concrete
Greater than 4570	Excellent
3655–4570	Good
3050–3655	Doubtful
2130–3050	Poor
Less than 2130	Very poor

on the Basis of the Pulse Velocity. Equation 1 is used to obtain the ultrasonic pulse velocity, where L is the distance between the two transducers in meters and T is the travel time in seconds.

$$\text{UPV} = L/T \tag{1}$$

2.3.2 Compressive Strength Test

The holed concrete cube filled with different cement paste is tested for compressive test to determine the effectiveness of the paste in repairing and retrofitting work. This test is also conducted in accordance ASTM C109.

3 Result and Discussion

Based on the data obtained, pulse velocity and compressive strength values for all types of specimens were illustrated. The relationship between pulse velocity and compressive strength is also illustrated.

3.1 Ultrasonic Pulse Velocity Test

Figure 1 accesses the quality of concrete cube filled with different specimens. It shows the reading of pulse velocity of each concrete cube filled with different types of paste.

As observed, cube filled with OPC paste shows the highest pulse velocity value with reading of 3945.46 m/s. Cube filled with 5P5.2H shows a pulse velocity of 3912.11 m/s which is in between cube filled with OPC and 5P. Cube filled with 5P tends to have the lowest pulse velocity among the three with value of 3889.29 km/s. Based on the table of Classification of Quality of Concrete on the Basis of the Pulse Velocity, all cubes with different cement paste fall on the good quality category.

However, a higher velocity reading shows that the cube is filled with cement paste with less defects and more compact. Therefore, OPC seems to be fill the cube better with less defect. Whereas, 5P is not able to fill the cube properly resulted in defects. Other than that, it is probably due to the porous surface of POFA particle. This might be the cause of low pulse velocity reading since porous structure is less compact. However, cube filled with 5P5.2H tends to have moderate defects since it can fill the cube slightly better than 5P. This is probably due to the reaction sodium hydroxide to disperse the cement and POFA particle. Dispersion can make the paste more workable; hence, it is easier to trap air bubble during the cube filling process. Thus,

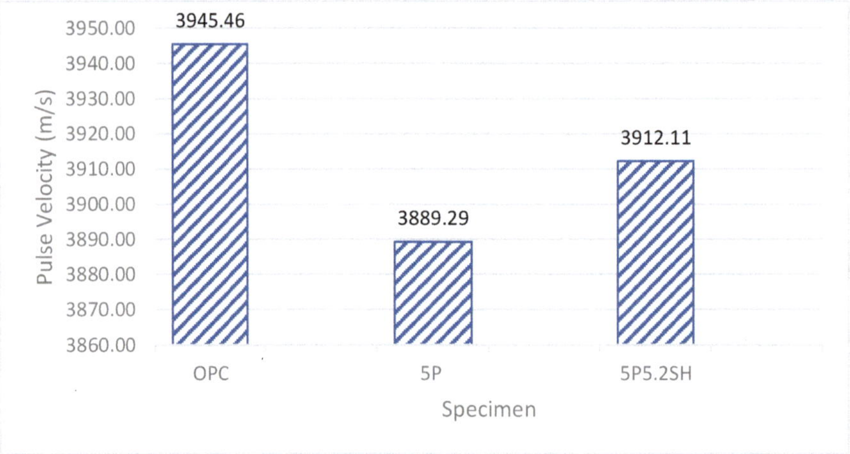

Fig. 1 Pulse velocity for tested specimens

this could cause defects in the filling resulting in lower velocity reading compare to OPC filled cube.

3.2 *Compressive Strength Test*

Figure 2 shows the compressive strength of concrete cube filled with different types of specimens.

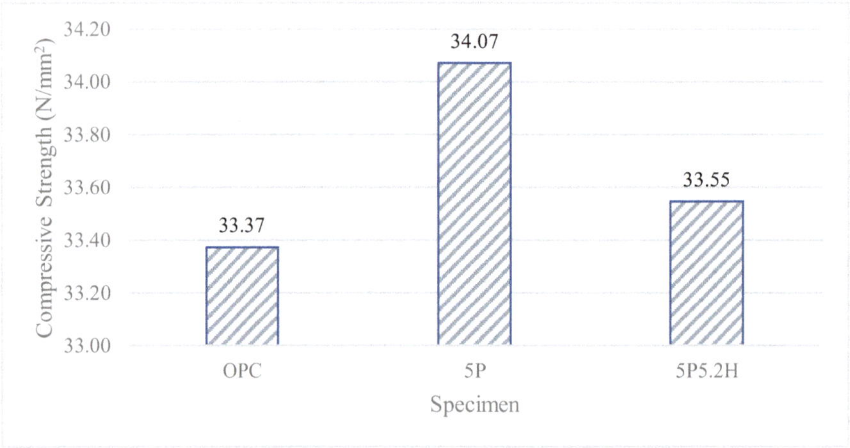

Fig. 2 Compressive strength for tested specimens

As seen in Fig. 2, the highest compressive strength is achieved by cube filled with 5P, with reading 34.07 N/mm^2. Followed by cube filled with 5P5.2H and OPC with reading 33.5 N/mm^2 and 33.37 N/mm^2 respectively. This shows that 5P filling is able to achieve the best overall compressive strength of concrete cube. POFA has been known to provide enhancement in concrete; therefore, by incorporating it into paste, it can definitely improve the strength properties of paste as well. Therefore, when 5P acts as filling material, it is able to strengthen the hole on the concrete cube, which is basically the core of the cube, hence strengthens the whole cube.

However, when sodium hydroxide is incorporated in POFA cement paste as demonstrated by 5P5.2H, the concrete cube tends to have lower compressive strength. This is probably due to the reaction of sodium hydroxide toward the POFA particle. As we know, cement produce CSH gel and calcium hydroxide. When POFA is used as partial replacement to cement, it generally produces more CSH gel since the reactive silica from POFA react with the rest of the calcium hydroxide produced. However, when there is too little calcium hydroxide for the reactive silica to reacted to, then plenty reactive silica are left unreacted. But, by incorporating sodium hydroxide, it can react with the reactive silica to produce a stable compound known as sodium metasilicate and water. The production of water by sodium hydroxide with reactive silica may be the main reason which leads to dispersion ability. However, the excessive water can also reduce the strength of paste especially when it is leached out, causing pores on the hardened cement. This explains the lower compressive strength of cube filled with 5P5.2H.

This shows that POFA replacement of cement does improve the compressive strength of the paste. However, when the POFA reacted with sodium hydroxide, it tends to slightly reduce the compressive strength. Hence, the overall compressive strength of the cube is influenced by the type of cement paste that fills the hole.

3.3 The Relationship Between Ultrasonic Pulse Velocity and Compressive Strength

Based on Figs. 1 and 2, a relationship graph is derived and is shown in Fig. 3. Figure 3 shows the relationship between the pulse velocity and the compressive strength of concrete cube filled with different types of cement paste.

As observed in Fig. 3, OPC filled cube tends to have the lowest compressive strength despite high pulse velocity reading. 5P tends to have the highest compressive strength but lowest pulse velocity reading. Whereas, 5P5.2H shows to have both compressive strength and pulse velocity reading between OPC and 5P. This shows that cube filled with OPC has less defects even though the cube it fills has the lowest early paste compressive strength compare to 5P and 5P5.2H. Although 5P paste has better compressive strength, the low pulse velocity reading indicates that there might be defects in the cube filling. Whereas, 5P5.2H fills the cube slightly better compared to 5P even though it has a slightly lower paste strength than 5P. Therefore, based on

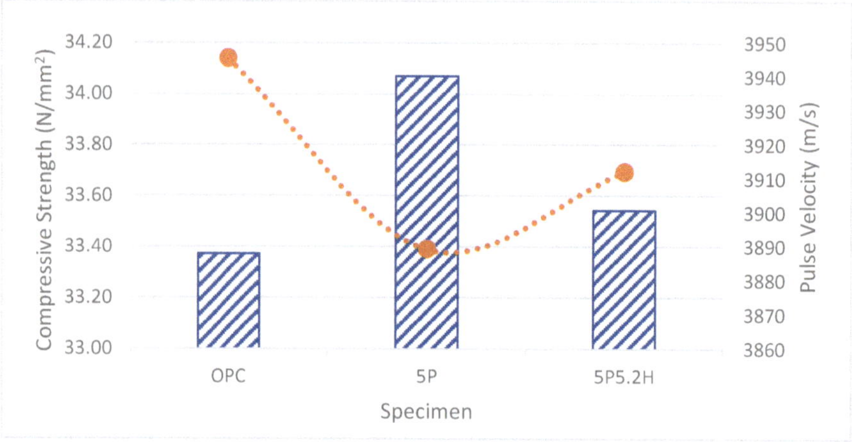

Fig. 3 Relationship between ultrasonic pulse velocity and compressive strength

this it can be concluded that the compressive strength of the filled cube is influence by the cement paste filling quality rather than the filling type.

4 Conclusion

This study stresses on the effectiveness of the cement paste in concrete repairing and retrofitting. Three different types of cement paste are ordinary Portland cement, POFA cement, and Activated POFA cement. The effectiveness in concrete repairing and retrofitting are accessed through ultrasonic pulse velocity test and compressive strength test. Therefore, based on the conducted experiment and analysis of data, the findings are as follows:

- OPC-filled cube tends to show the fastest ultrasonic pulse velocity compared to the others. This shows that paste has less defects when used to fill the cube.
- On the other hand, for compressive test, 5P-filled cube shows the highest compressive strength. This shows that 5P cement paste is able to improve overall strength of the cube compare to the other cement paste.
- The relationship between the pulse velocity and the compressive strength of filled cube can be concluded where the strength of the filled cube is not affected by the types of cement filling. Rather, it is influenced by the filing quality.

Based on the findings of this study, it is concluded that the uses of POFA does provide better strength performance to cement paste compare to cement paste alone. However, POFA cement paste has shown to have more defects compare to OPC. On the other hand, POFA reacted with sodium hydroxide tends to performed moderately in terms of quality and strength compare to both OPC and POFA cement. Even with moderate compressive strength, it has less defects which is a good attribute for repairing and retrofitting material since it can be used in long term.

Acknowledgements The author would like acknowledge the Research Management Institute for providing the "Geran Inisiatif Penyeliaan" (GIP) (600-IRMI/FRGS-RACER 5/3 (059/2019)) and (600-IRMI/FRGS 5/3 (178/2019)), Institute for Infrastructure Engineering and Sustainable Management (IIESM) and Faculty of Civil Engineering, UiTM Shah Alam for providing the accommodations to accomplish the experimental work for this study. The author would also like to thank the Malaysian Newsprint Industries Sdn Bhd and BASF Malaysia for providing the material used for this study.

References

Abdul Awal ASM, Warid Hussin M (2011) Effect of palm oil fuel ash in controlling heat of hydration of concrete. Procedia Eng 14:2650–2657. https://doi.org/10.1016/j.proeng.2011.07.333

Abdulkareem OA, Ramli M (2015) Optimization of alkaline activator mixing and curing conditions for a fly ash-based geopolymer paste system. Mod Appl Sci 9(12):61. https://doi.org/10.5539/mas.v9n12p61

Aydin, E. (2016) Influence of water content on ultrasonic pulse—echo measurements through high volume fly ash cement paste. Physmechanical Charact. October 2006. https://doi.org/10.1080/09349840600788004

Bahadure B, Naik N (2013) Effect of alkaline activator on workability and compressive strength of cement concrete with RHA. Int J Comput Eng Res3(5). http://scholar.google.com/scholar?hl=en&btnG=Search&q=intitle:Effect+of+Alkaline+Activator+on+Workability+and+Compressive+Strength+of+Cement+Concrete+with+RHA#0

Carrasco MT (2017) Alkaline activation of different aluminosilicates as an alternative to Portland cement : Alkali activated cements or ... Alkaline activation of different aluminosilicates as an alternative to Portland cement : alkali activated cements or geopolymers La ac. August, 4–12

Hamada HM, Jokhio GA, Yahaya FM, Humada AM, Gul Y (2018) The present state of the use of palm oil fuel ash (POFA) in concrete. Constr Build Mater 175:26–40. https://doi.org/10.1016/j.conbuildmat.2018.03.227

Huang Y, Lin Z (2010) Effect of sodium hydroxide on the properties of phosphogypsum based cement. J Wuhan Univ Technol, Mater Sci Ed 25(2):342–345. https://doi.org/10.1007/s11595-010-3342-7

Huseien GF, Tahir MM, Mirza J, Ismail M, Shah KW, Asaad MA (2018) Effects of POFA replaced with FA on durability properties of GBFS included alkali activated mortars. Constr Build Mater 175:174–186. https://doi.org/10.1016/j.conbuildmat.2018.04.166

Imran NF, Hasri MN, Fitri NAK, Hasan D, Ramli MZ (2020) Effects of palm oil fuel ash (POFA) towards consistency and setting time properties of concrete effects of palm oil fuel ash (POFA) towards consistency and setting time properties of concrete. 020035(October 2018). https://doi.org/10.1063/1.5062661

Joshi SV, Kadu MS (2012) Role of Alkaline activator in development of eco-friendly fly ash based geo polymer concrete. Int J Environ Sci Dev 3(5):417–421. https://doi.org/10.7763/IJESD.2012.V3.258

Mujah D (2016) Compressive strength and chloride resistance of grout containing ground palm oil fuel ash. J Clean Prod 112:712–722. https://doi.org/10.1016/j.jclepro.2015.07.066

Ridzuan ARM, Al Bakri Abdullah MM, Arshad MF, Mohd Tahir MF, Khairulniza AA (2014) The Effect of NaOH concentration and curing condition to the strength and shrinkage performance of recycled geopolymer concrete. Mater Sci Forum 803(August 2014):194–200. https://doi.org/10.4028/www.scientific.net/MSF.803.194

Wi K, Lee HS, Lim S, Song H, Hussin MW, Ismail MA (2018) Use of an agricultural by-product, nano sized palm oil fuel ash as a supplementary cementitious material. Constr Build Mater 183:139–149. https://doi.org/10.1016/j.conbuildmat.2018.06.156

A Review on Tin Slag Polymer Concrete as Green Structural Material for Sustainable Future

M. S. Manda, M. R. M. Rejab, Shukur Abu Hassan, and Ma Quanjin

Abstract Ordinary Portland cement concrete (OPC) is widely utilized in construction industry as structural material, but it has environmental issue due to natural resources consumption and Carbon emission. Therefore, polymer concrete (PC) with tin slag (TS) waste are introduced to replace aggregate and cement in OPC. Previous research on potential to apply TSPC as structural material has provided compressive strength data which shows that it can compete with OPC. PC using polyester and 100% TS aggregate with resin-aggregate ratio 30:70 consist of fine (<1 mm) uniformly graded aggregate has achieved compressive strength 58.21 MPa. After that, in another study, gap graded performance of TSPC using raw (4 mm) and coarse (2 mm) TS aggregate introduced and result in compressive strength 37.71 MPa, highest compared to other variation. By applying external FRP strengthening with two layers of CFRP increase strength to 125.07 MPa and finally uniformly graded TSPC with three layers of CFRP wrapping increased strength to 156.88 MPa. This discovery has contributed to the beginning of active study in TSPC as green structural material for sustainable future.

Keywords Green · Structural Material · Polymer Concrete · Tin Slag · Compressive Strength · FRP Confinement · TSPC · Sustainable

---

M. S. Manda
Politeknik Sultan Haji Ahmad Shah, Kuantan, Pahang, Malaysia

M. S. Manda · M. R. M. Rejab (✉) · M. Quanjin
Structural Performance Material Engineering (SUPREME), Faculty of Mechanical & Automotive Engineering Technology, Universiti Malaysia Pahang, Pekan, Pahang, Malaysia
e-mail: ruzaimi@ump.edu.my

M. Quanjin
e-mail: neromaquanjin@gmail.com

S. A. Hassan
Centre for Composites, Universiti Teknologi Malaysia, Skudai, Johor, Malaysia
e-mail: shukur@utm.my

M. R. M. Rejab
School of Mechanical Engineering, Ningxia University, Ningxia, China

R. Hassan et al. (eds.), *Green Infrastructure*,
https://doi.org/10.1007/978-981-16-6383-3_4

1 Introduction

Concrete and steel have been fully manipulated in the design and development of a structure, especially in building construction industry. Concrete characteristics are mainly classified as high strength and brittle, and while steel, on the other hand, is generally considered high strength and ductile. However, to compensate both brittleness of concrete and the ductility o f steel, the new combine properties of concrete and steel are introduced, resulting in new structural material known as steel-reinforced concrete structure. In the building construction industry, concrete has played the most dominant role over steel as concrete poses better strength to weight ratio, better insulator and more economical and environmentally friendly compared to steel. According to Hosamo et al. (2019), buildings made of concrete are more environmentally friendly than those made of steel because of higher energy consumption during build-up and manufacturing. Besides that, the production of steel is also more complex, which required large iron extraction and steel production plant compared to cement production, which is the main ingredient in concrete material. Thus concrete is the preferred structural material, and since the introduction of concrete as a structural material, it has evolved in many aspects such as types of aggregates, aggregates grading, curing conditions, fillers, admixtures, types of binders in concrete (Lahri and Dixit 2015) and also strengthening methods (Ueda 2014).

In terms of binders, conventional concrete use cement to hold the aggregates that forming a concrete. However, the production of cement affecting the environment as it involves natural resources such as limestones which are the main ingredient in the production of cement. In addition to that, the productions of cement are also causing air pollution due to the escaped cement particles from production plant and also Carbon emission in the forms of Carbon Dioxide (CO_2) and Carbon Monoxide (CO). This emission is caused by a large heating kiln in a sintering process to produce clinker, which is the main part in cement production where crushed limestones and other material burn at 1400 °C (Neville and Brooks 2010). Therefore, sole dependence on cement as binder in a concrete material has been studied, and new types of binders are introduced to minimize the environmental impact of cement for sustainable future. The new materials as concrete binder are known as Alternatives Cementitious Materials (ACMs). Among new types of binder or ACMs in concrete material are fly ash cement, geopolymer cement, blast furnace slag cement, and polymeric binders (Mueller et al. 2017).

Other than cement, natural aggregates in concrete materials are also affecting the environment because the natural aggregates resources may run out if it is continuously consumed. Aggregates variation in concrete can be considered as coarse and fine aggregates where gravels are a common natural coarse aggregate and river sand are common fine aggregate materials. These types of natural aggregates, such as gravels, granites, stones, and river sands, are widely used in concrete production, and the consumption of these aggregates has caused the increasing trend in the activity of aggregate mining process. The effects of natural aggregate mining are water pollution, sediment growth, mudflows, erosions, landslides or noise and dust due to stone blasting effects (Langer and Arbogast 2002, Ozcan et al. 2012; Ukpong 2012). Besides aggregates, fillers such as quartz or glass are also added, especially in high strength concrete (HSC) to fill the void, thus increasing the strength and density of a concrete material. As for alternative aggregates and fillers, there are many industrial by-products that have been used as alternatives for aggregates and fillers. Some of them are slag wastes from metal extraction smelting process. Other than that, recycle concretes, marble wastes, quarry wastes, recycle glasses, palm oil fuel ashes, fly ashes, and rice husks have also been introduced as alternative aggregate and filler replacement in concrete material production (Ismail et al. 2013).

According to Wang (2016), industrial waste as slags may be used as aggregate in concrete material. Therefore, as a rising concern in green material study, this article will focus on reviewing the potential of using tin slag as aggregate in concrete material for structural application. Other than by-products or slags of irons and steels production as alternative aggregates in concrete material, tin slag from tin smelting process also has the potential to become alternative aggregates in concrete material production. The reason is that tin slag waste volumes are increasing, and Malaysia, as tin producing country, is one of the tin slag wastes contributor in the world (Omar 2000). Currently, worldwide tin slags are dump on a paid land field with tight regulation as it contains naturally occurring radioactive (NOR) and heavy metal elements that could escape into underground water pool, thus polluted water resources (Sulaiman and Kamaruzaman 2016). A brief review has found that tin slag has been studied as alternative aggregates replacement in both cement concrete and polymer concrete. However, the suitability of using tin slag as aggregates in polymer concrete is higher compared to cement concrete. Therefore, this paper is intended to explore the potential and suitability to use tin slag as aggregates in polymer concrete for structural application and to identify gap in current research of tin slag polymer concrete.

2 Polymer Concrete as Replacement to Cement Concrete Material

Previous literature has shown that cement concrete has low tensile strength, low flexural strength, poor durability, high porosity, and vulnerable to acidic exposure. Therefore, to overcome the problems related to the undesirable properties of cement concrete material, new version of concrete material is needed. This finding is concluded based on Steinberg et al. (1968) and Dikeou et al. (1969) studies, which reported that by replacing water-cement with epoxy-hardener as aggregates binder, a new enhanced concrete material had been produced. These are believed to be the first articles that introduce a new composite material known as polymer concrete. Polymer concrete is a term used to represent the mixture of aggregates and polymeric binders to form a concrete material. Later in 1971, polymer concrete studies are well recognized with the establishment of design guideline and property database by the setup of American Concrete Institute (ACI) 548 Committee. For comparison, a report from Mebarkia et al. (1990) has shown that compressive strength of cement concrete is 10–60 MPa while polymer concrete has shown 40–150 MPa strength.

Polymer concrete is a composite material consists of aggregates and polymeric binders. Early development of polymer concrete uses natural resource as aggregates where gravels, crushed stones, and sands are utilized. However, today there are many types of aggregates from solid industrial wastes and recycle materials which have been utilized as natural aggregate replacement in polymer concrete as well as cement concrete alternative aggregates study that has been previously implemented. The difference is that in polymer concrete, the aggregate binder materials are polymer resin instead of cement as in cement concrete material. Polyester and epoxy resins are the most dominant binder used in polymer concrete material development. The reason is that polyester resins are relatively more economical with moderate strength, while epoxy resins pose higher strength but less economical, besides both resins are widely available compared to others.

According to Asif and Ansari (2013) and Bedi et al. (2014), polymer concrete has many advantages over cement concrete. They have rapid curing time, low cured shrinkage, excellent adhesion to most surfaces, resistance to chemicals and corrosions, excellent damping properties, low water absorbability, and also have the ability to be cast into a complex shape. Besides that, in the aspect of mechanical properties, polymer concrete also has high compressive strength, high specific stiffness, high flexural strength, high tensile strength, and high durability, which reduce the need for maintenance. Polymer concrete properties modification also has many variations compared to cement concrete. Some of them are binder content ratio, polymer hardener ratio, aggregate size distribution, aggregate grading, micro fillers characteristics, and curing condition. In fact, even with improved properties relative to cement concrete, polymer concrete strengthening is also available in many methods which have been studied in previous works where it can be internally strengthened with many types of fiber reinforcement as well as external reinforcement with metal tube or fiber-reinforced polymer confinement (Barrera et al. 2015).

3 Application of Conventional Polymer Concrete

The applications of polymer concrete are also preceded cement concrete as it poses superior properties. Due to better mechanical properties, polymer concrete can be applied as structural material to replace cement concrete structure. In term of chemical and corrosion resistance, polymer concrete is suitable in the application of acid tanks, hydraulic structures and geothermal energy process structure. Previous literature report has also shown that precast product from polymer concrete is already widely produced. Among them are manholes, drains, building claddings, drains, highway median barriers, sewer pipe linings, and overlays. Besides that, the excellent adhesion properties of polymer concrete have made it suitable to be applied as a repair material for cracked or damaged concrete structures. Because of that, polymer concrete may be classified as potential material to replace cement concrete for sustainable future, especially with the integration of industrial wastes and by-products as aggregates (Yeon 2010; Asif and Ansari 2013; Bedi et al. 2014). Figure 1 shows the available commercial application of pre-cast polymer concrete.

4 Tin Slag Polymer Concrete As Potential Structural Material

Tin slag is a by-product of tin smelting process that contributes significantly to mineral waste production in Malaysia. The volume of that waste is indicated by the refined tin production of 23,700 tons in the year 2019 based on a report by statista.com, (2019) in cooperation with World Bureau of Metal Statistics. In the smelting process, tin ore concentrates, which contain a significant amount of radioactive minerals such as monazite, zircon, and ilmenite, were used as raw materials. During processing, tin slag containing naturally occurring radionuclides (NOR) is generated as waste (Sulaiman and Kamaruzaman 2016). In addition to being a potential source of radioactive proliferation, tin slag currently presents an ominous increase in Malaysian industrial waste production. In contradistinction to wastes such as plastics and metals, disposing of tin slag seems to be economical and easy in practice, as reported by the Malaysian Nuclear Agency (MNA). This practice might increase the danger of leaching heavy elements in slag such as arsenic, lead, cadmium, and zinc to underground drinking water pools. An insightful study of inorganic pollutant presence in drinking water samples collected from all over Malaysia has already pointed out their presence (Omar 2000). All these potential hazards need the attention of policymakers and researchers to find and promote suitable alternative practices for dealing with tin slag and increase its utility.

According to the Institution of Engineers Malaysia, the government encourages the construction industry to promote the growth of sustainable development by environmentally friendly and sustainable resources (Samad et al. 2018). In addition to that, Brito et al. (2016) also reported that the introduction of recycle aggregates in concrete is important to the sustainable solution of future needs. Therefore, the introduction of tin slag polymer concrete as structural material application is intended to meet the growth of sustainable development by environmentally friendly and sustainable resources by recycling tin slag. Tin slags are industrial wastes which are also hard to be disposed because of their heavy metal element and naturally occurring radionuclides. The fundamental study of integrating tin slag in cement and polymer concrete has already proven it capability to become an alternative structural material with required properties, safely containing heavy metal element and reducing the naturally occurring radionuclides level to a safe level (Hashim et al. 2018). Other than recycling industrial waste, the consumption of tin slag as aggregates may also preserve natural aggregates such as stones, gravels, and sands. However, according to Yusof (2005), scarcity of information regarding its performance in a variety of

Fig.1 Commercial application of pre-cast polymer concrete (Yeon 2010) **a** Communications manholes, **b** Reinforced plastic composite pipe, **c** Polymer concrete manholes, **d** Installation of polymer concrete flume, **e** The installed of polymer concrete flume, **f** HIC (High Integrity Container), **g** Polymer concrete pre-slope trench

concrete mixes is the chief cause of its unpopularity. Thus, the previous studies are further reviewed to collect the information which presents the performance of TSPC as structural material.

5 Highlight on Previous Study in Tin Slag as Aggregates in Polymer Concrete

Based on the articles that have been reviewed, there are limited articles that describe TSPC in details. The literature search that has been performed has found what can be considered as the oldest Tin Slag as aggregate study, but it is not specifically for polymer concrete structures. In fact, the study is also not a published article but a thesis by Yusof (2005). In the study, Yusof has investigated the potential of using tin slag as aggregates in road pavement. The tin slags used in the study are from Malaysian Smelting Corporation, Penang Island, Malaysia. Yusof has measured the basic physical and mechanical properties of the tin slag specimens. Table 1 provides the summary of the properties while Fig. 2 shows raw tin slag particle size and shape as report by Yusof.

In terms of mechanical properties, Yusof (2005) has measured several properties such as particle density, aggregate crushing value (ACV), aggregate impact value (AIV), and 10% aggregate finest value (TFV) to assess initial potential to apply tin slag as aggregate. ACV give the relative measure of aggregate resistance to crushing under gradually applied load, while AIV gives the relative measure of aggregate resistance to impact load. Results from Table 1 indicate that tin slag has shown brittleness material property. According to BS 812–110:1990 for testing aggregate, if the ACV are >30%, AIV and TFV data must be performed to assess suitability of an aggregate. Because AIV value is <30% with 24.9%, tin slag is considered satisfactory to be applied as aggregate according to specification for highway work (SHW) standard. The overall results of the study concluded that it is possible to use tin slag in road pavements, but further studies are warranted for particular applications. However, there was no continuation found regarding Yusof (2005) study in conjunction as there was also no report where tin slag are commercially applied as aggregates in road pavement work.

Then, Sulaiman et al. (2007) have performed microforous XRF analysis on Malaysian tin slag. Tin slag sample was also collected from Penang Island. In this

Table 1 Physical and mechanical properties of tin slag (Yusof 2005)

Density (kg/mm^3)	Particle size (mm)	Particle shape	Color	Raw condition	ACV (%)	AIV (%)	TFV (kN)
3520	1.18–10	Round, Angular Elongate and Flaky	Blackish	Wet and Sticky	31.4	24.9	22.1

Fig. 2 Raw tin slag particle size and shape (Yusof 2005)

study, the intention is to observe the distribution of the elemental oxide of tin slag composition. The sample is directly analyzed without prior treatment, and the results show that among high-intensity element found in tin slag samples are Iron, Silica, Calcium, and Stannum. Before the study of tin slag as aggregates in polymer concrete, there were several studies found in using tin slag as cementitious material by Rustandi et al. (2017a, b) and Yusof (2005). The reason is that the compositions of tin slag are almost similar to ordinary OPC composition by the presence of various oxides such as calcium, silica, Alumina, Iron, Sulphur, and magnesium. However, the result shows decreasing strength with the increase of tin slag content. The conclusion states that ground tin slag is not suitable to be used as cementitious material.

Besides the study to evaluate the suitability of tin slag as cementitious material, Rustandi et al. (2017a, b) also performed another study to observe corrosion behavior of carbon steel in concrete material composed of tin slag waste in aqueous chloride solution. After that, Riastuti et al. (2018) also performed a similar study with different steel specimens and testing specification. Knowledge in corrosion resistance of reinforced steel in concrete containing TS is important as concrete use in structural material application is all steel-reinforced to produce an improved and combine property for structural load-bearing application. In Rustandi study, it is concluded that the presence of tin slag in concrete material results in more active corrosion on embedded steel. However, with the addition of 50% CaO in the concrete mixture, the corrosion rate decreases. Rustandi et al. (2017a, b) suggested that CaO are required to improve corrosion resistance of concrete material with the presence of tin slag. However, in Riastuti work, the findings are not obvious as corrosion rate of steel in the concrete specimens only show a small increase with 10% TS (2.93mpy) and 30% TS (1.81mpy) addition compared to 100% OPC (1.32mpy). In fact, with the addition of 20% TS (1.28mpy), the corrosion rate decrease compared with OPC alone. Even though both studies did not share the same finding, there are possibilities to assume that corrosion rate increase as tin slag presence in the concrete material considering different test specification and error possibility in Riastuti et al. (2018) work. These findings provide initial data to aid in strengthening design of concrete composed of tin slag by using nonmetallic material preferably.

According to available data on aggregate evaluation, tin slag may be applied as aggregate in concrete material with enough AIV (24.9%) but still not as strong as granite with twice AIV (10.4%). In term of ACV and TFV evaluation, tin slag is considered not competitive enough to be applied as aggregate in concrete material. Hashim et al. (2018) have performed a preliminary study of tin slag concrete mixture to assess its potential in terms of strength and the concrete capability to contain tin slag radiation in a safe limit which is 1 mSv/year. The strength of concrete, which composed of OPC, TS, and river sand at three different proportions, has shown acceptable strength. For normal concrete with river sand and OPC, the strength is 20 MPa, while for TS concrete with TS and cement, the strength is 20.8 MPa. But if the concrete composition is modified by mixing river sand, 20% TS and OPC, the strength increased to 24.8 MPa, which show that TS has a potential as aggregate in concrete material. In term of radiation, raw tin slag has higher naturally occurring radionuclides with the highest of 7.3 Bq/g from Th-232, 6.7 Bq/g from U-238, 3.4 Bq/g from Ra-226 and 1.2 Bq/g from Ra-228, as report by Omar 2008 in the tin slag composition. However, in term of radiation exposure, Atomic Energy Licensing Regulation (AELB, 2011) has set that the annual limit public dose is 1 mSv/year. However, this exposure is subjected to time of occupancy and distance from radiation source. Therefore, the occupancy factor for outdoor is 0.2 and for indoor is 0.8. The radiation exposure from concrete with tin slag mixture is also measured 20 and 100% TS which shown 0.8 μSv/hr and 1.5 μSv/hr. Actually, normal concrete also has naturally occurring radionuclides measured at 0.5μSv/hr. Thus, when measuring concrete with 20% tin slag, the reading shown 0.8μSv/hr equivalent to 0.3μSv/hr increment from tin slag addition. Therefore, with 20% tin slag addition in concrete mixture M30 with river sand, the radiation exposure is 0.3μSv/hr with 0.2 occupancy factor, which equivalent to 0.525 mSv/year. The results indicate that for 20% TS mixture in concrete, the radiation exposure is still below the safe limit. Based on calculation, Hashim et al. (2018) also suggested that, in term of radiation exposure, TS can be used as aggregate in concrete mixture up to 40% from overall composition.

According to Yusof (2005), during testing to evaluate tin slag potential as aggregate using ACV (31.4%), AIV(24.9%), and TFV(22.1kN), it is known that tin slags are more brittle compared to other aggregates. Therefore, using tin slag as aggregate in cement concrete seems unsuitable as both tin slags and cement are brittles. The reason is that in concrete material, aggregates and binders role are expected to produce a combined property which is higher strength with reducing brittleness, especially for a structural material application. Because of that, combining the brittleness of tin slags with a brittle-ductile property of polymer seems more suitable, and higher compressive strength is expected to be achieved. Tin slag polymer concrete research starts with the study of tin slag particle size effect on optimal mixture ratio of TSPC under compression by Faidzal et al. (2018). In the study, two parameters which are particle size and aggregate-resin ratio are investigated in their relationship with mechanical properties of TSPC. It is reported that the finest (<1 mm) size of tin slag particles with uniform graded had contributed to higher mechanical strength and modulus compared to larger sized tin slag particles. The reason is that in the finest sized tin slag particle, bond area between the aggregates particle surface and polymeric binder is higher, resulting in increased bond strength, thus better stress distribution throughout the TSPC specimen. In term of aggregate-resin ratio, the optimum ratio reports in the study are 70:30 by using unsaturated polyester as polymeric binder with 1% MEKP as initiator. Increasing resin content above optimum value reduces the TSPC strength as too much resin can form soft zones in inter-particle spaces. However, reducing resin content below optimum value can result in voids formation and not enough resin to coat the entire tin slag particle, thus reducing TSPC strength by the weakening of aggregate-resin bond strength. In Faidzal et al. (2018) study, it can be concluded that with fine (<1 mm) size tin slag aggregates and 70:30 aggregate-resin mixture ratio, the highest strength of 58.21 MPa is obtained.

Based on the study from Faidzal et al. (2018), two main conclusions are accepted, which are finer size tin slag aggregate and 70:30 aggregate-resin mixture ratio will result in highest strength TSPC. Then, Shakil and Hassan (2020) have performed a further study in TSPC by introducing two main approaches but still based on Faidzal study as reference. First is to formulate gap graded using tin slag aggregates with mixture of coarse and fine size particles. This approach is expected to increase the compressive strength of the TSPC by the interlocking action between fine and coarse aggregates during load bearing application. The concept of cement concrete strengthening by gap aggregate grading is introduce using tin slag aggregates. Conventionally, in cement concrete strengthening by gap grading aggregates, gravels are used as coarse aggregate, and river sand is used as fine aggregate to strengthen the concrete material. Second approach in Shakil study is by introducing fiber-reinforced polymer (FRP) confinement which is already proven its capability to increase the strength of other commercial concrete structure includes cement and polymer concrete. FRP strengthening is preferred because, based on Rustandi et al. (2017a, b) study, it is concluded that steel reinforcement strengthening in concrete composed of tin slag can increase the corrosion rate of the steel. Therefore, external FRP strengthening of TSPC is introduced by Shakil instead of steel reinforcement.

In Shakil and Hassan (2020) study, the researcher has performed an experiment to analyze the compression behavior of TSPC under one and two-layer glass fiber reinforced polymer (GFRP) and carbon fiber-reinforced polymer (GFRP) confinement. Upon compressive test on unconfined TSPC specimen, the highest strength is specimen with 4 mm + 2 mm tin slag particle size with 37.71 MPa. The highest strength of gap graded TSPC specimen in Shakil experiment are less than Faidzal et al. (2018) findings with 58.21 MPa. It can be concluded that uniformly graded tin slag aggregate with finer particle size has better performance compared to gap graded aggregates. Therefore, after evaluating TSPC study for structural application up to Shakil study, it can be summarized that uniform graded with finer size tin slag particle aggregates and CFRP confinement has shown the best performance in term of compressive strength.

After the study by Shakil, Hassan et al. (2020) have reported another study regarding compressive behavior of TSPC confined with CFRP. In the study, Hassan has used the optimum specification for TSPC specimen from previous studies by Faidzal and Shakil to further investigate the potential of TSPC specimen as a structural material. Therefore, Hassan has cast the TSPC specimen with fine size (<1 mm) aggregates and mixing 70 part of the aggregate to 30 part of polyester resin to form the optimum strength TSPC short column specimen as report in the literature. In addition to that, by referring to Shakil study, which reports that FRP confinement using CFRP has increased the specimen strength compared to other variation as well as controlled TSPC specimen, Hassan has re-applied the FRP confinement strengthening to the specimen using CFRP with addition of 3 ply confinement. In this experimental study, the results show that maximum compressive stress significantly increases to 103% from unconfined to 1 ply of CFRP confinement of TSPC, meanwhile for 1 ply to 2 ply only increase 14.9% and same case for 2 ply to 3 ply only gain about 13.6%. Confinement effectiveness increases twice for 1 ply confined TSPC from the unconfined TSPC, but further confinement only shows a slight increase in the TSPC specimen strength.

6 Discussion and Suggestion

According to the literature, TSPC previous study is first initiated by Faidzal et al. (2018), followed by Shakil and Hassan (2020), and the most recent by Hassan et al. (2020). Because of that, TSPC is considered new material and there was a wide area to be explored to further investigate the potential of TSPC by referring to the research trend in conventional PC study. The gap of this study based on available recent findings are identified and summarized in Fig. 3.

To help in predicting the structural performance, FEM of conventional OPC and PC structures are widely available. FEM in simulation software may also facilitate the optimizing process of a concrete structure quickly and at a lower cost compared to experimental process. In the case of TSPC, there was no available FEM found in the previous literature, which make the process of TSPC modeling become troublesome as there were no baseline parameters to be referred. Therefore, there was a potential

Previous study in Tin Slag as Aggregate		
Study	Investigation Focus	Findings
Hashim et al. (2018)	Mixture of tin slag in cement concrete as fine aggregate	Very little strength improvement (24.8MPa)
Faizal et al. (2018)	Introduce TSPC with particle size variation	Uniform fine TS particle increased strength (58.21MPa)
Shakil et al. (2020)	Introduce gap-graded TSPC and FRP Confinement	Uniform graded better than gap-graded but confinement enhanced strength (118.42 Mpa)
Hassan et al. (2020)	Introduced multilayers CFRP confinement effect on TSPC	Increased CFRP confinement layers enhanced strength, with 3 ply CFRP achieve maximum strength (156.88 MPa)

Research gap in TSPC study	
Study	Potential study focus
FEM Model of TSPC under compression	FEM model of cement concrete available in literature but no literature found on TSPC. FEM can help predict strength performance.
Other FRP material confinement	Previous study has perform GFRP and CFRP confinement on TSPC. There was no available literature found on other FRP material (BFRP and AFRP)
Metallic material confinement on TSPC	In cement concrete external strengthening, metallic material has improve strength drastically. So, there was potential to apply steel tube confinement on TSPC to investigate the response.
Effect of different FRP orientation on TSPC	In cement concrete structure, external strengthening using FRP material in spiral orientation has enhanced the compressive strength compared to confinement in hoop direction. So there was potential to perform spiral FRP wrapping on TSPC specimen.
Effect of other resin of TSPC	In previous study, TSPC is prepared using Polyester resin only. So, there was potential to investigate effect of other type of resins consumption on TSPC strength.

Fig. 3 Research gap in TSPC study

study in developing a finite element model of TSPC by focusing on compression behavior.

Previous study on cement concrete structures has reported the confinement of concrete structure using aramid fiber (Yang et al. 2015), basalt fiber (Thorhallsson and Snaebjornsson 2016; Nabi et al. 2019; Yang et al. 2018), and steel tube (Kedziora and Anwaar 2019; Tao et al. 2013). All of these studies have provided new fundamental data contributing toward the concrete structure design for strengthening purpose. Thus, by referring to these trends in concrete strengthening research, confinement material of TSPC specimen may also be explored by using other materials such as aramid fiber, basalt fiber, and steel tube confinement. Therefore, to start a study on the effect of other material confinement on TSPC structure is a novel way to contribute in TSPC research community worldwide. Besides that, hybrid confinement has also increase the concrete strength base on study by Xu et al. (2018) and Chellapandian et al. (2017). Therefore, in TSPC confinement study, hybrid confinement study may be initiated by focusing on GFRP and CFRP hybrid confinement to understand TSPC behavior under compression further.

According to similar study on cylindrical concrete column strengthening using FRP confinement by Cheng et al. (2014), spiral confinement in angular direction of 0.6° has shown an increase in compressive strength. The study begins by observing that the measured rupture strain of FRP jackets, which is considerably lower than the rupture strain obtained from the flat coupon test. The reason is that lateral strain expansions are not uniform throughout the entire cylindrical concrete column specimen upon uniaxial load application. The highest lateral strain expansions are in the middle of the specimen and then gradually decrease in both directions toward the top and bottom of the specimen which reduces the confinement effectiveness of horizontal confinement. Based on this study, spiral wrapping on TSPC column specimen may also be introduced to investigate its potential in strengthening of TSPC specimen.

In polymer composite materials, other than polyester resins, there are other resins which better in strength such as epoxy resins (Bedi et al. 2014). Therefore, epoxy resins may be used in preparing TSPC specimen to observe the specimen behavior under compression. Besides that, resins curing condition such as hardener ratio, temperature, curing periods, and post curing treatment may also be further studied to be applied to enhance TSPC specimen strength as these parameters increased bond strength between aggregates and polymeric binders. The findings in previous literatures regarding bond strength increment between aggregates and resin may be applied in preparing TSPC specimen with improved properties thus increasing the potential o f TSPC application as structural material.

7 Conclusion

The introduction of TSPC as green structural material replacement for natural aggregates cement concretes are proven as it ensure no emission during production of its resins compared with cement production. Furthermore, using tin slag as aggregates will preserve natural aggregates and may reduce quarrying activity for gravels, stones or stones that could affecting natural ecosystem. From the literature, researches regarding the utilization of tin slag as aggregates in polymer concrete for structural application are still in the first stage. Current state of TSPC for optimum compressive strength can be concluded as 30:70 resins to aggregate ratio with uniformly graded fine aggregates. In addition to that, the CFRP confinement layers are also proportional with compressive strength gain. The gap in this study is widely applicable in exploring other structural strengthening potential using experimental study by referring to the chronology of strengthening study that has been experienced by cement concrete and conventional polymer concrete. Among them are other material confinement, hybrid confinement, wrapping with different fiber orientation, other polymeric resins application, and optimum curing condition. Finally, based on current report, it can be briefly state that TSPC has shown its potential in becoming green material for structural application.

Acknowledgements This review is to explore gap and research potential in TSPC study in the aspect of external strengthening to be applied as structural material. The funding of this research comes from a collaborative research grant from University Technology Malaysia (CRG UTM—Tin Slag) under Vot. CRG: RDU192311 and also from Postgrad Research Scheme 2021(PGRS), UMP under Vot. Project PGRS210339.

References

Asif A, Ansari A (2013) Polymer concrete as innovative material for development of sustainable architecture. In: 2nd international conference on emerging trends in engineering & technology

Barrera GM, Santiago EV, Lahri A, Dixit D (2015) Alternatives to cement in concrete–a review. Int J Sci Eng Res 6(10):50. ISSN 2229-5518

Bedi R, Chandra R, Singh SP (2014) Reviewing some properties of polymer concrete. Indian Concrete J 88(8):47–68

Brito JD, Ferreira JDS, Pacheco J, Soares D, Guerreiro M (2016) Structural, material, mechanical and durability properties and behaviour of recycled aggregates concrete. J Build Eng 6:1–16

Chellapandian M, Prakash SS, Rajagopal A (2017) Analytical and finite element studies on hybrid FRP strengthened RC column elements under axial and eccentric compression. Compos Struct 184:234–248

Cheng JF, Peng SC, Deng J, Chen GM, Tao Y (2014) A technique for reducing strain concentrations in FRP wrapped concrete columns. In: Proceedings of the 7th International conference on FRP composites in civil engineering (CICE 2014)

Dikeou JT, Kukacka LE, Backstrom JE, Steinberg M (1969) Polymerization makes tougher concrete. ACI J Proc 66(10):829–839

Faidzal MMY, Hassan SA, Omar B, Zakaria K, Zaharuddin MFA (2018) Particle size effect on optimal mixture ratio of tin slag polymer concrete under compression. J Built Environ Technol Eng 5. ISSN 0128-1003

Hashim MJ, Mansor I, Ismail MP, Sani S, Azmi A, Sayuti S, Ibrahim MZ, Anuar AA, Rahim AAA (2018) Preliminary study of tin slag concrete mixture. In: Agency nuclear Malaysia: engineering division, industry technology division and waste technology development centre. IOP Conf. Series: Materials Science and Engineering, vol 298, 012014p

Hassan SA, Hanan UA, Yahya MY, Wahit MU (2020) Behaviour of tin slag polymer concrete confined with carbon fibre reinforced epoxy. J Penelitian dan Karya Ilmiah Lembaga Penelitian Universitas Trisakti 5(2). ISSN: 0853-7720, ISSN (e): 2541-4275

Hosamo H, Sarwari P, Aljoulbek MO (2019) Comparative study of concrete and steel buildings in China. Technical Report

Ismail S, Hoe KW, Ramli M (2013) Sustainable aggregates: the potential and challenge for natural resources conservation. In: Proceeding of AMER international conference on quality of life, pp 100–109

Kedziora S, Anwaar MO (2019) Concrete-filled steel tubular (CFTS) columns subjected to eccentric compressive load. In: AIP conference proceedings 2060, 020004

Lahri A, Dixit D (2015) Alternatives to cement in concrete—a review. Int J Sci Eng Res 6(10):50. ISSN 2229-5518

Langer WH, Arbogast BF (2002) Environmental impacts of mining natural aggregate. In: Deposit and geoenvironmental models for resource exploitation and environmental security, pp 151–169. U.S. Government

Mebarkia S, Vipulanandan C (1990) Aggregates, fibers and coupling agent in polyester PC. In: Materials engineering congress, ASCE Denver, Colorado, vol 2

Mueller HS, Haist M, Moffat JS, Vogal M (2017) Design, structural properties and structural performance of sustainable concrete. Sustain Civil Eng Struct Constr Mater SCESCM 2016):22–32
Nabi P, Petkova D, Donchev T, Costas G, Khorsanditalab MR (2019) Behaviour of BFRP reinforced concrete element under axial loading. J Fiber Reinf Polym Reinf Concr Struct:14
Neville AM, Brooks JJ (2010) Concrete technology, 2nd edn. Pearson Education Limited, England UK. ISBN: 9780273732198
Omar M (2000) NORM waste management in Malaysia. In: Proceeding of international conference on the safety of radioactive waste management.IAEA spain, pp 89–92
Ozcan O, Musaoglu N, Seker DZ (2012) Environmental impact analysis of quarrying activities established on and near a river bed by using remotely sensed data. Fresenius Environ Bull 21(11)
Riastuti R, Cahyadi A, Pratesa Y, Siallagan ST (2018) The study of corrosion resistance of reinforcement steel embedded in concrete composed of commercial portland cement and final tin slag against chloride environment. In: International conference on concrete engineering and technology, CONCET, 431(5)
Rustandi A, Nawawi FW, Pratesa Y, Cahyadi A (2017a) Evaluation of the suitability of tin slag in cementitious materials: mechanical properties and leaching behavior. In: International conference on chemistry and material science (IC2MS) 2017, vol 299, 012046p
Rustandi A, Cahyadi A, Sonia ST, Nawawi FW, Pratesa Y (2017b) Corrosion behavior of carbon steel in concrete material composed of tin slag waste in aqueous chloride solution. In: International conference on chemistry and material science (IC2MS) 2017, vol 299, 012047p
Samad AAA, Theng CWY, Ching TE, Mohamad N, Tambichik MA, Bosro MZM (2018) Effect of palm oil fiber (POF) to strength properties and fracture energy of green concrete. J Instit Eng Malaysia 79(1)
Shakil UA, Hassan SA (2020) Behavior and properties of tin slag polyester polymer concrete confined with FRP composites under compression. J Mech Behavior Mater 29(1)
Steinberg M, Dikeou JT, Kukacka LE, Backstrom JE, Colombo P, Rubenstein S, Kelsch JJ, Manowitz B (1968) Concrete-polymer materials—first topical report, BNL 50134 (T-509) and USSR General Report No. 41. Brookhaven National Laboratory/U.S. Bureau of Reclamation, Denver, 83p
Sulaiman MYM, Muslimin M, Amin L (2007) Microfocus XRF analysis of Malaysian tin slag. J Nuclear Relat Technol 4:227–232
Sulaiman MYM, Kamaruzaman NS (2016) Standardless EDXRF analysis methods of U and Th in Malaysian tin slag waste. Mater Sci Forum 840:410–415, Trans TechPublications
Tao Z, Wang ZB, Yu Q (2013) Finite element modelling of concrete-filled steel stub columns under axial compression. J Constru Steel Res 89:121–131
Thorhallsson ER, Snaebjornsson JT (2016) Basalt fibers as new material for reinforcement and confinement of concrete. J Solid State Phenomena 249:79–84. ISBN 978-80-906097-0-9
Ueda T (2014) Material conditions necessary for strengthening concrete structures. In: Proceeding of 2nd international conference on sustainable civil engineering structures and construction materials 2014 (SCESCM 2014) (pp 321–334)
Ukpong EC (2012) Environmental impact of aggregate mining by crush rock industries in Akamkpa local government area of cross river state. Nigerian J Technol (NIJOTECH) 31(2):128–138. ISSN 1115-8443
Wang GC (2016) Slag use as an aggregate in concrete and cement-based materials. In: The utilization of slag in civil infrastructure construction (pp 239–274). Woodhead Publishing Series in Civil and Structural Engineering
Xu L, Wei C, Li B (2018) Damage evolution of steel-polypropylene hybrid fiber reinforced concrete: experimental and numerical investigation. Adv Mater Sci Eng:23 (Article ID 1719427)
Yang H, Song H, Zhang S (2015) Experimental investigation of the behavior of aramid fiber reinforced polymer confined concrete subjected to high strain-rate compression. J Constru Build Mater 95:143–151

Yang LV, Wu X, Zhu Y, Liang X, Cheng Q, Gao M (2018) Compression behavior of basalt fiber-reinforced polymer tube-confined coconut fiber-reinforced concrete. J Adv Mater Sci Eng 2018:10
Yeon KS (2010) Polymer concrete as construction materials. Int J Soc Mater Eng Resour 17(2)
Yusof MAW (2005) Investigating the potential for incorporating tin slag in road pavements, Ph.D. Thesis. University of Nottingham, School of Civil Engineering, USA

The Effect of Waste Paper Sludge Ash Addition to the Fresh and Hardened Properties of Ultra-High-Performance Concrete

Muhamad N. A. Kamaruddin, Adiza Jamadin, Sakhiah Abdul Kudus, and Nur Kamaliah Mustaffa

Abstract Waste paper sludge ash (WPSA) is a waste material that came from the paper industry. WPSA is used as a cement replacement in producing concrete and was investigated on its chemical, physical, and mechanical properties. The increase in air pollution and environmental problem due to the construction material has limited the use of natural resources now. Therefore, the WPSA became a new innovation material that can be used for the construction industry to support the green technology due to less presence of sulfate at only 0.57% of the total weight. Carbon dioxide (CO_2) and sulfur dioxide emissions also can be reduced since less cement productivity is involved. The objective for this study is to identify the workability of ultra-high-performance concrete (UHPC) with three different percentages of WPSA by conducting the slump test and also to identify the compressive strength and the moisture content between the WPSA and UHPC. Result shows that the maximum compressive strength measured was at 5% more than the reference mix at 28 days.

Keywords Compressive strength · Workability · Waste paper sludge ash · Ultra-high-performance concrete · Green material

1 Introduction

Concrete which is the most frequently used material in the construction industry and is widely used not only in Malaysia, but worldwide. Building towers, highways, and bridges, for instance, are concrete buildings that can be seen in everyday life. Usually, concrete mixture consists of cement, coarse aggregate, fine aggregate, and

M. N. A. Kamaruddin · A. Jamadin (✉) · S. A. Kudus · N. K. Mustaffa
School of Civil Engineering, College of Engineering, Universiti Teknologi MARA Shah Alam, 40450 Shah Alam, Selangor, Malaysia
e-mail: adiza@uitm.edu.my

S. A. Kudus
e-mail: sakhiah@uitm.edu.my

N. K. Mustaffa
e-mail: nurkamaliah@uitm.edu.my

R. Hassan et al. (eds.), *Green Infrastructure*,
https://doi.org/10.1007/978-981-16-6383-3_5

water. Hard concrete will form the mixture of these materials. The development of Ultra-high-performance concrete (UHPC) has begun during the 90s. Use of UHPC in the developing world beats the issue of restricted burden and sturdiness properties of typical concrete and high strength cement (HSC). Issue emerges utilizing normal cement and HSC when applied to exceptionally extraordinary conditions and remarkable plan structures. Fortification, dam, atomic plant, and other extraordinary structures must be planned with uncommon kind of concrete which can oppose load more than 100 MPa and are strong when open to those outrageous conditions. This is the place UHPC can give the solutions.

To make concrete production cost-effective, the best approach is probably to use waste materials instead of natural resources. Based on research by Ahmad et al. (2013), the paper and board industries, paper plant sludge is a significant financial and ecological issue. A large quantity of recycle sludge is being produced worldwide. In India, 0.7% of total residential waste generated includes paper debris. The UK provides more than 1.5 thousand tons of recycling paper slop every year. For the paper and board industry, paper factory sludge is an important economic and natural issue. The material is the product of paper being de-increased and re-pulped. The paper sludge reuse and conversion courses are land-spreading as rural compost, burning at the paper factory in combined heat and power (CHP) plants, transporting paper sludge ash, or landfill transfer. The humidity content is normally up to 40%.

Sani et al. (2011) reported that Malaysia Newsprint Industry (MNI) generates about 80 tons of burned wastepaper effluent a day. The largest component of waste paper effluent derives from tree canning and paper manufacturing in which about 80 tons of paper waste is dumped regularly to a Pahang wasteland. As population expansion and request increase significantly, existing waste activity continues to increase resulting in environmental pollution problems. It is important to solve the problem of waste management and reduce the cost, reuse those resources and develop them through commercial products, and that had been in line with the efforts of the Malaysian community to improve green modern economy development. Therefore, it is necessary to improve the use of WPSA. Furthermore, it is not only possible to avoid the environmental impacts by using WPSA as a supplementary material, but also this study proposed on how the use of WPSA paper at specific percentages can bring advantages in terms of the characteristic of ultra-high-performance concrete as the standard concrete mix design.

This study is aimed to determine the effect of WPSA in percentage to the UHPC. The concrete was tested on the level of its strength and hardness criteria to maintain the quality of concrete. The objective was to guarantee that concrete containing WPSA could be used in Malaysia's construction projects and to obtain the optimum design mix of UHPC concrete added with WPSA. The percentages of WPSA admixture were handled within the range of 5, 10, and 15% of additional material. Concrete cube that been taken out from cube test will be weighed before and after the drying process. After that, this concrete is tested by mechanical and hardness properties. The study included focus on experimental work and analytical investigation where the concrete designs based on concrete mix design are properly cured until the compressive strength exceeds 100 MPa and above. Furthermore, the compression experiment will test nine reference concrete cubes without any internal or external confinement with dimension of 100 × 100 × 100 mm. This study includes analysis on the cube sample for compressive strength where the concrete cube that contains different percentages of WPSA (5, 10, and 15%) with dimensions of 100 × 100 × 100 mm will be tested under compression test.

2 Literature Review

Paper mill sludge consists of mineral fillers, small cellulose fibers, water, inorganic salts, and organic matter. Paper sludge ash is generated from the paper production process by burning waste paper sludge. Mirza et al. (2015) found the pH range was thoroughly studied from an almost neutral level to around × 12 based on its structure, and the particle density was approximately × 2.2 to 2.9 g/cm^3, the optimum dry density was × 0.65 to 0.95 g/cm^3. Monosi et al. (2010) found that the paper ash bulk density was found to be equivalent to 1200 kg/m^3 and was lighter than normal sand, predictably between × 2500 and 2600 kg/m^3 and 2100–3100 kg/m^3 binders, generally used in mortar production. Tay (1987) found that sludge ash's normal specific gravity is 2.81. Ahmad et al. (2013) found the paper sludge ash's specific gravity is 2.6. Khanbilvardi et al. (1995) described the sludge debris is inorganic, 28 percent water content, and 1.83 bulk specific gravity. Ishimoto et al. (2000) examined the paper sludge containing 60% water and 40% solids. Nevertheless, solids produce 30% ash, others are beginning to lose. Liaw et al. (1998) observed 75.40% average moisture content in paper sludge and 70.11% ignition loss and 19.63% ash start misfortune.

Lime and silica are the main constituents found in WPSA. The other major elements are small in quantities which is less than 2%, with the exception of being 4%. It also includes trioxide of aluminum, ferrous trioxide, magnesium oxide, sulfate, potassium oxide, etc. Dewatering at low temperatures which is less than 200 °C and disposal at high temperatures will reduce the volume weight. Organic compounds are burned at temperatures of around 350 to × 500 °C during combustion, while mineral fillers and inorganic salts at higher temperatures, which is more than 800 °C, are converted into the corresponding oxides. The most common oxides in Paper Sludge Ash (PSA) are CaO, Al_2O_3, MgO, and SiO_2. The most damaging factor is the enlargement due to the hydration of CaO to Ca_xOH_2, which happens when it occurs after settling in unsoundness. The rapid hydration of lime would give the solution high alkalinity, which in the case of WPSA could be used to produce pozzolanic materials or ground granulated blast furnace slag (GGBFS).

The significant factor in developing UHPC is to enhance the micro and macro properties of its mixture design to ensure mechanical homogeneity, optimum density of particle packing, and minimum fault size (Schmidt et al., 2005). For ultra-high-performance concrete (UHPC), a high percentage of cement is used compared to normal-strength (NS) and high-performance concrete (HPC) (Schmidt et al., 2005; Ghafari et al., 2015). The rise in cement capacity has been observed to increase UHPC compressive strength, however, compressive strength appears to decline beyond the ideal cement content around × 1700 kg/m^3 due to limited aggregate participation (Talebinejad et al., 2004). Cement of moderate Blainexfineness and tri-calcium aluminum content is favored by less than 6 percent due to its lower water demand (Wille et al., 2011). Also used for the development of UHPC were special micro-fine cements with particle size smaller than standard Portland cement (Strunge et al., 2008).

Due to UHPC's very low water/binder ratio, only part of the total cement hydrates and cement anhydrate can be replaced with crushed quartz, fly ash, or blast furnace slag. For example, up to 30, 36, and 40% by volume of cement in UHPC mixtures can be substituted separately with squashed quartz, impact heater slag, or fly ash, without negotiating compressive strength (Ma and Schneider 2002; Soutsos et al. 2005). Using silica fume as a binder can also enhance UHPC's usability by filling voids between coarser particles due to a much finer particle size and ideal spherical shape. In addition to this, micro filler impact, through its pozzolanic reactions, silica fume also enhances UHPC's strength properties (Ma and Schneider 2002; Richard and Cheyrezy 1995). Various studies (Ma and Schneider 2002; Matte and Moranvillex 1999) have prescribed silica fume calculations of 20–30% of the total binder material to achieve denser particle packing and pozzolanic reactivity in UHPC, resulting in higher resistance properties. For example, UHPC suggested an optimal dosage of 25% by cement weight of low carbon content 0.5% silica fume (Wille et al., 2011).

For UHPC mixtures, a very low water/binder ratio (w/b) is used. Richard and Cheyrezy (1995) accounted for the lowest w/b of 0.08; in any case, this ratio did not guarantee thick molecule pressing. In past studies (Richard and Cheyrezy 1995; Larrard and Sedranx 1994; Gao et al. 2006; Wenyux et al. 2004; Shix et al. 2015), an optimal w/b ratio of 0.13–0.20 has been suggested in order to achieve optimum relative density and distributed flow. Nevertheless, researchers (Wille et al. 2011; Drollx 2004) also accomplished compressive strength by using 0.25 w/b above 150 MPa. It therefore appears to be claimed that w/b is not the main force controlling UHPC factor. Other important parameters are the properties of mixing materials, mixing procedures, and the form of mixer.

The reduced workability of UHPC can be solved by the addition of effective super plasticizers (SP). Compatibility between the mixture ingredients and the form of SP used relies entirely on the correct SP measurements. Improved compatibility may result in lower measurements of SP. In addition, gradual or delayed addition of SP instead of adding the SP at once has been found to strengthen the workability of UHPC mixtures due to an improved dispersing effect (Tue et al. 2008). Various studies (Schmidt et al. 2004, 2012; Fehling et al. 2008) used cement-weighing SP measurements to improve the functionality of UHPC mixtures somewhere between 1 and 8 percent. SP measurements by cement weight of 1.4–2.4% are generally recommended (Wille et al. 2011).

3 Methodology

Using the required grading curves, the size of fine aggregates for concrete mixing can be calculated. Nevertheless, in comparison to the limit defined in BS 882, this method requires the use of fine aggregates in a small range of grading. Table 1 shows the fine aggregate grading limits. Kong et al. (1987) reported that the use of fine aggregate for the blending of a size 100 × 100 × 100 mm concrete cube specimen shall have the gradation within the M grading range. Nevertheless, given the difficulty of identifying the fine aggregates that have limited grading, some researchers only suggest the use of fine aggregates which pass through the sieve test of 600 μm. Nevertheless, in this report, based on BS 410, the natural sand that passed through 5 mm sieve and retained at 600 μm is used. To protect it from humidity, the sand is contained in a dry and airtight jar. The fine aggregate's fineness component was found to be 2.75 by taking a 1000 g dry sand test. The fine aggregate is filtered into sieve sizes of 4.75, 2.36, 1.18, 0.6, 0.3, and 0.15 mm. Finally, by adding all the cumulative percentage values and dividing it by 100, the value of the fineness module can be obtained.

The coarse aggregates are separated by three nominal maximum sizes, which are 40, 20, and × 10 mm, based on the BS 882:1992. Suitable selection of optimum coarse aggregate size is essential in concrete mix design as it affects concrete strength. In addition, the coarse aggregate should be cleaned and free of chemicals that can lead to concrete degradation. Crushed stones are typically coarse aggregates used for concrete mixing. In addition, coarse aggregate grading in compliance with BS 882 allows the percentage size of coarse aggregates to be uniform to the maximum limits as shown in Table 2. Based on the BS 882, the grading limit and total size of coarse aggregates are calculated because the aggregates could influence the water and cement needs in the concrete mix design, hence affecting the concrete strength. However, this could be avoided by choosing the water-to-cement ratio correctly, which will give a wide range of grading that can be used without affecting the

Table 1 Grading limits for fine aggregates (BS 882: 1992)

Sieve size	Percentage by mass passing BS sieve			
	Overall time	Additional limit of grading		
		Coarse	Medium	Fine
10.0 mm	100	–	–	–
5.00 mm	89–100	–	–	–
2.36 mm	60–100	60–100	65–100	80–100
1.18 mm	30–100	30–90	45–100	70–100
600 um	15–100	15–54	25–80	55–100
300 um	5–70	5–40	5–48	5–70
150 um	0–15	–	–	–

Table 2 Grading limits for coarse aggregates (BS 882: 1992)

Standard sieve (mm)	Percentage by weight passing the standard sieves		
	Nominal size of aggregate		
	40–5 mm	20–5 mm	14–5 mm
20.0	100	–	–
37.5	90–100	100	–
20	35–70	90–100	100
14	–	–	90–100
10	10–40	30–60	50–85
5	0–5	0–10	0–10

concrete strength. Marsh (1988) mentioned that the maximum size of the coarse aggregate is commonly taken as 20 mm–5 mm.

The coarse aggregates included in this analysis are those retained on a 9 mm sieve and those passing through a 20 mm sieve which provides a range of coarse aggregate sizes from × 10 to × 20 mm. The fineness module gained for the coarse aggregates is 7.17 by sieving 5000 g dry weight sample and sieving it through the sizes of 80, 40, 20, 10, 4.75, 2.36, 1.18, 0.6, 0.3, and 0.15 mm. The value of the modulus of fineness shows that the average size of the sample of coarse aggregates falls between 10 and 20 mm.

3.1 Mix Design

Concrete mix design commonly uses dimensions or ratios such as 1:2:4, respectively, representing the proportions of cement, fine aggregates, and coarse aggregates. The ratio is either determined by weight or volume. Although in terms of its simplicity of expression, this concrete mix design system has an advantage. However, it will cause a disruption when describing the effect of the mixing measurement on the concrete features. This is related to the importance in defining the amount of cement needed to cast a given concrete size. Thus, using the typical mix concrete layout sheet to measure the concrete mixing percentage is the most ideal way to specify mixing parameters for each individual material engaged in concrete mixing in terms of volume. Moreover, by measuring specifically the mixing percentages required for concrete mixing, the wastage of materials typically occurring will be minimized (Marsh 1988). The reason is that the mix design sheet can produce one meter cubic of concrete for each individual material, namely, concrete, water, fine aggregate, and coarse aggregate, with designated quantities in terms of mass. The design mix proportion for this study is depicted in Table 3. Upon continuing to the casting process, the fresh concrete is then tested to assess its workability using the slump test. In addition, the vibrating table is being used to vibrate the mold to ensure that

Table 3 Design mix proportion

Casting mix design UHPC control
Size: 100 × 100 × 100 mm
Total cube: 12

Cement (800 kg /m^3)	Fine 433 (kg /m^3)	Coarse (800 kg/m^3)	Water (160 kg/m^3)	Admixture (16 kg/m^3)	WPSA 0%
7.92 kg	4.29 kg	7.92 kg	1.58 kg	0.16 kg	–

Casting mix design UHPC 5% WPSA
Size: 100 × 100 × 100 mm
Total cube: 12

Cement (800 kg /m^3)	Fine 433 (kg /m^3)	Coarse (800 kg/m^3)	Water (160 kg/m^3)	Admixture (16 kg/m^3)	WPSA 5%
9.36 kg	5.07 kg	9.36 kg	1.87 kg	0.19 kg	0.47 kg

Casting mix design UHPC 10% WPSA
Size: 100 × 100 × 100 mm
Total cube: 12

Cement (800 kg /m^3)	Fine 433 (kg /m^3)	Coarse (800 kg/m^3)	Water (160 kg/m^3)	Admixture (16 kg/m^3)	WPSA 10%
9.36 kg	5.07 kg	9.36 kg	1.87 kg	0.19 kg	0.94 kg

Casting mix design UHPC 15% WPSA
Size: 100 × 100 × 100 mm
Total cube: 12

Cement (800 kg /m^3)	Fine 433 (kg /m^3)	Coarse (800 kg/m^3)	Water (160 kg/m^3)	Admixture (16 kg/m^3)	WPSA 15%
9.36 kg	5.07 kg	9.36 kg	1.87 kg	0.19 kg	1.40 kg

the concrete is evenly compacted while minimizing the air void produced in the concrete.

3.2 Slump Test

Slump test is a guide for measuring the strength of the concrete, which is the degree to which the concrete was either too wet or too dry. The mix must be adjusted to achieve greater consistency by changing the quantities of sand and coarse aggregate to correct admixture. However, there can be no change to the total amount of water used in the mix specified for each bag of cement. Commonly, a 300 mm tall concrete cone is prepared and allocated for settling in compliance with BS 1881: Part 102: 1983, therefore increasing the concrete cone's height is used as a measure of the concrete's consistency. The basic method for performing the slump test is to fill the cone with concrete mix in three layers with a steel rod cylinder at each layer tamped for 25 cycles. After that, raise the cone up vertically immediately and set aside the cone mold next to the cement. At last, the slump can be calculated by the type of slump that is the real slump, shear slump, or collapse slump that has been obtained. The equipment for slump test are slump cone, tamping rod, slump test base, scoop, scrub brushes, slump cone filing funnel, trowels, measuring tape.

3.3 Compressive Strength Test

The concrete cube test's compressive strength gives an overview of all the concrete characteristics. A concrete cube is a test cube of 100 × 100 × 100 × 100 mm per dimension. The cube form of the mold is filled with cement used in the works. The cube cast is cured for 3, 14, and 28 days and the cube's compressive strength is overcome by measuring it on the compressive testing machine at the end of 3 days, 14 days, and 28 days. The strength of the concrete used in works defines the compressive strength. The concrete cube's compressive strength complies with BS 1881–116: 1983 and is taken as the total compressive load per unit area that it can withstand.

3.4 Water Absorbtion Test

The water absorption test is a way to measure the consistency of the concrete that plays a key role in monitoring the concrete's reliability. It is in accordance with the BS 1881–122: 1983. The permeability is the control measure which influences the quantity of water to flow through the microstructure of the concrete. The concrete cube sample containing different percentage quantity of waste sludge paper is set to dry until testing is started by placing it under the sunlight until it is dry and weighs as W_1. In addition, the sample must be weighted as W_2 before the cube is set to dry. The water absorption of the specimen can be derived from the given equation:

$\text{Water Absorption } (\%) = [(W_2 - W_1)/W_2] * 100$

W_1 = Mass of dried specimen (g)

W_2 = Mass of wet specimen (g)

4 Results and Discussion

Based on the main objectives of this study, the experimental work is categorized into two, which are fresh concrete test and hardened test. The testing for hardened concrete test involved compressive strength test and determines the moisture content of the cube sample, while for fresh hardened concrete test involved the slump test only. The analyses of the results started with the slump test with the purpose of evaluating the workability of concrete containing WPSA. Besides that, the evaluation in terms of strength performance of the concrete are also analyzed to get the optimum

Table 4 Result of slump value

Percentage of waste paper sludge ash as an additional material	Slump value (mm)	Degree of workability
0% (control)	250	Very high
5%	280	Very high
10%	230	Very high
15%	230	Very high

percentage of WPSA as an additional material. The results presented in this chapter are the average results where more than one sample for each mixture were tested.

4.1 Slump Test

The slump test was performed in accordance with the BS 1881: Part 102: 1983 to investigate the degree of workability of a fresh concrete containing different percentages of WPSA. The results of slump test on fresh concrete UHPC of different mix proportions are displayed in Table 4. The results obtained show that as the addition material of WPSA, it will result the increasing of the workability of the concrete at 5% and then the slump decreased with the increase in WPSA content. The workability is defined as high degree of workability when the slump value is higher than 100 mm. WPSA particles absorbed more water as compared to cement and thus decreasing the workability of concrete mix. Slump was maximum for the concrete mixture containing 5% WPSA in place of cement. Although the workability of the concrete gradually decreased, the slump value for all the mix proportions still has a very high workability because of the physical properties of UHPC concrete. Based on study conducted by Sajad et al. (2013), the percentage water absorption increased with increase in WPSA content where the lowest value of water absorption was found for concrete mix with 5% WPSA content. In conclusion, the presence of WPSA in the concrete will increase the water absorption, therefore it will affect the workability of the slump test value.

4.2 Water Absorption

The result of the water absorption is illustrated in Tables 5, 6, 7 and 8. Water absorption tests were carried out at 3, 7, 14, and 28 days. An increase in water absorption was observed with increase in WPSA. The increment of water absorption capacity of the concrete will lead to a lesser durability property due to high porosity. The higher porosity will affect the concrete surface to weathering action, abrasion from traffic loads, and chemical attack more dominantly. The study from Sani et al. (2011) stated

Table 5 The result of water absorption for day 3

Sample (%)	Day 3			
	mass of dried specimen (kg)	Mass of wet specimen (kg)	Water absorption (%)	Avg of water absorption (%)
0	2.55	2.65	3.7735	4.0076
	2.43	2.53	3.9525	
	2.45	2.56	4.2968	
5	2.51	2.67	5.9925	4.6844
	2.50	2.61	4.2145	
	2.50	2.60	3.8461	
10	2.52	2.67	5.6179	4.7895
	2.46	2.59	5.0193	
	2.58	2.68	3.7313	
15	2.26	2.36	4.2372	5.5045
	2.30	2.45	6.1224	
	2.44	2.60	6.1538	

Table 6 The result of water absorption for day 7

Sample (%)	Day 7			
	Mass of dried specimen (kg)	Mass of wet specimen (kg)	Water absorption (%)	Avg of water absorption (%)
0	2.4	2.51	4.3824	4.3399
	2.42	2.52	3.9682	
	2.45	2.57	4.6692	
5	2.47	2.58	4.2635	4.5110
	2.45	2.58	5.0387	
	2.49	2.6	4.2307	
10	2.47	2.58	4.2635	4.7147
	2.49	2.61	4.5977	
	2.51	2.65	5.2830	
15	2.49	2.6	4.2307	5.0962
	2.45	2.64	7.1969	
	2.49	2.59	3.8610	

that the mix of WPSA on the concrete cube was very high in water absorption. The result from his study stated that the mean value of eight number of sampled mixes is 27.05% and the maximum water absorption was 29.26% compared to the normal concrete.

Table 7 The result of water absorption for day 14

Sample (%)	Day 14			
	Mass of dried specimen (kg)	Mass of wet specimen (kg)	Water absorption (%)	Avg of water absorption (%)
0	2.5	2.6	3.8461	4.3516
	2.46	2.61	5.7471	
	2.51	2.6	3.4615	
5	2.49	2.6	4.2307	4.4269
	2.54	2.67	4.8689	
	2.75	2.87	4.1811	
10	2.42	2.55	5.0980	4.7596
	2.47	2.61	5.3639	
	2.52	2.62	3.8167	
15	2.46	2.6	5.3846	5.2902
	2.45	2.58	5.0387	
	2.43	2.57	5.4474	

Table 8 The result of water absorption for day 28

Sample (%)	Day 28			
	Mass of dried specimen (kg)	Mass of wet specimen (kg)	Water absorption (%)	Avg of water absorption (%)
0	2.5	2.6	3.8461	3.4143
	2.57	2.67	3.7453	
	2.57	2.64	2.6515	
5	2.45	2.55	3.9215	4.1496
	2.48	2.59	4.2471	
	2.46	2.57	4.2801	
10	2.51	2.61	3.8314	4.5977
	2.49	2.62	4.9618	
	2.47	2.6	5.0000	
15	2.49	2.62	4.9618	5.0898
	2.48	2.59	4.2471	
	2.48	2.64	6.0606	

4.3 Compressive Strength Test

Based on Table 3, there were 48 concrete cube samples casted for all the mix proportions. Although, the mix design for this study requires a target mean strength of 100 N/mm^2 at 28 days, the concrete samples only started to show a positive result

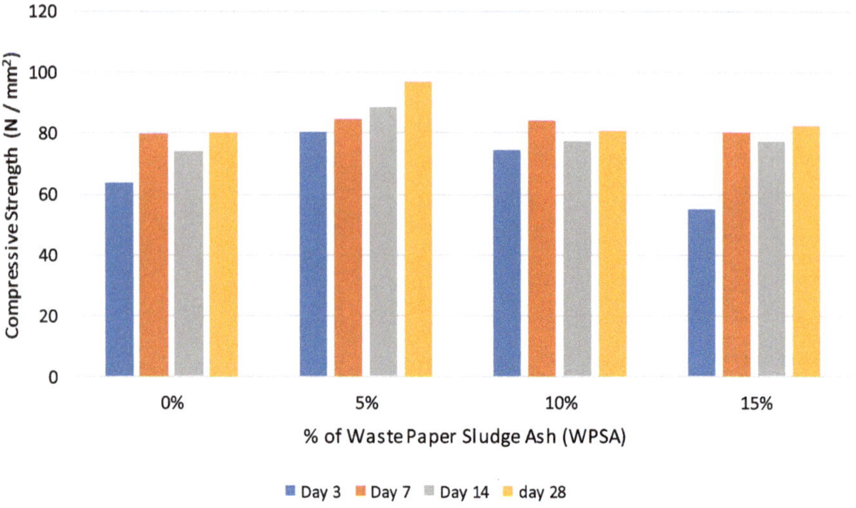

Fig. 1 Compressive strength for 3,7,14, and 28 days concrete cube samples

when replacement of WPSA was done at 5%. Figure 1 shows the results of compressive strength of all mix proportions for all 3, 7, 14, and 28 days. An increase in compressive strength was observed at 5% replacement of cement by WPSA and immediately decrease at 10 and 15%. The maximum compressive strength measured was at 5% more than the reference mix at 28 days corresponding to concrete mix containing 0% WPSA as an additional material for UHPC. Compressive strength for concrete mix with 10, and 15% WPSA content was found to be decreasing in value of compressive strength compared to the control sample concrete. Study by Sajad et al. (2013) stated that the compressive strength increases at 5% of concrete that contain the WPSA and decrease in next percentage of WPSA which is 10 and 15%. He stated that the maximum compressive strength was 15% more than the reference mix at 28 days compared to concrete mix containing 0% WPSA. Furthermore, the compressive strength for concrete mix with 10, 15, and 20% WPSA was found to be less than the reference mix.

4.4 Relationship Between Water Absorption and Compressive Strength

Based on the testing result shows that the increase in water absorption was observed with increase in WPSA. With an increasing water absorption capacity of the concrete, it will lead to a lesser durability property due to high porosity. The higher porosity will affect the concrete surface to weathering action, abrasion from traffic loads, and chemical attack more dominantly. Furthermore, the result of the compressive strength is related to the age of the concrete, where the increasing number of days for curing will increase the compressive strength of concrete. The maximum compressive strength measured was at 5% more than the reference mix at 28 days corresponding to concrete mix containing 0% WPSA as an additional material for UHPC. Compressive strength for concrete mix with 10, and 15% WPSA content was found to be decreasing in value of compressive strength compared to the control sample concrete. The optimum percentage to improve the concrete strength when adding WPSA in UHPC was at 5%.

Helal et al. (2015) stated that the concrete test result is affected by many factors such as surface smoothness, test specimen geometric properties, test specimen age, concrete surface and internal moisture condition, type of cement and coarse aggregate, type of mold and surface concrete carbonation. Study by Sajad et al. (2013) stated that the percentage water absorption increased with increase in WPSA content. The lowest value of water absorption was found for concrete mix with 5% WPSA content. He also stated that the density decreased with increase in WPSA content. The results showed the decrease in dry weight of concrete cube specimens for concrete mix with 20% WPSA content as compared to reference mix. This reduction in density can be attributed to lesser specific gravity of WPSA as compared to cement. Thus, WPSA concrete is light weight in nature.

5 Conclusion

In conclusion, the experiment of the study was to focus on the objective of the effect of the WPSA in percentage to UHPC. The objective for this is to determine the workability of UHPC with different percentages of WPSA by conducting the slump test and to determine the compressive strength and the water absorption of UHPC with different percentages of WPSA. For the first objective, it shows that as the addition material of WPSA will result in the increasing of the workability of the concrete at 5% and then the slump value decreased with the increase in WPSA content. Although the workability of the concrete gradually decreased, the slump value for all the mix proportions still has a very high workability because of the physical properties of UHPC concrete.

Furthermore, for the second objective, an increase in water absorption was observed with increase in WPSA. The results showed 49.07% increment in water absorption compared to the control sample with the absence of WPSA at 28 days. With an increasing water absorption capacity of the concrete, it will lead to a lesser durability property due to high porosity. Moreover, for the compressive strength test, an increase in compressive strength was observed at 5% replacement of cement by WPSA and immediately decrease at 10 and 15%. The maximum compressive strength measured was at 5% more than the reference mix at 28 days corresponding to concrete mix containing 0% WPSA as an additional material for UHPC. Finally, the optimum percentage to improve the concrete strength when adding WPSA in UHPC was at 5%.

Acknowledgements The authors would like to express their gratitude to the Ministry of Higher Education Malaysia and Universiti Teknologi MARA, Shah Alam for supporting this publication via research grant 600-RMC 5/3/GRR (004/2020).

References

Ahmad S, Malik MI, Wani MB, Ahmad R (2013) Study of concrete involving use of waste paper sludge ash as partial replacement of cement. IOSR J Eng 3(11):06–15

Bin Mohd Sani MS, Muftah F, Ab Rahman M (2011) Properties of waste paper sludge ash (WPSA) as cement replacement in mortar to support green technology material. In: 2011 3rd international symposium & exhibition in sustainable energy & environment (ISESEE). IEEE, p 94

British Standard Institution BS 882:1992, Specification for aggregate from natural source for concrete

British Standard Institution BS 1881–111:1983, Testing concrete, Method of normal curing of test specimens

British Standard Institution BS 1881–102:1983, Testing concrete, Method for determination of slump

British Standard Institution BS 1881–122:1983, Testing concrete, Method for setermination of water absorption

British Standard Institution BS 1881–116:1983, Testing concrete, Method for determination of compressive strength

Fehling, E, Schmidt, M, Stuerwald S (Eds) (2008) Second International Symposium on Ultra High Performance Concrete, Kassel, Germany, p 902

Ghafari, E., Costa, H., & Julio, E. (2015). Critical review on eco-efficient ultra-high performance concrete enhanced with nano materials. Constr Build Mater 101(1):201–208

Ishimoto H, Origuchi T, Yasuda M (2000) Use of papermaking sludge as new material. J Mater Civ Eng 12(4):310–313

Khanbilvardi R, Afshari S (1995) Sludge ash as fine aggregate for concrete mix. J Environ Eng

Liaw CT, Chang HL, Hsu WC, Huang CR (1998) A novel method to reuse paper sludge and co-generation ashes from paper mill. J Hazard Mater 58(1):93–102

Ma J, Schneider H (2002) Properties of ultra-high performance concrete. Leipzig Annual Civil Engineering Report (LACER), 7:25–32

Matte V, Moranville M (1999) Durability of reactive powder composites: influence of silica fume on leaching properties of very low water/binder pastes. Cem Concr Res 21(1):1–9

Mirza MI, Malik MI, Ahmad S, Tariq T, Bashir M, Chowdhary U (2015) Prospect of partial utilization of waste glass powder and waste paper sludge ash in concrete. Int J Civ Eng Technol (IJCIET) 9(6):160–217

Monosi S, Sani D, Tittarelli F (2010) Used foundry sand in cement mortars and concrete production. The Open Waste Manag J 8:3(1)

Richard P, Cheyrezy M (1995) Composition of reactive powder concretes. Cem Concr Res 25:1501–1511

Schmidt M, Fehling E, Geisenhanslueke C (Eds) (2004) International Symposium on Ultra High Performance Concrete, Kassel, Germany, p 868

Schmidt M, Fehling E (2005) Ultra-high-performance concrete: research, development and application in Europe. In: 7th International Symposium on Utilization of High Strength High Performance Concrete Vol 1, pp 51–77

Soutsos M, Millard S, Karaiskos K (2005) Mix design, mechanical properties, and impact resistance of reactive powder concrete (RPC). In: International Workshop on High Performance Fibre-Reinforced Cementitious Composites in Structural Applications, pp 549–560

Strunge T, Deuse T (2008) Special cements for ultra-high performance concrete. In: Proceedings of the Second International Symposium on Ultra High Performance Concrete, Kassel, Germany, pp 61–68

Tay JH (1987) Bricks manufactured from sludge. J Environ Eng

Talebinejad I, Iranmanesh A, Bassam S, Shekarchizadeh M (2004) Optimizing mix proportions of normal weight reactive powder concrete with strengths of 200–350 MPa. In: Proceedings of the International Symposium on UHPC, Kassel, Germany, pp 133–141

Tue N, Orgass M, Ma J (2008) Influence of addition method of superplasticizer on the properties of fresh UHPC. In: Proceedings of the 2nd International Symposium on Ultra-High Performance Concrete, Kassel, Germany, pp 93–100

Wille K, Naaman A, Montesinos G (2011) Ultra-high performance concrete with compressive strength exceeding 150 MPa (22 ksi): a simpler way. ACI Mater J 108(1):46–54

Flexural Performance of Strengthened Glued Laminated (GLULAM) Timber Beam Using Glass Fibre-Reinforced Polymer (GFRP)

A. O. A. Zamli, R. Hassan, M. N. M. Sidek, A. Awaludin, B. Anshari, N. H. A. Hamid, and S. M. Sapuan

Abstract Several reinforcement applications using glass fibre-reinforced polymer (GFRP) as strengthening material show greater structural strength without an increase of its dimensions. Though reinforcement is proven to be workable, there is minimal understanding of the optimum area should the GFRP be used in strengthening. This research aims to investigate this issue by testing the GFRP strengthening of glulam beams with a dimension of 80 mm x 140 mm x 3000 mm under four-point flexural loading. Three (3) specimens of each made for control, and three different areas covered by GFRP, are 30% (G30), 40% (G40), and 50% (G50), respectively. Results

A. O. A. Zamli
School of Civil Engineering, Universiti Teknologi MARA, 40450 Shah Alam, Selangor, Malaysia

R. Hassan (✉) · M. N. M. Sidek · A. Awaludin
Institute for Infrastructure Engineering and Sustainable Research (IIESM), Universiti Teknologi MARA, 40450 Shah Alam, Selangor, Malaysia
e-mail: rohan742@uitm.edu.my

M. N. M. Sidek
e-mail: muhdnorhasri@uitm.edu.my

A. Awaludin
e-mail: ali.awaludin@ugm.ac.id

A. Awaludin
Civil and Environmental Engineering Dpt, Gadjah Mada University, Grafika Street # 2, UGM campus, Sleman Yogyakarta 55281, Indonesia

B. Anshari
Department of Civil Engineering, Universitas Mataram, Majapahit 62 Mataram, Nusa Tenggara Bar, Indonesia
e-mail: buan.anshari@unram.ac.id

N. H. A. Hamid
Institute for Infrastructure, Engineering and Sustainable Management (IIESM), Universiti Teknologi MARA, 40450 Shah Alam, Malaysia
e-mail: norha454@uitm.edu.my

S. M. Sapuan
Laboratory of Biocomposite Technology Institute of Tropical Forestry and Forest Products (INTROP), Universiti Putra Malaysia UPM, 43400 Selangor, Serdang, Malaysia
e-mail: Sapuan@upm.edu.my

R. Hassan et al. (eds.), *Green Infrastructure*,
https://doi.org/10.1007/978-981-16-6383-3_6

show that adding 50% of the GFRP area increases 4.73% of the load-carrying capacity of the control beam; however, it reduced 3.84% the Modulus of Elasticity (MoE) and increased 2.16% of Modulus of Rupture (MOR) of the beam. Since the average maximum bending capacity of the G50 is 4.74% greater than that of the G30 and G40 strengthened area; therefore, G50 is considered to be more effective. While analysis of the damage pattern of the glulam beam and GFRP strips indicated a severe effect of delamination due to the difference of material behaviour when placed under load.

Keywords Tropical glulam · Light Red Meranti · GFRP reinforcement and flexural resistance

1 Introduction

Glued laminated (glulam) timber being an engineered component manufactured explicitly under pressure, makes it a solidly predictable structural member. Within the timber construction method, it is without any doubt to consider as the most robust and most reliable method of designing an engineered timber component, especially for complicated structural design or large buildings. Glulam is one of the timber products that contribute to the green development in the building industry through carbon dioxide retention and the less energy consumption used in the production of glulam compared with other construction materials. In common practice, to further increase the strength parameters of the glulam component, the method is usually by selecting a higher timber grade or increasing its dimension.

Recent studies have found that glass fibre-reinforced polymer (GFRP) reinforcement could apply to glulam for supporting strength used during manufacture and applicable for later structural repairs (Arriaga et al. 2011). While those studies show the viability of further reinforcing the timber or its connections with GFRP, minimal experimental data is available covering the wide selection of tropical timber. One of the tested species, solid Meranti (Shorea spp) beam with density 851 kg/m3 and 17% moisture content, was tested using artificial crack reinforced with carbon fibre-reinforced plastic (CFRP) by Awaludin and Marianti (2016). They found that moment capacities of beam increase as the length of CFRP were increased. Hassan et al. (2015) reported strengthening Yellow meranti and Bintangor using small clear samples on the effect of GFRP strengthening based on 45° and 90° orientation. They found that aligning GFRP in diagonal (45°) to the grain significantly produced a higher maximum load-carrying capacity than the perpendicular (90°) to the wood grain.

It is also questioned whether the flexural capacity of the glulam beam will be affected by laminating GFRP within the laminate with the selected strengthened area. A full cover of reinforcement may be beneficial, but the cost of the material may outweigh the necessity to put it into practice. Less coverage of the tensile surface area could mean more in any sense within the scope of the construction industry if there is an understanding of the optimum balance in the cost–benefit equation.

Documentations and studies related to the application of glulam using local tropical timber and the usage of GFRP as glulam reinforcement, unlike that of steel as concrete reinforcement, are very limited. Up to this writing, no publications reported on the strengthening of glulam beams using Malaysian timber. Few related Malaysian glulam beam studies include the finite element modelling of glulam tropical timber beam in bending (Shakimon et al. 2016) and bending strength of selected tropical glulam (Wan Mohamad et al. 2014). In contrast, Moya et al. (2013) and Campos et al. (2012) tested solid wood for I-beam.

Thus, this study aims to determine the performance of GFRP structural properties as reinforcement for glulam beam made of Light Red Meranti, a Malaysian timber species. Subsequently, this research seeks to determine the flexural behaviour and physical character of the control glulam beam compared with samples reinforced with GFRP and compare the delamination characteristics between layers of the glulam lamination with and without the GFRP reinforcement layer.

2 Flexural Strength of Timber Member

Variability of properties in timber is the main problem that creates an unpredictability factor in timber designs. This variability is because every tree matures in different environments and seasonal conditions, which greatly influence the strength factor of the timber itself. According to Saribiyik and Akgul (2010), the joints reinforced with GFRP bars have higher bending strengths than the adhesively bonded connection sample with the ratio of 300% indifference, while Zakaria (1997) mentioned that bending stress is lower than tensile strength. Thus, it requires an increase in bending strength.

By comparing the characteristics of timber, glulam, and GFRP reinforced glulam, the GFRP reinforced glulam provides less scatter in mechanical properties. The design value is also improved, which means that it is possible to construct wooden structures with GFRP reinforced glulam that can resist higher loads. The solid timber reinforced with GFRP can achieve the same level of strength as the GFRP reinforced glulam. The modulus of rupture (MoR) obtains the breaking strength of a purely hypothetical beam for points beyond the elastic limit. For the four-point bending test, the MoR and MoE were calculated based on the following Eqs. (1) and (2);

$$MOR = \frac{aP_{\max}}{2\left(\frac{bh^2}{6}\right)} \tag{1}$$

$$MOE = \frac{aL^2(F_2 - F_1)}{16\left(bh^3/12\right)(w_2 - w_1)} \tag{2}$$

where,

P_{max}—maximum load (N),

L—load span (mm),
a—distance between a loading position and the nearest support (mm),
b—width of beam (mm),
h—depth of beam (mm),
$F_2 - F_1$— increment of load on the straight line portion of the load deformation curve (N) and
$w_2 - w_1$— increment of deformation corresponding to $F_2 - F_1$(mm)

2.1 Glued Laminated Timber

Glued laminated timber or glulam is an engineered timber component. Multiple smaller timber pieces are joined together and manufactured under pressure to make it into a solid and highly predictable timber structural piece. Although it is not a rule for glulam laminations wood grain to run parallel with each other, it may be designed otherwise as needed by the structural requirements.

Finger-joints connect each continuous piece in a lamella, so this will make it stronger, straighter, and more uniformly solid rather than regular solid timber and also overcome some natural timber limitations such as strength reducing knots, allowing them to support heavier loads and spanning longer distance than normal timber (André 2006). This increase of strength factor could be understood why the need for discontinuous timber parts to be linked together into a continuous lamella utilizing connections such as finger-joints to reduce the defect variable of timber within a glulam component. In conclusion, manufacturing the structural element using the glulam method would allow it t o bear greater strength than the conventional method or bear a similar load with smaller design dimensions (André 2006).

2.2 Glass Fibre-Reinforced Polymer (GFRP)

GFRP is one category of fibre reinforcement polymer (FRP). FRP is a composite material that composes two or more material components (Abu Bakar et al. 2006). As reported by Abu Bakar et al. (2006, Isa 2005) found that GFRP has tension strength similar to CFRP, but the elasticity modulus is less than three or four times with CFRP. Harvey and Ansell (2003) reported that GFRP is cheaper than CFRP.

GFRP may also be applied in different forms and shapes and have also been used before as reinforcement and repair material, especially for concrete material as an external wrapping layer (Arduini et al. 1996) or as reinforcing bars embedded in the concrete, similar to steel rebars (Bank et al. (1997; Cosenza et al. 1997).

A study by Hadi et al. (2016) has shown clear evidence that the GFRP-wrapped column performed better again with concentric and eccentric load compared to the ordinary steel reinforced column without GFRP wrap. By looking at the layout of the reinforcement design from a two-dimensional perspective, it could hypothesize that GFRP may provide sound reinforcement against lateral load by applying it in a similar structure in a two-dimensional pattern onto the glulam surface. GFRP is a spiral alignment that can resist both deflection and shearing force of the structural component that applies to a glulam design.

GFRP as timber repairs by Yusof (2010) and Gentile (2000) and Mamlouk and Zaniewski (2011). Many other similar studies are structural under dynamic load by Chan et al. (2011). Detailed observation on the shearing effect between layered timber beam and solid timber beam has also been studied by Kim et al. (1997). According to Kim et al. (1997), both de-bonding and delamination significantly influence a laminate's stiffness, strength, and stability (Kim et al. 1997).

GFRP as for the timber reinforcement reports by few researchers as Gardner (1989); Dagher and Bragdon (2004); Hassan et al. (2009); Fava et al. (2012); Saribiyik and Akgul (2010); Alhayek and Svecova (2012); Gentry (2011); Davids et al. (2004) and Yang et al. (2016). The mechanical properties of timber strengthened with GFRP are higher than un-strengthened timber, as Arriaga et al. (2011) demonstrated that GFRP is preferable for maintaining deteriorated timber structures. Radford et al. (2002) found that GFRP would improve the ductility of timber. Furthermore, Saribiyik and Akgul (2010) explained that the bending strength of the connection points strengthened with GFRP bar was higher than un-strengthened adhesively bonded connection by about 300%. While in timber construction, GFRP has either been applied as a composite sheet layer between glulam laminations (Dagher and Bragdon 2004).

Arriaga et al. (2017) present a simplified analysis and design method to repair deteriorated timber beam-ends using glass fibre-reinforced product (GFRP) bonded rigid plates. Pinus radiata tested on the effect of temperature (−20 to 50 °C) and moisture content (9–30%) changes were analysed in two groups of six short beam specimens and in bonding shear test specimens using two different epoxy formulations. They found that the theoretical model predicts conservative values of load-carrying capacity. Temperature and moisture changes cause a reduction of mechanical properties, but to different degrees depending on the epoxy formulation.

In addition, Chew et al. (2016) have shown a Mengkulang glulam block with the GFRP sheets as reinforcement placed between the laminations, managed to acquire the difference of strength increase of 12.6% from the control samples. They put an entire layer of GFRP over the whole surface of the laminations between the first and second layers and between the third and the fourth layer of the laminations. In terms of variations of GFRP alignment and applied surface area, the Italian Research Council (2004) has performed studies on the increasing properties of the shear strength of structural beam components with fibre reinforcements. The alignments for FRP application by wrapping and increment in the surface area applied to the structural surface. Previous studies also touched on the issue where the presence of the reinforcement fibre could remedy the effects of the delamination between different layers of the structural component. However, the cause of delamination is related to the difference in moisture and temperature during curing, which leads to the outer layer breaking and peeling off, revealing the inner reinforcing steel to corrosive effects of the environment.

Hay et al. (2006) studied reinforcing sawn timber stringers for existing bridges in Manitoba, Canada. It focused mainly on repairing the horizontal splits that appear at each end of the stringer, which is a common defect due to age. Hay et al. (2006) design schemes were to deal with critically damaged areas. From their observation, the fibres arranged in the longitudinal direction (zero direction) had the highest capacity of strength and along with findings by John and Lacroix (2000) that utilized U-shaped GFRP sheets to provide both shear and flexural reinforcement. Basterra et al. (2017) tested the possibility of partial reinforcement of a glulam beam where the GFRP layer was placed over the tension zone by a low ratio of 1.07 and 1.6%. Their test results showed an average improvement of 12.1 to 14.7% in stiffness, while the moment capacity increased to 23%. GFRP reinforcement also reduces the influence of various timber defects like knots or cracks, which would affect the glulam strength.

Balmori et al. (2020) presented an experimental structural-scale test campaign to analyse the flexural behaviour of low-grade maritime pine (Pinus pinaster Ait.). The duo timber beams were reinforced with an internal glass fibre-reinforced polymer (GFRP) sheet. Unreinforced duo beams and internally reinforced with a unidirectional GFRP sheet with an areal mass of 1200 g/m^2 were produced and tested. They found that bending behaviour improved, and the internal GFRP reinforcements seemed to decrease the influence of wood singularities and wood heterogeneity on mechanical properties. The addition of a low GFRP reinforcement ratio (1.07%) in the tension zone of the duo beams provided an average improvement of 8.37% in bending stiffness (EI) and an increase of up to 18.45% in ultimate moment capacity.

3 Materials and Methods

3.1 Glulam Beam Design Details

The glulam samples made of Light Red Meranti were fabricated into two design groups with similar dimensions of 140 mm x 80 mm x 3000 mm (with five lamella layers each with a thickness of 28 mm). One set of the control sample was made as a standardized glulam beam. In contrast, the other set with four lamellas and GFRP reinforcement strips were applied before being fully assembled into a complete five-layered glulam beam. Figures 1 and 2 show the detailed dimension o f he glulam beam and the layout of the GFRP strips layered inside the beam located on the surface of the first lamella counting from the bottom of the beam.

The type of GFRP used is a sheet in woven form and uni-direction. Three (3) sets of GFRP strips were prepared for the glulam reinforcement. Figure 3 shows the GFRP sheet that have been cut into the required strips width and installed in between the glulam laminations. The thickness of the GFRP was approximately 0.6 mm. Each was cut to the length of 3300 mm which are longer than the length of the beam itself to compensate for the stretching and stapling of the strip end to prevent creeping from shrinkage. The glulam GFRP strips were done and installed by the manufacturer at

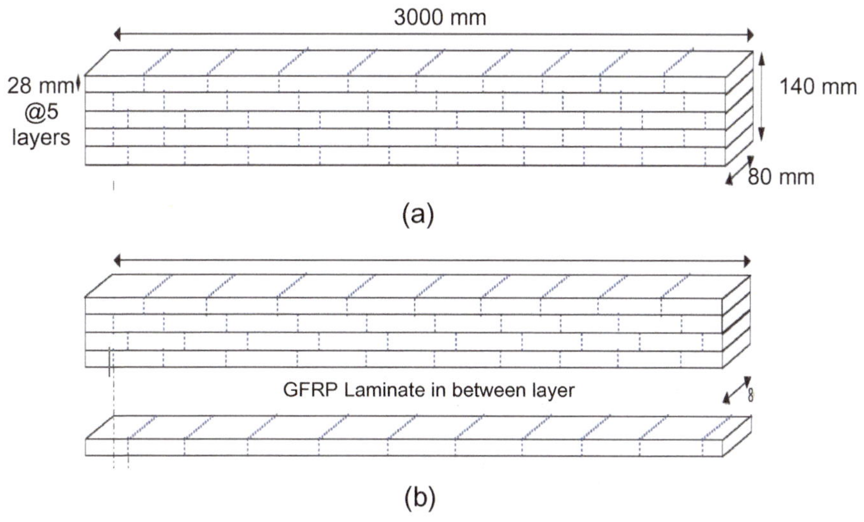

Fig. 1 Lay up of glulam beam **a** Five lamellas and **b** location of GFRP layered on the surface of first layer from the bottom

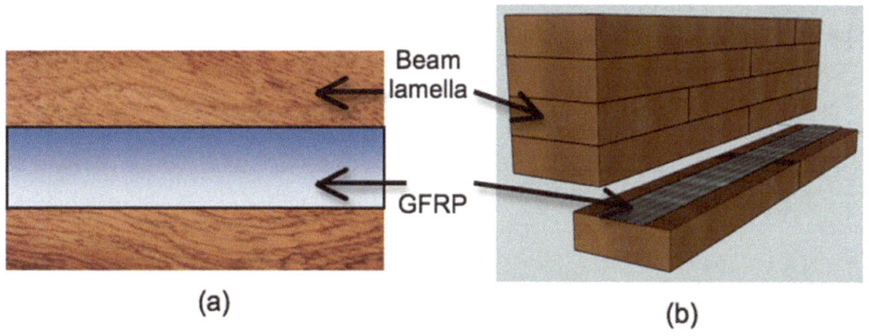

Fig. 2 Layout of the GFRP sheet over the glulam surface area **a** Plan view of the GFRP laid out parallel to the wood grain **b** expanded isometric view with the GFRP strips positioned inside

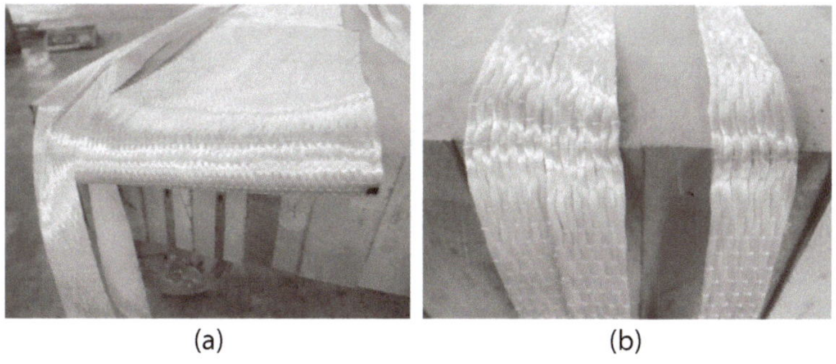

Fig. 3 GFRP as reinforcement material **a** sheets and **b** strips

Table 1 Specimens preparation

Sample criteria	Sample code	Size (mm)	GFRP area (mm_2)
Control	A	80 X140 X 3000	–
30% GRFP	G30	23 X 140 X 3000	72,000
40% GFRP	G40	32 X 140 X3000	96,000
50% GFRP	G50	40 X 140 X 3000	120,000

the glulam factory. Each set of strips was cut to the width of 24 mm (30% of lamella surface area), 32 mm (40% of the lamella surface area), and 40 mm (50% of the lamella surface area) of the three GFRP strips, respectively. Each beam specimen was coded for identification (Table 1).

As shown in Fig. 4, the strips were stapled to the glulam beam end surface during glueing step to ensure a straight uniform layout and adhered firmly to prevent structural defects due to slippage from shrinking creep.

The adhesive used for permanent structural application was manufactured with type A, bonded using phenolic resin-based adhesive, which shows black glue lines. Type A sealant will not deteriorate under heat, freeze, or high moisture conditions being fully durable for timber components exposed to the weather. It is a material with high strength and structural reliability while being a cost-effective and sustainable construction material. The epoxy resins and hardener were used, which are suitable for bonding wood to wood and wood to GFRP. The primer coat was applied by using

(a) (b)

Fig. 4 Installation of the GFRP reinforcement; **a** GRFP strip laid between glulam lamellas **b** Glulam beam compressed after the GFRP strip installed

Epo bond primer. The primer coat is essential to ensure that the surface is clean from dust, better bonding between wood and GFRP, and provide additional protection to the material.

As a bonding agent, the epoxy resin could cure at room temperature and provide solid adherents. The curing time required was 24 h. A total of 20% was allowed for wastage. The mixing ratio of adhesive for Part A and hardener (Part B) is 2.5 : 1 by weight. The adhesive was manually mixed until well blended for about 5 min at a constant rate. The adhesive spread should be uniformed and with sufficient quantity according to the adhesive manufacturer's recommendations.

3.2 *Flexural Test*

For the requirement of the four-point flexural test using the Universal Testing Machine (UTM) UTS-348, each of the glulam component was measured and cut accordingly to requirement of flexural test BS EN 408:2010 + A1:2012 where the length of the test member should be 19 times of its depth which in this experiment was about 2660 mm. It also specifies that the rate of movement of the load applied is to be no greater than (0.003 h) mm/s. So, the rate of movement to the average depth of the samples at 140 mm would be at the constant load applied at the displacement rate of 0.4 mm/s until the sample suffered complete structural failure. The test samples set up with deflection change were measured in the horizontal and vertical direction as shown in Fig. 5.

Two identical loads were applied at 1/3L, and three LVDTs (linear variable displacement transducer) were mounted in the horizontal and vertical direction, respectively. Measurement of beam deflection and magnitude of the applied load was continuously recorded and saved. Visual observation was also performed during the test to understand the failure development, and the test was carried out until failure (Fig. 6).

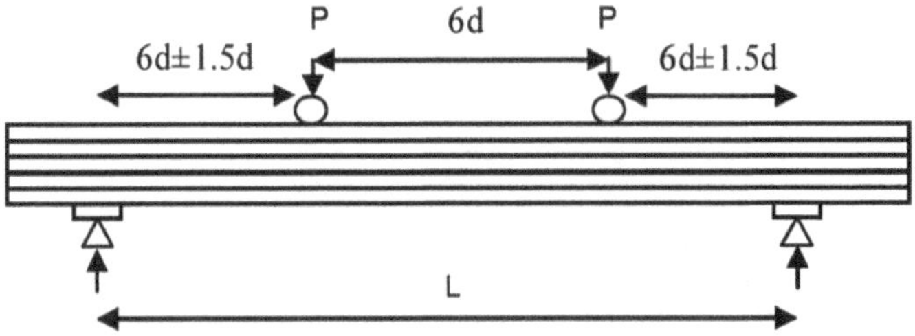

Fig. 5 Layout for the four-point flexural test (d: depth of beam, P: applied load)

Fig. 6 Test sample laid out ready for the flexural test with LVDT mounted at each end and in the middle of the beam span

4 Result and Discussion

4.1 Pattern of Failure

The failure pattern of the specimens was observed where each of the control specimens reached the highest load before an abrupt failure from delamination and splitting of the finger-joints. However, the abrupt termination of load-bearing capability observed was only valid for the specimens without any reinforcement. The specimens with GFRP reinforcement failure patterns occurred more or less the same as the control sample. Still, the point of failure is gradual, where the reinforcement had prevented a total structural disintegration (Fig. 7).

The failure of the strengthened glulam is more gradual compared to the control beam. This failure was due to the transfer of the load interchanges between the timber laminate and GFRP fibres until the point where both timber and GFRP fibres ultimately failed to maintain structural integrity. The load transfer process brought heavy shear damage, delamination, and shattering splinters of the timber components. Loud creaks were also heard, indicating the gradual shear and splintering of the laminations and GFRP fibres breaking apart. The damage pattern was found similar to the study done by Davids et al. (2004).

Fig. 7 Glulam beam sample A (Control) after failure at maximum load

Fig. 8 Glulam beam A (Control) sample failure **a** failure at finger joint and **b** splintering of the lamella

Inadvertently, the result has indicated a loss of strength for each of the samples that were reinforced with GFRP. All these findings point to a straightforward conclusion: GFRP reinforcement area must not be less than equal to the area of the timber lamination that they bonded to in between the layers. The difference of material bonded (timber-to-GFRP and timber-to-timber) as discovered from the delamination and the flexural test shows that GFRP has a more excellent bond to the timber than timber-to-timber bonding. In addition, from observation during the flexural test, the GFRP fibres have more stretching percentage properties. GFRP more significant difference in the rate of dimensional distortion than timber forces the glue line to shear away, leading to excessive delamination, shear across the grain and splintering of the timber lamination. This finding has also reflected the statement by Kim et al. (1997) that both de-bonding and delamination significantly influence the stiffness, strength, and stability of a laminate (Kim et al. 1997).

Two samples were rejected during the experimental work as the failure was due to defects in the beam production. The beam finger joint had separated, and delamination between layers occurred and formed a wide gap at the earlier experimental setup stage. The physical failure was not at the maximum tensile position (mid-span) and before the total rupture of the beam (Fig. 8). Therefore, from all 12 specimens prepared, only ten specimens were successful. The splintering or shearing of the timber lamination from observation was gradual as the timber component began to fail; the majority share of the load transferred to the GFRP reinforcement (Fig. 9). In practice, this would serve as a visible and audible early warning for the failure risk of the timber structure.

4.2 *Load Capacity of Light Red Meranti Specimens*

The percentage difference of G30 and G40 reinforced average maximum flexural capacity to the control specimen is reduced by 7.42%, 5.40%, while G50 has increased by 4.73%, respectively (Table 2).

One particular pattern of failure is as shown for specimen No. 3 (G50). The sample has a high load of 56.14kN at a corresponding mid-span displacement of 70.23 mm.

Fig. 9 Glulam beam G50 sample failure

Table 2 Maximum flexural capacity of control and GFRP strengthened beam

Specimen	A (Control)	GFRP reinforcement		
		G30	G40	G50
	Max Load (kN)	Max Load (kN)	Max Load (kN)	Max Load (kN)
1	34.09	54.17	NA	55.61
2	60.4	55.93	47.09	NA
3	65.57	38.09	53.85	56.14
Average	53.35	49.4	50.47	55.88
% Different	0	−7.42	−5.40	4.73

This sample shows an interesting result, as shown in Fig. 10, where the curve behaved multi-linearly, showing the interaction pattern between the timber laminates and the GFRP reinforcement. The load decreases as the timber began to fail. The GFRP reinforcement became taut and held on before the bond failed and transferred the load-bearing back to the timber.

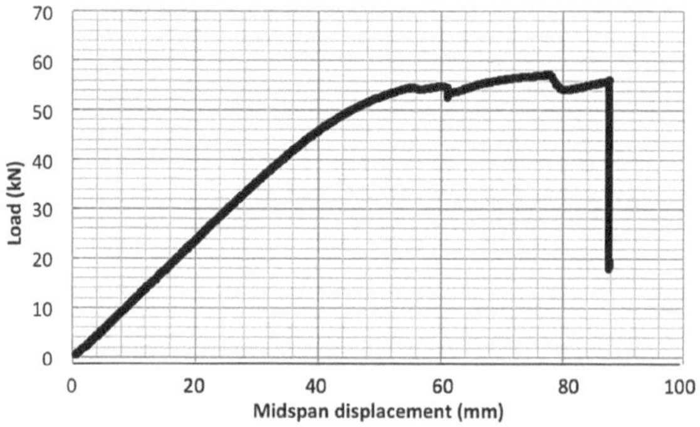

Fig. 10 Glulam beam sample 3 (G50); load versus displacement failure at maximum load

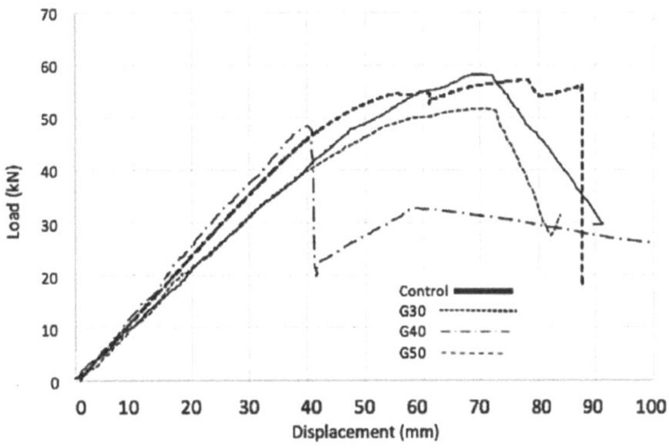

Fig. 11 Typical Load versus displacement of three different reinforcement areas

Figure 11 shows the typical load versus displacement failure for the control and three different reinforcement areas. The G50 specimen at 55.88 kN acquires the maximum average load. While beam A (control), G30, and G40 achieved the load average of 54.69 kN, 49.40 kN, and 50.47 kN, respectively.

The results analysis and observation of the typical failure show that instead of reinforcing the glulam beam, the stretching of the G30 and G40 strips delaminated the bottom layer of lamination that weakens the whole beam onset of load applied upon it. Figure 12 shows that only G50 shows the maximum load capacity higher than the control specimens. It shows that the alignment and layering of the GFRP strips have seriously affected the glulam strength detrimentally, hinting and reducing the surface engagement between the timber laminate. It suggests that the GFRP reinforcement layer is best aligned as maximum area as possible and placed at the outermost layer to reduce tensile failure due to the difference in structural properties between timber and GFRP as the reinforcement material.

Based on the lower strength of the glulam beam shown for 30% and 40% GFRP, reinforcement strips indicate that only 50% or more timber layers or GFRP strip coverage is beneficial to the glulam beam strength.

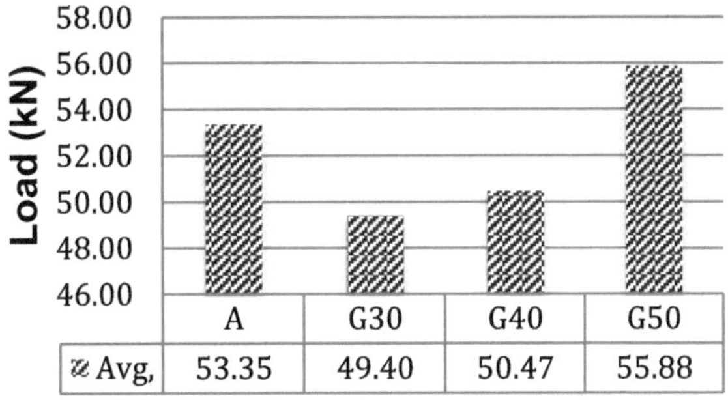

Fig. 12 The average load-carrying capacity for glulam beam specimen

4.3 Modulus of Elasticity (MoE)

The average modulus of elasticity results is summarized in Table 3. The MoE for G30 is increasing from the control sample, decreasing for G40 reinforcement, and further decreases for G50 reinforcement. The highest average of MoE is 20,644 MPa for the G30 sample and the lowest average of MoE is 19,426 MPa for the sample with G50 reinforcement. The difference in the percentage of decrement is hardly significant since the higher the GFRP percentage, the lower the MoE of the beam.

Figure 13 also shows G30 has the highest MoE that reached 2.19% higher than control and decreases with increase of the GFRP area for G40 and G50 by 2.15% and 3.84%, respectively. G50 reinforcement has less elasticity value than G30 reinforcement. Specimens with the mixed area of timber and GFRP reinforcement surface between the lamination have less elasticity to recover themselves from deformation. From the observations, the stretch and slip of the GFRP strip where it delaminated from the timber lamination impart a permanent deformation.

Table 3 Modulus of Elasticity (MoE) of glulam beam specimen

Specimen	A (Control)	GFRP reinforced		
		G30	G40	G50
	MoE (Mpa)	MoE (Mpa)	MoE (Mpa)	MoE (Mpa)
1	18,932	18,103	NA	18,766
2	20,932	18,837	20,255	NA
3	20,739	249,918	19,280	20,087
Average	20,201	20,644	19,767	19,426
% Different	0	2.19	-2.15	-3.84

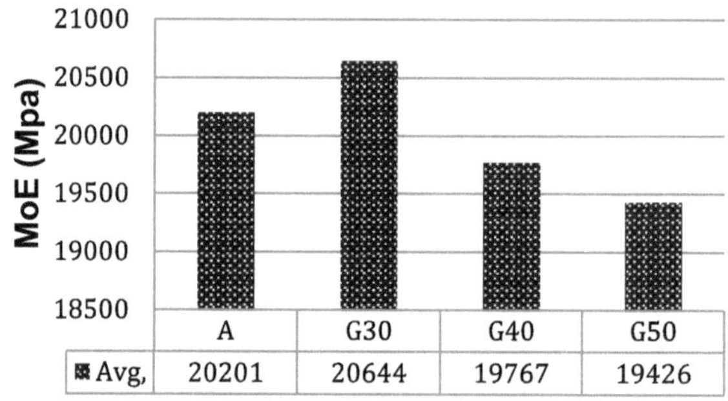

Fig. 13 The average modulus of elasticity (MoE) comparison

Table 4 Summary of Modulus of Rupture (MoR) of Light Red Meranti glulam beam specimen

Specimen	A (Control)	GFRP reinforced		
		G30	G40	G50
	MoR	MoR	MoR	MoR
	Mpa	Mpa	Mpa	Mpa
1	54.80	87.00	NA	89.4
2	97.10	89.90	75.7	NA
3	111.80	61.20	86.5	90.2
Average	87.90	79.40	81.1	89.8
% Different	0%	−9.67	−7.74	2.16

4.4 *Modulus of Rupture (MoR)*

The results of the average modulus of rupture are summarized in Table 4. It shows that the MoR average for the G50 sample is higher than any other sample with reinforcement. The G50 sample has the highest MoR average at 89.8 Mpa. However, a decrease in MoR average is observed for the G30 and G40 compared to the control, A specimens.

While the difference of MoR is negligent, the average percentage in decrement of MoE of glulam beam for different GFRP strips area observed to be quite significant. As shown in Table 4 and Fig. 14, the average percentage decrement of MoR for G30 and G40 are -9.67%, -7.74%, and increment of 2.16% for G50. It shows that the highest area of reinforcement at G50 area has the highest increment in strength. The difference percentage between G30 and G50 reinforcement support the theory that more of either kind of material (i.e. fibres, timber) is a must for the optimized performance of reinforcement. It indicates that 50% or more than coverage of GFRP is expected to give the best performance.

The test results showed that specimens with GFRP strips along the wood grain for G30 and G40 possessed less strength than the G50 reinforced specimen. In contrast, G50 shows an increment of strength compared to unreinforced beam. These results support using reinforcement and findings from previous research that GFRP fibre will

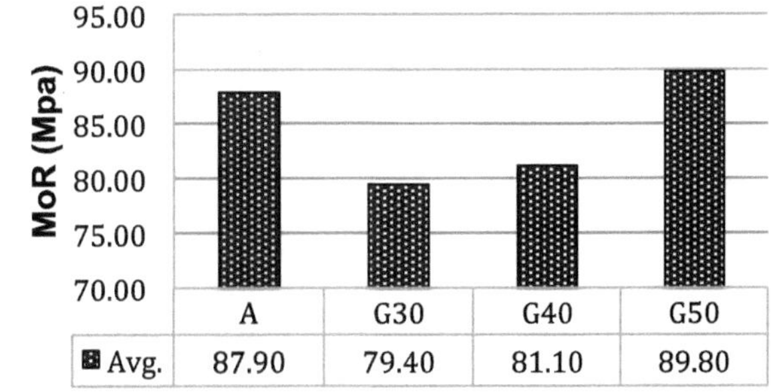

Fig. 14 The average modulus of rupture (MoR)

improve flexural capacity of wood. This experiment discovered that a reinforcement area that is either less than 50% to the lamination bonded surface area instead is detrimental to the overall improvement of the glulam beam. Therefore, it is suggested that the optimum reinforcement of GFRP in between layers is more than 50% area of coverage. However, due to the limitation of test specimens in this study, further research with more specimens is suggested for more accurate results.

5 Conclusion

The glulam beams' flexural tests with varied G30, G40, and G50 area of reinforcement over the lamination surface has an average maximum load capacity of 49.40 kN, 50.47 kN, and 55.88 kN, respectively, compared with the control samples that had an average maximum load at 53.35kN. This result clearly shows that an increase of GFRP strengthened area will increase the maximum load capacity, decrease the MoE, and increase the MoR, respectively. While analysis of the damage pattern of the glulam beam and GFRP strips indicated a severe effect of delamination due to the difference of material behaviour when placed under load. These findings conclude that the alignment and layering of the GFRP strips could seriously affect the glulam strength detrimentally, hinting, and the underlying mechanics at work. It suggests that the GFRP reinforcement layer is best aligned as more than 50% of the area and placed at the outermost layer to reduce tensile failure due to the difference in structural properties between timber and the reinforcement material.

Acknowledgements The authors wish to acknowledge financial support from Universiti Teknologi MARA through MITRA PERDANA grant (600-IRMI/PERDANA 5/3/MITRA (001/2018)-2) and special assistance by Faculty of Civil Engineering laboratory technical staff.

References

Abu Bakar S, Mohd Amir W, Mohd Sam AR, Abdullah R, Saleh AL (2006) Mechanical properties of timber-carbon fibre reinforced polymer composite under tensile loading. In: Proceedings of the 6th asia-pacific structural engineering and construction conference (APSEC 2006), 5–6 June 2006, Kuala Lumpur, Malaysia

Alhayek H, Svecova D (2012) Flexural stiffness and strength of GFRP-Reinforced timber beam. J Compos Constr 16:245–252

André A (2006) Fibres for strengthening of timber structures. Civil and Environmental Engineering. Struct Eng 3, Luleå tekniska universitet, Luleå University of Technology. ISSN 1402-1528

Arduini M, Di Tommaso A, Manfroni O, Nanni A (1996) Failure mechanisms of concrete beams reinforced with FRP flexible sheets. In: Proceedings of the 2nd International conference on advanced composite materials in bridges and structures, Acmbs-Ii, Montreal

Arriaga F, Íñiguez-Gonzáles G, Esteban M (2011) Bonding shear strength in timber and GFRP glued with epoxy adhesives. Wood Res 56(3):297–310

Arriaga F, Fernandez-Cabo JL, Jose R, Aira JR (2017) Timber beam bearing reinforcement with GFRP glued-in plates: strength and hydrothermal effects. J Mater Civil Eng 29(2)

Awaludin A, Marianti SA (2016) Repair of timber beams using CFRP: materials,technologies and components for sustainable buildings. Central Europe, https://www.researchgate.net/publication/331587065_Repair_Timber_Beams_using_CFRP. Accessed 03 Jul 2021

Balmori JA, Basterra LA, Acuña L (2020) Internal GFRP reinforcement of low-grade maritime pine duo timber beams. Materials (basel) 13(3):571. https://doi.org/10.3390/ma13030571. PMID:31991708; PMCID: PMC7040746

Bank LC, Frostig Y, Shapira A (1997) Three dimensional FRP grating cages for concrete beams: a pilot study. ACI J Struct 94(6):643–652

Basterra LA, Balmori JA, Morillas L, Acuña L, Casado M (2017) Internal reinforcement of laminated duo beams of low-grade timber with GFRP sheets. Constr Build Mater 154:914–920

BS EN 408:2010+A1 (2012) Timber structures: structural timber and glued laminated timber. Determination of some physical and mechanical properties, CEN National Members

Campos MBS, Menezzi CHSD, Souza D (2012) Flexural properties of wood i-beams flanged with tropical hardwoods. J Trop Forest Sci (JTFS) 24(3)

Chan R, Albermani F, Kitipornchai S (2011) Stiffness and strength of perforated steel wall. In: The proceedings of the twelfth east asia-pacific conference on structural engineering and construction (EASEC12). 26–28 Jan 2011, Hong Kong SAR, China, vol 14, pp 675–679

Chew AA, Rahim NSA, Mohd Yasim NHF, Hassan R (2016) Compression strength of composite glulam timber reinforced between first and second, third and fourth lamina. Jurnal Teknologi 78(5–4):49–55

Cosenza E, Manfredi G, Realfonzo R (1997) Behavior and modeling of bond of FRP rebars to concrete. J Compos Constr 1(2): 40–51

Dagher HJ, Bragdon M (2004) Advance FRP-Wood composites in bridge applications. J Transp Res Board. https://doi.org/10.3141/1814-28

Davids WG, Richie M, Gamache C (2004) Flexural fatigue of glulam beams with fiber-reinforced polymer tension reinforcing. Bridges 10(40700):48

Fava G, Parisi MA, Poggi C, Tardini C (2012) Fibre reinforced connection for timber members: an experimental study. Proc CICE

Gardner GP (1989) A reinforced glued laminated timber system. In: 2nd pacific timber engineering conference 1989, vol 2. Auckland, New Zealand

Gentile CJ (2000) Flexural strengthening of timber bridge beams using FRP. Thesis for Master of Science, University of Manitoba

Gentry TR (2011) Performance of glued-laminated timbers with FRP shear and flexural reinforcement. J Compos Constr ASCE

Hadi MN, Karim H, Sheikh MN (2016) Experimental investigations on circular concrete columns reinforced with GFRP bars and helices under different loading conditions. J Compos Constr 20(4):04016009

Harvey K, Ansell MP (2003) Improved timber connections using bonded-in GFRP rods. Thesis for Doctor of Philosophy, University of Bath

Hassan R, Ismail S, Ali Siblamurisi A, Abu Bakar I, Mat Isa C (2009) Effectiveness of glass fibre reinforced fabric (GFRP) for dowel type timber connection. UniversitiTeknologi MARA (UiTM), Malaysia Research Management Institute

Hassan R, Ghazali N, Zamli AOA (2015) Bending strength performance of selected timber species with different GFRP strips pattern. In: Hassan R, Yusoff M, Alisibramulisi A, Mohd Amin N, Ismail Z (eds) InCIEC 2014. Springer, Singapore. https://doi.org/10.1007/978-981-287-290-6_100

Hay S, Thiessen K, Svecova D (2006) Effectiveness of GFRP sheets for shear strengthening of timber. Constr Build Mater 106:550–563

Isa SM (2005) Sifat Mekanikal bagi rencam kayu-cfrp di bawah pembebanan lenturan, Final Year Project Report, UTM

Italian National Research Council (2004) CNR-DT 200/2004. Guide for the design and construction of externally bonded FRP systems for strengthening existing structures. Rome, Italy

John KC, Lacroix S (2000) Composite reinforcement of timber in bending. Can J Civ Eng 27(5):899–906. https://doi.org/10.1139/cjce-27-5-899

Kim Y, Davalos JF, Barbero EJ (1997) Delamination buckling of FRP layer in laminated wood beams. J Compos Struct 37(3–4):311–320

Mamlouk MS, Zaniewski JP (2011) Materials for civil and construction engineers. Pearson/Prentice Hall, Boston

Moya R, Tenorio C, Carranza M, Camacho D, Quesada-Pineda H (2013) Structural performance of i-beam fabricated from a fast-growing tree, *Gmelina Arborea.* J Trop Forest Sci (JTFS) 25(1)

Radford DW, Van Goethem D, Gutkowski RM, Peterson ML (2002) Composites repair of timber structures. Constr Build Mater 16:417–425

Saribiyik M, Akgul T (2010) GFRP bar element to strengthen timber connection systems. Prod J Fac Technol. Sakarya University, Turkey

Shakimon MN, Malek NJA, Hassan R, Ahmad Z (2016) The finite element modelling of glulam tropical timber beam in bending. Jurnal Teknologi 78(5–5). https://doi.org/10.11113/jt.v78.8579

Wan Mohamad WH, Ahmad Z, Abd Jalil A (2014) Bending strength of glulam from selected malaysian hardwood timber. Adv Mater Res 879:237–244. https://doi.org/10.4028/www.scientific.net/amr.879.237

Yang H, Liu W, Lu W, Zhu S, Geng Q (2016) Flexural behavior of FRP and steel reinforced glulam beams: experimental and theoretical evaluation. Constr Build Mater 106:550–563

Yusof A (2010) Bending behavior of timber beams strengthened using fiber reinforced polymer bars and plates (pp 33 – 35). Thesis for Doctor of Philosophy (Civil Engineering), UTM

Zakaria ML (1997) Bahan dan Binaan. Dewan Bahasa dan Pustaka, Kuala Lumpur. ISBN: 9789836228208, 9836228209

Overview on Bending and Rolling Shear Properties of Cross-Laminated Timber (CLT) as Engineered Sustainable Construction Materials

W. C. Lum, M. B. Norshariza, M. S. Nordin, and Z. Ahmad

Abstract Cross-laminated timber (CLT) is one of the preferred engineered timber products (ETP) used in the construction and building industry. CLT is an orthogonal and laminar structure that can be used as full-size load-bearing structural elements such as wall, floor element as well as a linear timber member. The timber pieces are visually graded before being manufactured into CLT panels. As a result, CLT is able to provide enhanced stability and performance compared with its solid timber counterparts. In order to use CLT products in the construction of timber structures, characteristic bending and shear properties are important values in design. Thus, this chapter provides an overview of the bending and shear properties of CLT made from different timber species with various densities including timbers such as softwood, temperate hardwood as well as tropical hardwood. The factors influencing their properties are summarized and discussed. Their respective failure modes are also reported and discussed. This chapter also covers the general manufacturing process and the applications of CLT in the construction and building sector. Limitations, challenges dealing with the applications of CLT in the construction industries and the future of CLT are also discussed.

Keywords Engineered timber products · Cross-laminated timber · Sustainable construction materials · Bending properties · Rolling shear properties · Failure modes

---

W. C. Lum · M. B. Norshariza (✉)
Institute for Infrastructure Engineering and Sustainable Management, Universiti Teknologi MARA Malaysia, 40450 Selangor, Shah Alam, Malaysia
e-mail: nshariza@uitm.edu.my

M. B. Norshariza · M. S. Nordin · Z. Ahmad
School of Civil Engineering, College of Engineering, Universiti Teknologi MARA Malaysia, 40450 Selangor, Shah Alam, Malaysia
e-mail: shaiful.nordin@mtib.gov.my

Z. Ahmad
e-mail: zakiah@uitm.edu.my

M. S. Nordin
Malaysian Timber Industry Board, Menara PGRM, Level 13-17No 8, Jalan Pudu Ulu, 56100 Kuala Lumpur, Cheras, Malaysia

R. Hassan et al. (eds.), *Green Infrastructure*,
https://doi.org/10.1007/978-981-16-6383-3_7

1 Introduction

Cross-laminated timber (CLT) is an engineered timber product (ETP) produced from sawn timber planks bonded together by the appropriate adhesive. CLT panels for construction consist of at least three layers of lamella made from either softwood or hardwood. Each timber lamination is arranged 90° (crosswise oriented) to each other. CLT is considered as a relatively new green construction material and was first developed in Switzerland and then further improved in Austria at the starting of twentieth century as an alternative building material to the conventional materials such as concrete, steel, masonry as well as solid timber (Zhou et al. 2014). Due to its orthogonal and laminar nature, CLT can be used as full-size load-bearing structural elements such as walls, floor structures, roofs elements as well as a linear timber member in a large number of innovative residential and commercial buildings (Wang et al. 2015).

There are many advantages in using CLT as preferred construction materials. Some notable advantages are summarized by Santoni et al. (2017). As summarized by the authors, CLT panels are commonly known to have high structural strength and can provide decent structural stability. CLT panels can be manufactured to fulfil specific safety requirements needed with cost-competitive to conventional building materials such as concrete, masonry and steel. Furthermore, the various construction elements can be prefabricated using CLT panels and are more time and labour efficient during onsite assembly. Recently, CLT is gaining traction worldwide as it is proven to be an environmentally friendly and sustainable engineered timber product. The CLT panel used in construction stores CO_2 in the timber in a process known as bio-sequestration. Another benefit mentioned is the ability to convert lower grade or lower class timber into a higher grade CLT panel. The lower grade timber can be used in the neutral axis or the transverse layer where the effect of the tension and compression is minimal.

Until recently, the manufacturing and applications of CLT are mostly confined to its region of origin in Central Europe. In the early twenty-first century, the applications of CLT as structural elements became widespread in Austria, Germany, Norway, Sweden, Switzerland and the United Kingdom (Crespell and Gagnon 2010). However, CLT is rapidly gaining traction outside these regions. Countries like the United States of America, Australia and New Zealand are starting to realize the potentials and benefits of CLT. CLT panels have excellent strength and stiffness as well as improved dimensional stability, which allow it to be used as a wide range of structural elements in various types of buildings. Buildings constructed using CLT are structurally simple and yet provide better design versatility at the same time. Other benefits of the incorporation of CLT in the construction industry are fast installation, decreased waste, less carbon emission, improved thermal performance and high seismic performance (Liao et al. 2017; Viguier et al. 2015).

It is necessary to understand the various properties of CLT products, as well as the entire building system incorporating CLT panels as structural elements for better standardization. In order to better introduce CLT products in the construction of timber structures, characteristic bending and rolling shear properties are important values in design. Therefore, this chapter covers an overview of the bending and rolling shear properties of CLT made from different timber species and also factors affecting the bending and rolling shear properties. The bending and rolling shear failure modes are also reported and discussed. The manufacturing process of CLT is briefly outlined as well.

2 Manufacturing Process Of CLT

CLT panels are generally manufactured in three or more layers with the same thicknesses of laminations. The layers are arranged in a 90 degree of traverse pattern. The manufacturing process of CLT panels, which generally involves the following steps, is illustrated in Fig. 1:

(i) Lamella selection, grading and classification
(ii) Lamella finger jointing
(iii) Lamella cutting to length and planning
(iv) Panel lay-up
(v) Adhesive application
(vi) Assembly pressing
(vii) CLT online quality control, machining and cutting
(viii) Product marking, packaging and shipping.

The effectiveness of the CLT manufacturing process depends on the consistency of the timber quality and the control of the parameters that affect the quality of adhesive bond.

3 Bending Properties Of CLT

Bending properties such as modulus of rupture (MOR) and modulus of elasticity (MOE) of CLT are normally determined using a four-point bending test. EN 408 and EN 16,351 are generally referred in the determination of bending properties, especially in Europe. In the literatures given below, some modified testing methods were also been used based on the specimens and specified criteria being studied and determined. The bending properties from various literatures are summarized in Table 1.

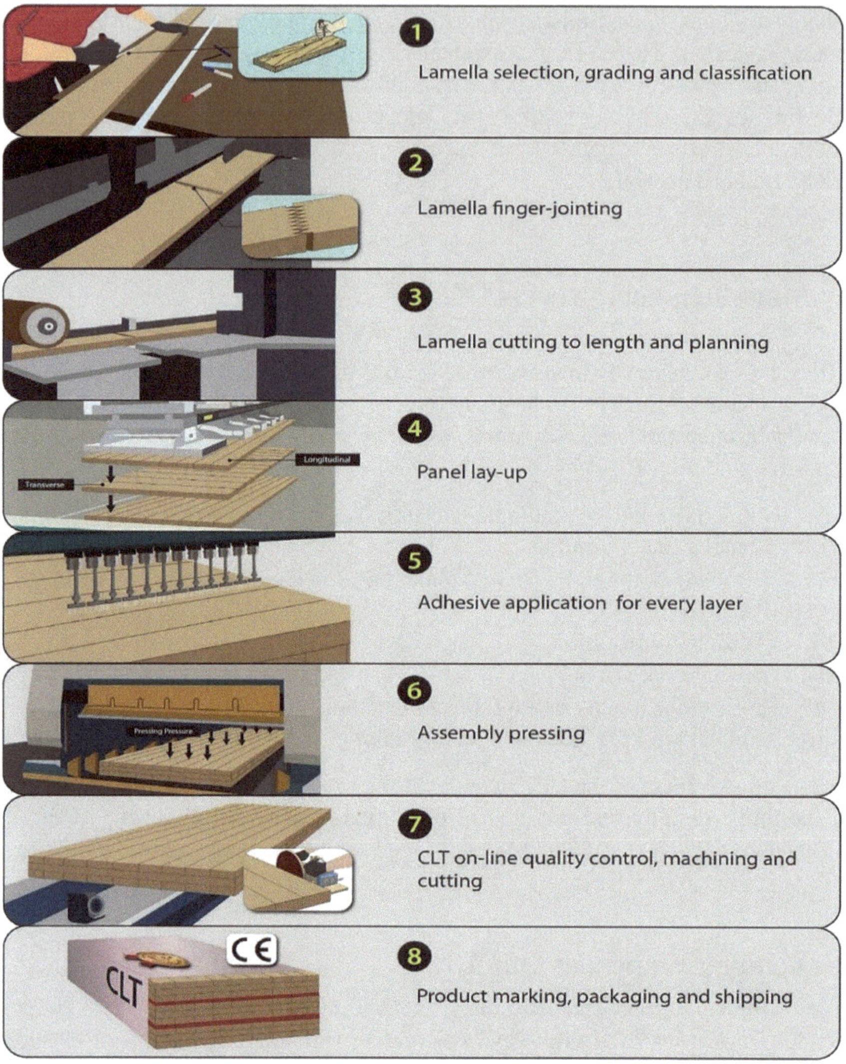

Fig. 1 Schematic diagram for typical CLT manufacturing processes

3.1 Factors Affecting Bending Properties

3.1.1 Timber Species and Density

Several researchers have studied the effects of timber species and density on the bending properties of CLT panels. In the study carried out by Franke (2016), beech (*Fagus* spp.) CLT showed better bending properties than spruce (*Picea* spp.) CLT. The result was expected as Beech (690 kg/m^3) having a higher density than spruce

Table 1 Bending properties of CLT panels correspond to the timber species used, testing methods and other related parameters

Timber species	Parameters	Testing method/comments	MOR (N/mm^2)	MOE (N/mm^2)	Reference
Eucalyptus urograndis, *Pinus taeda*	Species	Three-point bending	–	11,740 5461	Pereira and Calil (2019)
Eucalyptus nitens, *Eucalyptus globulus*	Species	Four-point bending	34.58–55.50 52.01–64.16	8900–12,100 11,000–13,700	Pangh et al. (2019)
Acacia mangium	Species, Type of adhesive	Four-point bending	27.78–36.55	10,740–12,693	Yusof et al. (2019)
Irish Sitka spruce (*Picea sitchensis*)	Species, thickness of panel	Four-point bending −60 mm −100 mm	35.71–35.98 34.08–34.43	7319–9552 6310–8404	O'Ceallaigh et al. (2018)
European Norway spruce (*Picea abies*)	Layer configuration	Four-point bending - CLT 90° - CLT 45°	35.2 47.5	8243 9517	Buck et al. (2016)
Beech (*Fagus* sp.)	Species	Four-point bending	43.8	12,306	Franke (2016)
Sesenduk (*Endospermum malaccense*)	Species	Four-point bending	47.14	13,069	Hamdan et al. (2016)
Irish Sitka spruce (*Picea sitchensis*)	Species, lamella thickness	Four-point bending −40 mm	24.56–37.67	–	Sikora et al. (2016)

(470 kg/m^3). The result was in agreement with the results obtained by Pereira and Calil (2019). Pereira and Calil (2019) tested CLT panels made from *Eucalyptus* timber and *Pinus* timber for bending properties. Their results showed that CLT panels produced from *Eucalyptus* timber exhibited higher strength and stiffness values compared to the panel made from *Pinus* timber. This result was well anticipated as the *Eucalyptus* panels produced in the study had a higher density than the *Pinus* panels. It is worth noting that *Eucalyptus* specimens showed higher drying defects after manufacturing and testing, however, the specimens did not show cracking. On the other hand, *Pinus* specimens include more natural growth defects, such as large knots, which may have reduced MOE of the *Pinus* specimens.

The study by Pangh et al. (2019) showed that two fibre managed plantation species, i.e., *Eucalyptus niten* and *Eucalyptus globulus* displayed higher flexural performance than other eucalyptus species reported in the literature. They also reported that the MOE and MOR of the CLT panels made from both species are positively correlated to the stiffness of the timber planks used in the top and bottom layers of the CLT specimens. It is generally known that, under flexural test, the top and bottom surfaces are subjected to compression and tensile load, respectively, and the middle layer acts as the neutral axis. This phenomenon is especially true in CLT panels where the middle layer is perpendicular to the top and bottom layers. This also explains the low rolling shear values in the middle layer during the bending test. Rolling shear properties of CLT will be further discussed in Sect. 4 of this chapter.

3.1.2 Type of Adhesives

The types of adhesives used also have an influence on the bending properties of CLT panels, as shown by the study of Yusof et al. (2019). The researchers studied the bending properties of CLT panels made from *Acacia mangium* bonded using two types of adhesives, namely, phenol-resorcinol–formaldehyde adhesive (PRF) and Polyurethane reactive adhesives (PUR). As demonstrated by their research, the mean MOE of the PRF-bonded panels was 12,639 N/mm^2, while the mean MOE of PUR bonded panels was 10,740 N/mm^2. The PRF-bonded panels showed better MOE properties, which was an 8% increment compared with PUR-bonded panels. The MOR of both types of panels also showed a similar trend. The mean MOR values of PRF-bonded panels (36.55 N/mm^2) were 14% higher than those of PUR-bonded panels (27.78 N/mm^2). It should be noted that the CLT panels produced in this study exhibited higher MOE but lower MOR compared with the results from the literatures.

3.1.3 Lamella Thickness and Layers Configuration

Other factors that influence the bending properties of CLT include Lamella thickness, size effect and layers configuration. Sikora et al. (2016) designed an experiment to determine the MOR, MOE and failure mode for CLT panels made from Irish Sitka Spruce (*Picea sitchensis*). Different thickness of lamella was set as the parameters in the study. In the study, the mean MOE results for each group of specimens tested were calculated per meter width. The researchers reported that there-ply CLT panels with the thickest lamella of 40 mm exhibited the highest MOE values. They also found out that increasing the thickness of the lamella from 20 to 40 mm increased the MOE values as much as 687–698%.

Buck et al. (2016) conducted a study to determine the effect of layer configuration on the bending properties of CLT panels. CLT panels with 45° or 90° alternating transverse layers were produced in the study. Four-point bending was used to determine the desired bending properties. They reported that CLT panels with 45° transverse layers performed better in bending tests when compared with conventional 90° transverse layers. The MOR of CLT with the 45° transverse layer increased by 35% while the MOE increased by 16% when compared with the CLT panels made with transverse layers arranged at 90°.

Results from past studies suggested that the MOE of CLT significantly depends on its lamella thickness. The higher the lamella thickness the higher the MOE. However, the MOR was statistically unfaceted by the lamella thickness of CLT specimens. Also, the orientation degree of the transverse layer has a significant effect on the MOR and MOE. The 45° arranged transverse layer shows both higher MOR and MOE.

3.2 Failures Modes of Bending Test

Generally, three modes of failure were observed in all CLT panels in the literatures, i.e. tension failure (Fig. 2), rolling shear failure (Fig. 3) and combined tension and rolling shear failure (Fig. 4). These failures were often accompanied by glue line failure. Pangh et al. (2019) reported that *Eucalyptus niten* CLT panel made with utility-grade timber failed on the tensile side. On the other hand, the CLT panels

Fig. 2 Tension failure

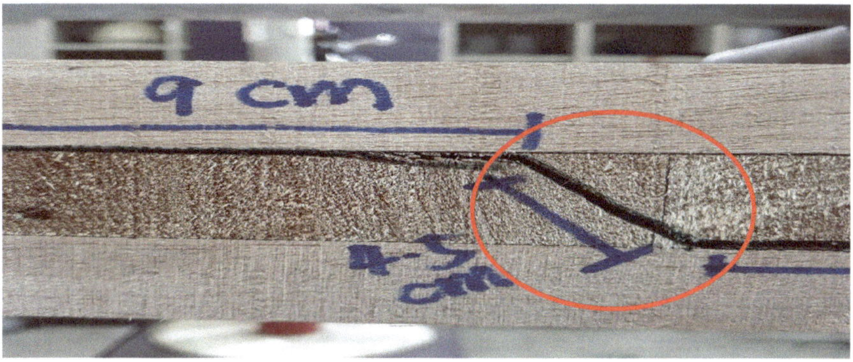

Fig. 3 Rolling shear failure

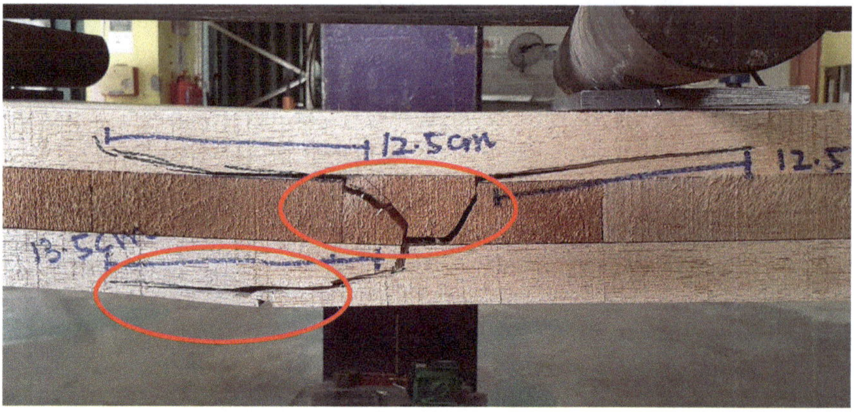

Fig. 4 Combined tension and rolling shear and failure

made with higher grade timber failed by rolling shear. Some degree of splitting was also observed in panels made with higher grade planks along with pure shear failure at the cross-section. However, CLT panels made from *Eucalyptus globulus* using higher grade planks failed under combined tension and rolling shear. The research also suggested that CLT panels produced from higher grade planks generally failed in the transverse layer due to rolling shear associated with splitting of the glue line. Based on the failure mode that occurred, the researcher concluded that the rolling shear strength of the CLT panels in the transverse layer together with the adhesion properties governs the overall bending performance of CLT panels.

Yusof et al. (2019) also found out that the CLT specimens they tested displayed similar modes of failure. The first type of failure mode is failure due to rolling shear stress in the transverse layer. This failure mode is more prevalent in PRF-bonded CLT than PUR-bonded CLT. The second failure mode occurred in the glueline, when the adhesive failed to sustain the subjected load. This type of failure mode was more commonly found on PUR-bonded CLT. The third failure mode is tension failure. This failure mode mostly happened in the tension zone of the lowest outer layer. Only PRF-bonded CLT showed this kind of failure mode. Sousa et al. (2013) also reported similar findings. The CLT specimens tested have mostly failed in the tension zone in the outermost layer and glue line failure in the middle layer. This finding is in agreement with the finding by Mohamad et al. (2011), which reported that the initial failure of the tested specimens often occurs in the glue line rather than in woods. Because of the existence of finger joint and adhesive in the CLT panels, most specimens tested exhibited brittle failure mode when compared with solid timber without finger joint and adhesive (Sikora et al. 2016).

Lim et al. (2020) reported that CLT panels treated and untreated with preservatives generally showed the same shear failure mode under four-point bending test setup. The load–deflection curves of the both CLT specimens were linear up to approximately 70% of their maximum loads. Then, the curves became nonlinear as shear cracks formed in the core layers at inclined angles. The nonlinearity became more severe as the shear cracks propagated towards the glue lines. Eventually, the wood fibres surrounding the cracks fractured in a brittle manner, which caused the load to abruptly drop.

4 Rolling Shear Properties of CLT

Rolling shear (RS) stress in CLT panel is defined as the shear stress acting on the radial–tangential plane perpendicular to grain, in other words, the shear stress of the cross-layers or transverse layers in CLT (Fellmoser and Blaß 2004). Rolling shear property of CLT panels is the decisive factor influencing the other mechanical properties of the panels such as MOR and MOE (Aicher et al. 2016). RS values of timber are rather low, normally in the range of 1.33–6 N/mm^2 (Table 2) when compared with its longitudinal shear values. As a result, precisely measuring the cross-layer's rolling shear properties is critical for CLT product design and application (Zhou et al. 2014). Rolling shear properties of CLT panels correspond to the timber species used, testing methods and other related parameters are summarized in Table 2. Typical rolling shear failure is shown in Fig. 5.

Rolling shear properties of CLT panels are determined by using either bending test method or two-plate shear test method. EN 408 and EN 16,351 are commonly used. However, it was commonly regarded that the two-plate shear test was a more appropriate test method for assessing the rolling shear modulus of a cross-layer in CLT. On the other hand, the bending test is more appropriate to determine the shear strength of CLT. It is because the bending test method could produce a failure mode more similar to that when CLT panels are subjected to bending load, which is a common loading configuration in the construction of building that produces rolling shear failure.

4.1 Factors Affecting Rolling Shear Properties

4.1.1 Influence of Species, Density and Sawing Pattern

According to Aicher et al. (2016), the rolling shear modulus of the tested CLT specimens was weakly correlated to their corresponding density regardless of the sawing pattern of the timber planks. However, when comparing CLT panels made with flat-sawn and pith boards, the difference in rolling shear modulus was about 20%, which was rather significant. The researchers also reported that, among the flat-sawn, quarter-sawn and semi-quarter-sawn specimens, semi-quarter-sawn specimens exhibited the highest rolling shear values. It can be concluded from the research that sawing pattern exerts more influence on the rolling shear modulus than the board density. Another study by Franke (2016) compared the rolling shear strength of CLT made of spruce and beech, respectively. The researcher found out that the rolling

Table 2 Rolling shear properties of CLT panels correspond to the timber species used, testing methods and other related parameters

Timber Species	Parameters	Testing method/comments	Rolling shear strength (N/mm^2)	Rolling shear modulus (N/mm^2)	Reference
United States grown southern yellow pine (*Pinus* spp.)	Species, Chemical treatment (micronized copper azole type C)	Four-point bending test (untreated) Four-point bending test (treated)	1.70–2.79 1.33–2.51	100–181 94–260	Lim et al. (2020)
Southern yellow pine (*Pinus* spp.)	Species, Presence of knots	Two-plate shear tests -No knot -With knot Short span bending test -No knot -With knot	1.95–2.05 2.33–2.62 1.77 1.86 –1.92	–	Cao et al. (2019)
Irish Sitka spruce *(Picea sitchensis)*	Thickness of panel	Four-point bending -60 mm -100 mm	2.14–2.22 1.39–1.40	–	O'Ceallaigh et al. (2018)
New Zealand Radiata pine (*Pinus radiata*)	Species, Testing method	Short-span bending tests Modified planar shear tests	1.97–2.45 1.99–2.33	–	Li (2017)
European beech wood (*Fagus sylvatica*)	Species, The presence of pith	Two-plate shear tests -with pith -without pith	4.5 6.0	370 370	Aicher et al. (2016)
Beech (*Fagus* sp.)	Species	Short-span bending tests	3.8	–	Franke (2016)
Irish Sitka spruce (*Picea sitchensis*)	Species	Four-point bending	1.09–2.09	–	Sikora et al. (2016)
Black spruce (*Picea mariana*)	Species, testing method	-Two-plate shear tests -Variable span bending test	2.57–2.84 2.02–2.13	31.55–38.12 46.27–88.53	Zhou et al. (2014)

shear strength beech CLT is more than two times higher than spruce CLT. This was owing to the fact that beech wood has a higher density than spruce wood.

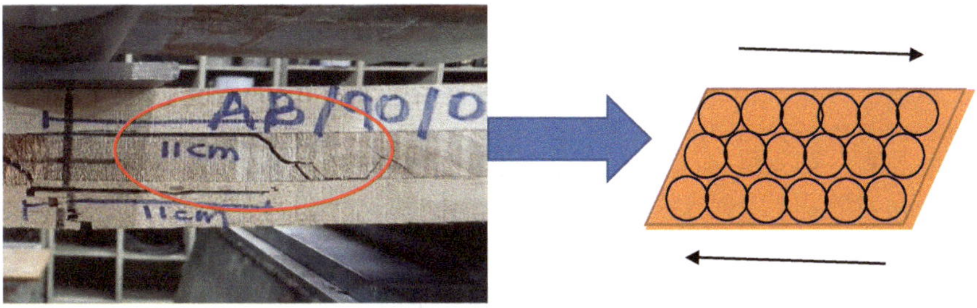

Fig. 5 Rolling shear phenomenon in CLT panels

4.1.2 Knots, Piths and Defects of Timber

In a study done by Cao et al. (2019), the RS strength of CLT specimens composed of cross-laminations with three knot conditions (without knot, intergrown sound knot and encased decayed knot) was determined. Surprisingly, the CLT panels made from the lamellae that contain knots exhibited better RS strength. The results obtained from this research suggest that the presence of knots in CLT panels did not adversely influence the RS strength properties. Furthermore, when comparing CLT panels made of lamellae with pith to CLT panels made of lamellae without pith, it was discovered that there was no substantial difference in RS strength between the two. The researchers also proposed that a two-plate shear test was more appropriate to be used to investigate the effects of particular cross-lamination features or conditions on RS strength. This is because the short span bending test method was more conservative, and the results obtained were not significantly affected by the heterogeneity of the cross-lamellae. Aicher et al. (2016) studied the effect of piths on the rolling shear strength of solid beech wood. However, the results were in contrast with those of Cao et al. (2019). Aicher found out that the pith in the specimens did indeed adversely affect its rolling shear strength. But it is worth noting that the specimens used were solid timber without taking the glue line of typical CLT specimens into consideration.

4.1.3 Lamella and Panel Thickness

Li (2017) studied the effect of lamella thickness (20 mm and 35 mm) on the RS strength properties of CLT panels made from Radiata pine. Short-span bending test and two-plate shear test were used to determine the RS properties. Both testing methods yielded similar results. The RS strength values of 20 mm thick lamella ranged from 2.33–2.45 N/mm^2. On the other hand, the RS strength values of 35 mm thick lamella ranged from 1.97 to 1.99 N/mm^2. In the bending tests, due to the short span, relatively high compressive stresses perpendicular to grain may be introduced in cross-layers thus affecting the evaluation of RS strength in cross-layers. In the modified planar shear tests, the specimens were loaded in a relatively "pure shear" mode and the minor compressive stress introduced by the small angle between the loading direction and the specimen major direction may not significantly affect the RS evaluation. Such a relationship between the lamination thickness and the RS strength can be also explained by the size effect on RS strength of wood. Besides the lamination thickness effect on RS strength, the width-to-thickness of laminations is also believed to affect RS strength evaluations. Sikora et al. (2016) also noted that RS strength was also adversely influenced by increasing CLT thickness. O'Ceallaigh et al. (2018) study the effects of panel thickness or the number of layers on the rolling shear strength of CLT. Two panel thicknesses of 60 mm (three-layered CLT) and 100 mm (five-layered CLT) were investigated. Four-point bending according to EN 408 was used to determine the material properties of CLT specimens. They found out that the rolling shear strength of CLT with 60 mm thickness ranged from 2.14 to 2.22 N/mm^2. On the other hand, the rolling shear strength of CLT with 100 mm thickness ranged from 1.39 to 1.40 N/mm^2. Thus, they concluded that rolling shear strength decreases as the panel thickness increases. However, the number of layers shows a negligible effect on the rolling shear strength.

4.1.4 Effects of Chemical Treatment

Lim et al. (2020) study the effect of chemically treated CLT panels on the rolling properties of the specimens. The specimens were subjected to a four-point bending test. From the results yielded, it was clear that preservative-treated CLT specimens had lower RS strength (1.87 N/mm^2) than the ones without treatment (2.16 N/mm^2). However, the treated CLT specimens showed higher RS modulus (147.72 N/mm^2) than the specimens without treatment (132.11 N/mm^2). It should be noted that the differences in the RS properties of the untreated and the treated CLT specimens were not statistically significant. Both treated and untreated CLT specimens exhibited rolling shear failure. Only untreated CLT specimens showed secondary bending failure modes near their loading points.

4.2 Failure Modes of Rolling Shear Test

Rolling shear failures from solid timber and CLT are discussed together in this subchapter to give a better understanding of the material properties. These failure modes are resulted from bending test and two-plate rolling shear test. This subchapter focuses on the failure mode of CLT tested using two-plate rolling test. In the two-plate shear tests, the specimens are subjected to a relatively "pure shear" loading. The failure modes of CLT specimens resulted from bending tests are discussed in Sect. 3.2.

In a study by Zhou et al. (2014), RS failure was studied using two-plate shear test and four-point bending test. When subjected to two-plate shear test (Fig. 6), the cracks started within the earlywood zone near the boundary between two growth rings, propagated in a zigzag pattern along growth rings and wood rays and accumulated in the

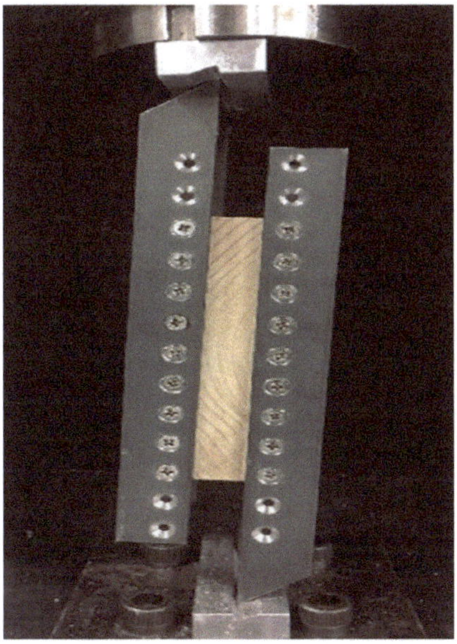

Fig. 6 Two-plate planar rolling shear test setup (*Source* Li 2017)

bonding area until delamination appeared (Fig. 7). Furthermore, rolling shear failure was observed in several locations along the cross-layer direction, demonstrating that the two-plate shear test was capable of producing nearly pure shear along the specimen. When subjected to four-point bending test, shear failure also initiated as a

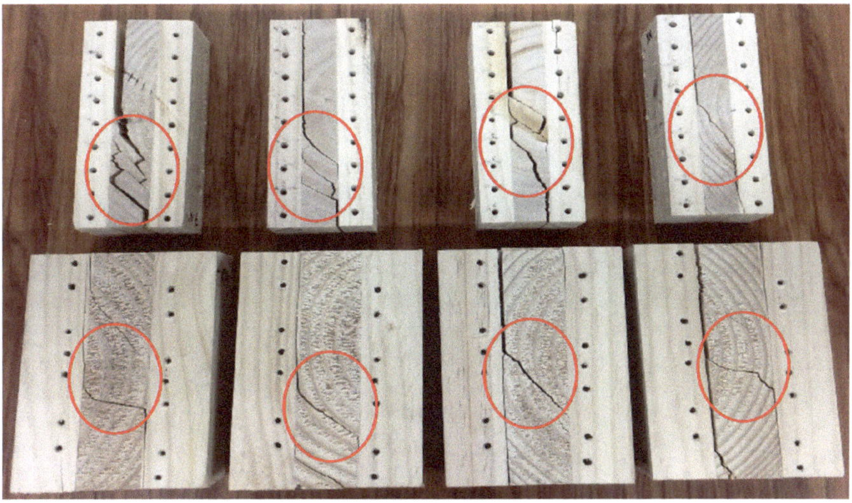

Fig. 7 Typical rolling shear failure of CLT specimens (*Source* Li 2017)

Fig. 8 Rolling shear failure of beech lamination (*Source* Aicher et al. 2016)

crack in the earlywood zones of the cross-laminate between the loading point and reaction support, mainly propagated along wood rays and secondly along a growth ring. Finally, specimens failed at the bonding surface or delaminated because of shear.

Lamella thickness and the presence of pith also affect the failure mode of CLT panels. Aicher et al. (2016) reported that solid beech lamination with piths exhibited pre-cracked and significant cracking under shear loading. Figure 8 shows the failure mode of beech lamination with the presence of pith. Cao et al. (2019) also found out that, for CLT made from yellow pine, initial shear cracks appear in the piths or in the earlywood/ decayed zone (brim) of the knots. Cracks propagated along or across the wood grain as shear stress increased, following the decayed knots' border. RS failures were found in all of the tested specimens, with cracks observed in the core layer at inclined angles to the glue lines. The failure patterns of the specimens showed that the sound knots prevented the fibres of cross-laminations to roll over each other, resulting in severe wood failure near the glue lines. The rolling shear failure mode of CLT made from yellow pine is illustrated in Fig. 9.

5 Applications of CLT in the Construction Industry

Structural components made from CLT are used primarily for walls and floor structures, but CLT panels may be used for a wide range of different applications, from small to large scale. Other applications of CLT include lift shafts and stairwells. Figure 10 illustrates the applications of CLT in the construction industry.

Fig. 9 Rolling shear failure of CLT specimens made with yellow pine (*Source* Cao et al. 2019)

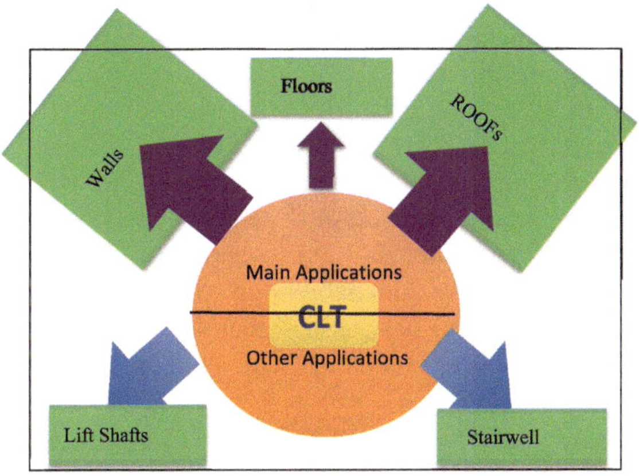

Fig. 10 Applications of CLT as load-bearing structural elements in the construction industry

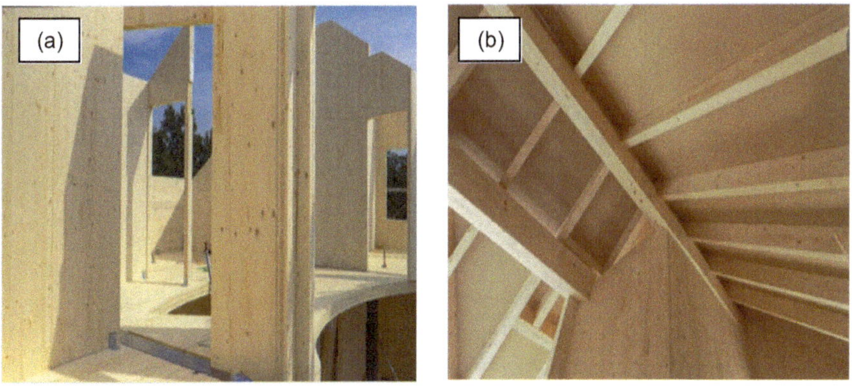

Fig. 11 Structural floor and wall structures (**a**) and structural roof truss made from CLT panels (**b**)

5.1 *CLT Panels as Floor and Walls Structures*

CLT panels can be used as structural floor and wall structures as shown in Fig. 11a. CLT panels can also be utilized as roof trusses of a building Fig. 11b. CLT as a floor structure must transfer vertical loads such as applied load and self-weight to the supports. The floor panel can also handle horizontal loads such as wind load. Panels are set on two supports in the most common form of a CLT structure. Load-bearing CLT panels may be loaded in one or more directions. It can be constructed as a simply supported panel strip if it is load bearing in one direction. It can be classified as a three or four-sided supported panel if it is equipped with two load-bearing directions. The principle for CLT floor structures is that the compression forces are absorbed by the top concrete block, and the majority of the tensile forces are absorbed by the underlying wooden frame in CLT floor structures. Rolling shear fractures can occur in CLT panels used for floor structures. Therefore, it is important that geometry and the quality of the manufactured CLT panels are taken into account when designing timber structure. CLT panels with no edge glueing or tongue and groove and with a width to thickness ratio of less than four are considered to have a bigger risk for shear failures.

Installations are a crucial aspect of a structure's design, and they commonly influence load-bearing component design. When wide holes are drilled in wood structures, reinforcement is often needed to guide forces past the installation holes. However, CLT panels have the advantage of being able to disperse and move forces to adjacent structures without the need for additional reinforcement, even with wide gaps. CLT wall panels usually have a high load-bearing capability. A linear load can be seen as the vertical load in a wall panel, and panels with a thickness of 80 mm can be built to take loads of over 100 kN/m.

Above are some examples of the application of CLT panels. It can be concluded that the mechanical behaviour of the CLT panel used in construction is complex. This is due to the nature of this engineered timber product as lamella of CLT is typically arranged in the perpendicular direction to each other. Besides that, the timber itself is also anisotropy in nature. Christovasilis et al. (2016) pointed out that this complexity of CLT actually works in favour of CLT in construction as that is usually the situations where perpendicular loads to the plane of the panel are applied, e.g., vertical or wind loads on floor or wall panels, loads in the vertical plane.

6 Challenges, Limitations and Future of CLT

The challenges and limitations faced by the CLT players in the construction industry include building regulations, the concern about raw materials and quality, the doubt on the performance of the CLT system, manufacturing technology, logistics and transportation etc. Various researches have been carried out to evaluate the performance of CLT construction against different building codes. The researches all suggested that

CLT structures meet the requirements stipulated, and, in some cases, even outperformed the requirements given (Hindman et al. 2012; Popovski and Gavric 2016). Nevertheless, there is still a concern about the performance of timber structures against fire, decay and earthquake resistance. More efforts and initiatives are needed in order to change the mindset and perspective of the public toward CLT panels and structures. Governments, universities and related industry players need to come together to provide a better information flow on the advantages and performance of CLT structures.

Another concern regarding the CLT panel and its structures is raw material supply and quality. Many consumers have negative feelings towards timber products and timber industry because they view it as the cause of deforestation and adversely impact the ecosystem. However, in order to ensure the feasibility of CLT products, most raw materials are sourced from sustainably managed forest and plantations. The timber industry has a positive effect on human lives, habitats and other natural resources that are interconnected with forests by developing sustainable practices for managing, harvesting and manufacturing forest products (Lippke et al. 2011). To minimize low timber yields, wildfires, water scarcity and severe animal impacts, native forests and plantations must be appropriately managed in all circumstances. Forest certification schemes like the Forest Stewardship Council (FSC) and the Program for the Endorsement of Forest Certification (PEFC) are well known for incentivising the private sector to adopt sustainable forestry practices.

CLT panels are commonly manufactured from softwood. The literatures suggested that most researches done were focused on the mechanical properties of CLT panels fabricated using softwood. Only limited studies on hardwood CLT panels were conducted. The lack of hardwood CLT panels might be due to some challenges faced when using hardwood as the raw materials for CLT. The challenges include longer drying time for hardwood compared to softwood of the same thickness and high shrinkage values of hardwood. Longer drying time will increase the cost of the end products, and high shrinkage value will interfere with the bonding performance of adhesive (Espinoza and Buehlmann 2018). However, in recent years, researchers have shown interest in hardwood CLT panels. This is a welcoming change that will promote and encourage the adoption of hardwood in the manufacturing CLT panels. Engineers and architects prefer hardwoods because they have higher mechanical properties than softwoods, allowing them to work with smaller cross-sections, longer spans and higher loads. Therefore, hardwood CLT has a bright future ahead.

7 Conclusions

CLT panels have great potential in the construction industry. It can be used as a green alternative to conventional materials. CLT panels can also be used to complement conventional building materials. Engineers, architects and designers have better flexibility and choices in selecting the desired building materials based on the required criteria. Nevertheless, despite the numerous benefits of CLT panels, the applications of CLT panels in the construction industry are still not gaining any desirable traction in developing countries mainly due to lack of exposure and financial constraints. Should CLT structures become a reality in the construction industry one day, such an industry would bring economic benefits to the participants along the CLT value chain and expand markets for underutilized timber species.

References

Aicher S, Christian Z, Hirsch M (2016) Rolling shear modulus and strength of beech wood laminations. Holzforschung 70(8):773–781. https://doi.org/10.1515/hf-2015-0229

BS En 16351 (2015) Timber structures—cross-laminated timber—requirements. BSI Standards Publication, London

Buck D, Wang XA, Hagman O, Gustafsson A (2016) Bending properties of cross laminated timber (CLT) with a 45° alternating layer configuration. BioResources 11(2):4633–4644

Cao Y, Street J, Li M, Lim H (2019) Evaluation of the effect of knots on rolling shear strength of cross laminated timber (CLT). Constr Build Mater 222:579–587. https://doi.org/10.1016/j.conbuildmat.2019.06.165

Christovasilis IP, Brunetti M, Follesa M, Nocetti M, Vassallo D (2016) Evaluation of the mechanical properties of cross laminated timber with elementary beam theories. Constr Build Mater 122:202–213. https://doi.org/10.1016/j.conbuildmat.2016.06.082

Crespell P, Gagnon S (2010) Cross laminated timber: a primer. FP Innovations 52p

Ehrhart T, Brandner R (2018) Rolling shear: Test configurations and properties of some European soft-and hardwood species. Eng Struct 172:554–572. https://doi.org/10.1016/j.conbuildmat.2019.06.165

EN standard (2012) 408 Timber structures–Structural timber and glued laminated timber–Determination of some physical and mechanical properties. European Committee for Standardisation, Brussels, Belgium

EN standard (2015) 16351 Timber structures–Cross laminated timber–Requirements. European Committee for Standardisation, Brussels, Belgium

Espinoza O, Buehlmann U (2018) Cross-laminated timber in the USA: opportunity for hardwoods? Curr for Rep 4(1):1–12

Eurocode 5 (2004) Design of timber structures. Part 1–1: general — common rules and rules for buildings EN 1995–1–1. CEN, Brussels

Fellmoser P, Blaß HJ (2004) Influence of rolling shear modulus on strength and stiffness of structural bonded timber elements. In: Larson HJ, Munch-Andersen J (eds) Proceedings of the CIB-W18 Meeting, 31 Aug-3 Sept 2004. Karlsruhe, Germany, Edinburgh, Scotland, UK

Franke S (2016) Mechanical properties of beech CLT. In: Proceedings of the WCTE 2016 world conference on timber engineering, Vienna, Austria (pp 22–25)

Hamdan H, Iskandar M, Anwar U (2016) Cross laminated timber: production of panel using Sesenduk timber species. Timber Technol Bull 59:1–6

Hindman D, Loferski J, Bond B, Kline E, Quesada-Pineda H J (2012) Development of low-grade hardwood cross-laminated timbers. Integr For Prod Grant NIFA (Number 2012-34638-20333)

Li M (2017) Evaluating rolling shear strength properties of cross-laminated timber by short-span bending tests and modified planar shear tests. J Wood Sci 63(4):331–337. https://doi.org/10.1007/s10086-017-1631-6

Liao Y, Tu D, Zhou J, Zhou H, Yun H, Gu J, Hu C (2017) Feasibility of manufacturing cross-laminated timber using fast-grown small diameter eucalyptus lumbers. Constr Build Mater 132:508–515. https://doi.org/10.1016/j.conbuildmat.2016.12.027

Lim H, Tripathi S, Li M (2020) Rolling shear modulus and strength of cross-laminated timber treated with micronized copper azole type C (MCA-C). Constr Build Mater 259:120419. https://doi.org/10.1016/j.conbuildmat.2020.120419

Lippke B, Oneil E, Harrison R, Skog K, Gustavsson L, Sathre R (2011) Life cycle impacts of forest management and wood utilization on carbon mitigation: knowns and unknowns. Carbon Manage 2(3):303–333. https://doi.org/10.4155/cmt.11.24

Mohamad WHW, Razlan MA, Ahmad Z (2011) Bending strength properties of glued laminated timber from selected Malaysian hardwood timber. Int J Civ Environ Eng 11(4):7–12

O'Ceallaigh C, Sikora K, Harte AM (2018) The influence of panel lay-up on the characteristic bending and rolling shear strength of CLT. Buildings 8(9):114. https://doi.org/10.3390/buildings8090114

Pangh H, Hosseinabadi HZ, Kotlarewski N, Moradpour P, Lee M, Nolan G (2019) Flexural performance of cross-laminated timber constructed from fibre-managed plantation eucalyptus. Constr Build Mater 208:535–542. https://doi.org/10.1016/j.conbuildmat.2019.03.010

Pereira MCDM, Calil Junior C (2019) Strength and stiffness of cross laminated timber (CLT) panels produced with Pinus and Eucalyptus: experimental and analytical comparisons. Matéria (Rio de Janeiro) 24(4). https://doi.org/10.1590/S1517-707620190002.0684

Popovski M, Gavric I (2016) Performance of a 2-story CLT house subjected to lateral loads. J Struct Eng 142(4):E4015006. https://doi.org/10.1061/(ASCE)ST.1943-541X.0001315

Santoni A, Schoenwald S, Van Damme B, Faust P (2017) Determination of the elastic and stiffness characteristics of cross-laminated timber plates from flexural wave velocity measurements. J Sound Vib 400:387–401. https://doi.org/10.1016/j.jsv.2017.04.018

Sikora KS, McPolin DO, Harte AM (2016) Effects of the thickness of cross-laminated timber (CLT) panels made from Irish Sitka spruce on mechanical performance in bending and shear. Constr Build Mater 116:141–150. https://doi.org/10.1016/j.conbuildmat.2016.04.145

Sousa HS, Branco JM, Lourenço PB (2013) Glulam mechanical characterization. In: Materials Science Forum (Vol 730, pp 994–999). Trans Tech Publications Ltd.

Viguier J, Jehl A, Collet R, Bleron L, Meriaudeau F (2015) Improving strength grading of timber by grain angle measurement and mechanical modeling. Wood Mat Sci Eng 10(1):145–156. https://doi.org/10.1080/17480272.2014.951071

Wang Z, Gong M, Chui YH (2015) Mechanical properties of laminated strand lumber and hybrid cross-laminated timber. Constr Build Mater 101:622–627. https://doi.org/10.1016/j.conbuildmat.2015.10.035

Yusof NM, Tahir PM, Lee SH, Khan MA, James RMS (2019) Mechanical and physical properties of Cross-Laminated Timber made from Acacia mangium wood as function of adhesive types. J Wood Sci 65(1):1–11. https://doi.org/10.1186/s10086-019-1799-z

Zhou QY, Gong M, Chui YH, Mohammad M (2014) Measurement of rolling shear modulus and strength of cross-laminated timber using bending and two-plate shear tests. Wood Fiber Sci 46(2):259–269. https://doi.org/10.1016/j.conbuildmat.2014.04.039

Delamination Test for Mengkulang Timber Species Using Methods A and C

N. J. Abd Malek, R. Hassan, A. Alisibramulisi, S. M. Abdullah Alesaei, and S. M. Sapuan

Abstract Malaysia is one of the tropical countries around the world, which is the pioneer in timber harvesting. Malaysian species such as Mengkulang is known for its strength and durability. The delamination test is considered a stimulation of the exposure to weather where it imitates the stresses applied to the wood during its life but in a much faster way and that result in the failure of the glue line between the wood sheets which is well known as delamination. The delamination test was conducted on Mengkulang timber species, testing using different methods. The total delamination percentage, maximum delamination percentage and moisture content before and after oven-dried were determined. Standards used in this research are Malaysian standard MS544, British standard BS EN 391:1995/2002 and British standard BS EN 301:1992.

Keywords Mengkulang · Delamination percentage · Maximum delamination · And moisture content

1 Introduction

Glulam timber is known as a type of structural engineered timber product. Glulam timber is made from several layers of dimensioned timber attached together. Through

N. J. A. Malek
School of Civil Engineering, Universiti Teknologi MARA, 40450 Shah Alam, Selangor, Malaysia

R. Hassan (✉) · A. Alisibramulisi
Institute for Infrastructure Engineering and Sustainable Research (IIESM), Universiti Teknologi MARA, 40450 Shah Alam, Selangor, Malaysia
e-mail: rohan742@uitm.edu.my

S. M. A. Alesaei
Faculty of Engineering and Built Environment, SEGi University Kota Damansara, Jalan Teknologi, 47810 Petaling Jaya, Selangor, Malaysia

S. M. Sapuan
Laboratory of Biocomposite Technology Institute of Tropical Forestry and Forest Products (INTROP), Universiti Putra Malaysia, 43400 UPM Serdang, Selangor, Malaysia

R. Hassan et al. (eds.), *Green Infrastructure*,
https://doi.org/10.1007/978-981-16-6383-3_8

lamination of smaller pieces of timber, a unit structural member of timber, which is big and tough, is formed. The new structural object can be used in several areas including horizontal beams, vertical columns, and arched, curved shapes. The ways these elements are being assembled provide the chance to have beams with many sizes and shapes. Nonetheless, these elements have their own defects as are subjected to the occurrence of delamination. This is resulted by assemblage errors such as an inadequate surface preparation of the lamellas or incorrect glue amount used to bond the lamellas amongst other errors or bad design of the structure. The delamination of glued laminated timber beams diminishes the capacity of these beams to carry load as well as their modulus of flexibility, which leads to greater deformations.

In Malaysia, these issues have led to flaws in the glued laminated timber beams, being necessary to repair the structure. Due to the fact that the removal of timber elements requites much time and cost a lot of money, a repair on-site is more preferable than any other option. The general belief in Malaysia is that timber is not a very effective material when being used in structure especially when it comes to locations that have a tropical climate like Malaysia. In this kind of climate, timber can suffer from insect and weathering harms. Malaysia is not alone in this belief, most of the European countries share the same type of philosophy and belief toward the usage of timber. Besides, there is an increasing awareness of the fact that timber is considered as higher fire hazard. Timber is more likely to cause fire compared with other structural materials like bricks and concrete.

In order to deal with this negative view of timber, a lot of studies have been conducted to determine the strength of the glued laminated timber. These studies determine the performance of delamination of the glued laminated timber for the purpose of ensuring that the glulam is appropriate and provide all the requirements just like any other materials used in the construction. Due to the limited usage as well as the small number of studies on glulam using tropical wood species, we do not have enough information about the topic. Delamination tests measure the integrity of glue lines by using it under fluctuating climate conditions. Therefore, these tests are used in the development and improvement of structural adhesives as well as in controlling the glued used in factory timber products designed uses in structures that would bear load. The test of delamination indicates in terms of Pass or Fail whether the adhesive bond can endure climatic pressures. Hence, this study reported on the delamination of Mengkulang timber species from SG5 group strength. Delamination of select timber species limited publications is available reporting on the delamination of tropical wood species. This study observed the delamination performance of glulam made from Mengkulang strength group timber SG5.

2 Literature Review

The beam made from laminas of wood was the first of engineering wood products mentioned by Hoadley (2000). Furthermore, those beams were made using sheets of wood spiked or dowelled with each other as it was found in the ships, old buildings, and constructions. Glued laminated timbers are applied widely in the construction of building. It is a timber product in which engineered structurally manufactured by sticking together individual segments of dimensional and strength graded timber parallel running with their grain under the conditions of manufacturing (Cheng and Hu 2011). Glulam has a great strength to weight ratio. It could be manufactured to any desirable shape and size, and span can be extended in length, while providing a beautiful natural look and effective utilization of the wood where it's lost in other substances. It can correspond to the different strength demands (Bakar et al. 2004). Figure 1 shows various standard shapes of glulam beams (Anonymous 1996).

Glulam can be defined as an engineered, stress-rated timber product assembled from chosen and prepared timber sheets glued jointly with adhesives with approximately parallel grain of the laminations to the longitudinal direction. Glulam is made by employ individual segments of great strength, kiln-dried timber, laminated jointly under pressure to shape massive timber components that keep the traditional charm of wood over with engineered intensity, elegant fire resistance, thermal efficiency, and dimensional consistency. Individual segments of timber are generally finger linked in the longitudinal direction in order to get laminates of demanded length (Hilbers et al. 2012). Figure 2 shows the process of manufacturing glulam (Anonymous 2003).

The highest thickness of each lamination is 50 mm. Horizontally glulam and vertically glulam are the two types of glued laminated member where the layers of the horizontal one are produced, and bond horizontally as presented in Fig. 3. However, the second type is vertically glulam in which the layers are produced and glued vertically in shape (Ozelton and Baird 2006).

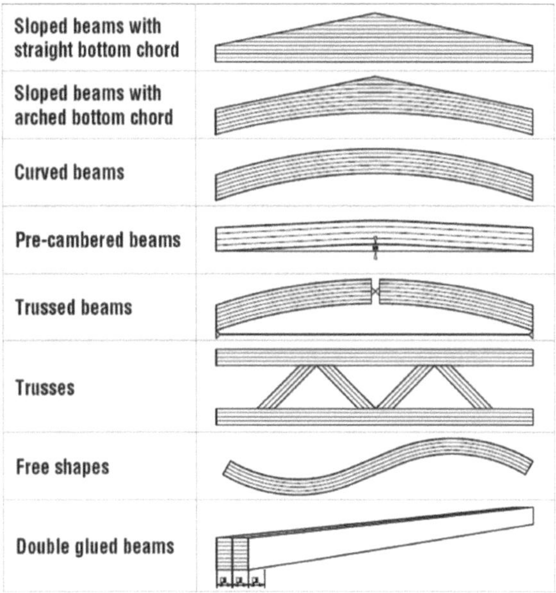

Fig. 1 Standard shapes of glulam beams (Anonymous 1996)

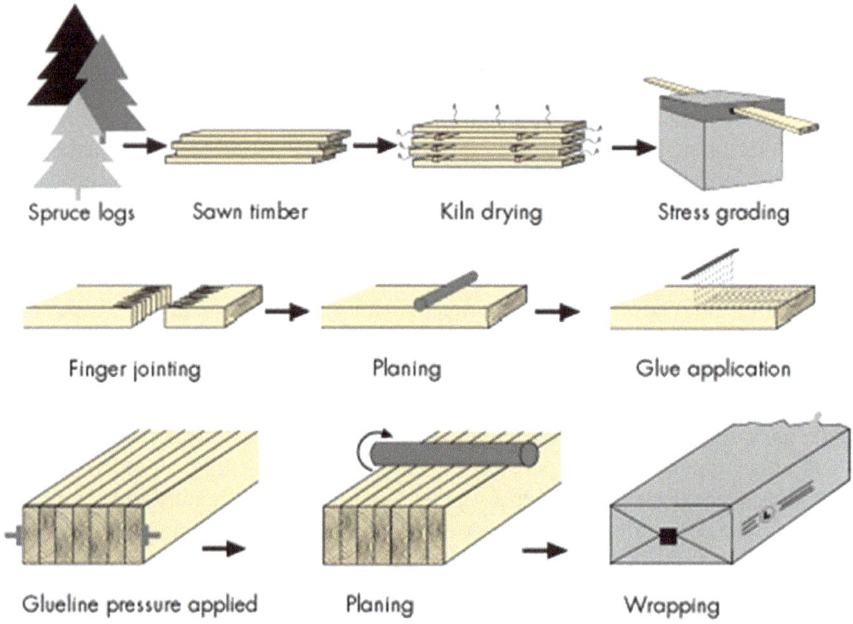

Fig. 2 Glulam manufacturing (Anonymous 2003)

Fig. 3 Glulam in different lamination and sizes (Ozelton and Baird 2006)

Glued laminated timber has unique production procedure, which permits the utilization of substance with desirable properties at desirable cross-section, along with complete quality control so it is considered as a highly engineered timber product. These large, laminated timbers can be made in nearly any straight or arched order for long-span case. Structural glulam members have been extensively used in developed countries especially Europe, Japan, and America (Wan Mohamad et al. 2014).

Regrettably, in Malaysia, the building construction industry did not utilize glulam as the building structure elements till recently, when Malaysian Timber Industry Board (MTIB) constructs an ideal glulam building by using Keruing and Resak glulam in Tebrau, Johor Bharu showing that local Malaysian timbers species can be used to manufacture glulam.

In Europe, typically use softwood to manufacture glulam where the most common types used are Norway Spruce and Silver where many kinds of buildings including gymnasiums, churches, recreational spaces, and auditoriums using long length glulam for full structural systems due to its suitability. Normally, utilizes in smaller buildings include ridge beams, doors, garage door headers and window headers, long-span girders, stair treads and stringers, and heavy timber trusses.

Singapore, Malaysia, Indonesia, Brunei, and Thailand are where equatorial timber normally can be found, for example; botanical name for Kempas is Koompassia Malaccensis and hail from Leguminosae family. It is categorized for treated timber, which appropriate for heavy construction while untreated timber suitable for structural use under cover. The classification of timbers with different strength group is presented in Table 1.

According to Wan Mohamad et al. (2014), Bintangor glulam is categorized under SG5 in term of group strength. Within the same group of Mengkulang, Bintangor glulam has passed the standard requirements according to MS758. While it gets less than 2% for total delamination after two cycles compared with woods from the group strength SG4 such as Resak which get 66.07% and prove that the wood from group strength SG5 able to withstand swelling and shrinkage in high temperature and humidity.

According to MS 544 (Part 2), Mengkulang timber is categorized into strength group 5 (SG 5). Figure 4 shows the Mengkulang texture timber species (Anonymous 2020).

Table 1 Strength group of timber (MS544:Part2:2001)

Strength group	Name of timber
SG 1	Balau, Bitis, Chengal, Penaga
SG 2/SG 3	Kempas, Kekatong, Belian, Mata Ulat, Petaling, Balau (red), Keranji, Kulim, Perah, Surianbatu, Bekak, Delek, Kembang Semangkok, Mertas, Penyau, Tualang
SG 4	Berangan, Kapur, Malabera, Meransi, Merpauh, Rengas, Dedali, Kasai, Merantibakau, Nyalin, Resak, Mempening, Merbau, Punah, Malaganggai
SG 5	Alan bunga, Brazil nut, Kungkur, Meranti (dark red), Meranti (white), Ramin, Tembusu, Babai, Gerutu, Kelendang, Mengkulang, Penarahan, Sengkuang
SG 6/SG 7	Ara, Damarminyak, Jenitri, Machang, Mersawa, Terap, Bayur, Jongkong, Meranti (light red), Meranti (yellow)

Fig. 4 Mengkulang timber (Anonymous 2020)

The typical name for lightwood Heritiera spp. in Malaysia is Mengkulang where it is considered as medium Harwood with cube air-dry density ranged from 625 to 895 kg/m^3. In terms of color, it is difficult to be differentiating between the red brown of sapwood and dark red brown of heartwood due to its light color. Mengkulang is a bit coarse when it comes to its texture with straight interlocked grain. In respect of shrinkage, the average of the tangential shrinkage is 1.5% where the average of the Radial shrinkage is 1.5%. Thus, shrinkage is considered high, so it is difficult to plane and saw, when it is uncovered to ecological conditions Mengkulang timber is not durable. It is also liable to lyctus beetle attacks in early stages. It is easy to treat Mengkulang timber with preservatives according to the Forest Research Institute Malaysia (Anonymous 2011).

Structural timber adhesives are one of the important elements for the process of manufacturing glulam. Thus, adhesives are in demand both for bonding the laminations together and for the finger joints in the individual laminations. Large finger joints utilize in construction a comparable adhesive is necessary.

Adhesive manufacturers involve a major industry, operating at a global scale and when it comes to structural timber products Adhesive manufacturers invest extraordinarily in developing and obtaining approvals. In addition to ensure favorable adhesive implementation, cleansing and operational safety adhesive companies supply precision equipment. Approved adhesives are defined and classified by internationally accepted standards. In connection with the Structural Eurocodes, these are developed through the European Standards Organization (CEN) and published by national standards bodies such as BSI.

Justo (2010) mentioned that Phenol Resorcinol Formaldehyde (PRF) is one of the most popular adhesives that have been used in laminated timber. Since it can cure at room temperature and has high humidity resistance making it sufficient for outdoor uses. The purpose of both the wood lamination to form glulam and finger joint it the adhesives family that have been utilized for a long period is the phenolic and amino plastic polycondensation range. In all situations, the adhesive manufacturers must explain that the specified trademark has been approved in accordance with BS EN 301 and the concerning standards. To meet these qualifications, general characterization such as simply naming the broad chemical family is not quite enough.

BS EN 301 covers PRF adhesive among the polycondensation types. In term of setting, they are divided into two hot and cold where the cold is the usual to be used in the process of glulam manufacture. In comparison with the *Melamine Urea Formaldehyde* (MUF) types, Phenol resorcinol formaldehyde PRF adhesives are somewhat more troublesome in terms of application, however, a number of PRFs trademarks have the highest potential moisture resistance, so regarding the Type I specification. They are yet recommended for utilization in external, hardwood glulam and in marine structures. In addition, there are many trademarks obtainable to fill the gap.

In general, the PRF adhesive considered one of the adhesive used in the structural wood which performed highly in exterior environment therefor the curing of the PRF adhesives most often happen in a alkaline environment especially in the wood species that have low rate of pH where the full treated wet and dry Norway Spruce samples glued with MUF adhesive shows a shear strength and percentage of wood failure clearly higher than the samples glued with the PRF adhesive according to Sernek et al. (2007).

Under this regulation, fundamental requirements must be accomplished in order to establish an initial type testing that all glulam adhesives are subjected to it. Also, to ensure the safety of the structure to be retained throughout its intentional lifetime, durability and creep resistance under the defined service class and hazard conditions must be included in the essential requirements. Structural timber adhesives are also employed for bonding-in threaded rods and identical connection components intended for ultimate installation during structure.

Glued laminated timbers were applied widely in the construction of building. It is a timber product in which engineered structurally manufactured by sticking together individual segments of dimensional and strength graded timber parallel running with their grain under the conditions of manufacturing as mentioned by Carlson et al. (2011). Glulam is made by employed individual segments of great strength, kiln-dried timber, laminated jointly under pressure to shape massive timber components that keep the traditional charm of wood over with engineered intensity, elegant fire resistance, thermal efficiency, and dimensional consistency. Individual segments of timber are generally finger linked in the longitudinal direction in order to get laminates of demanded length (Hilbers et al. 2012).

These large, laminated timbers can be made in nearly any straight or arched order for long-span case. Structural glulam members have been extensively used in developed countries especially Europe, Japan, and America (Mohamad et al. 2014).

Singapore, Malaysia, Indonesia, Brunei, and Thailand are where equatorial timber normally can be found, for an example; botanical name for Kempas is Koompassia Malaccensis and hail from Leguminosae family. It is categorized for treated timber which appropriate for heavy construction while untreated timber suitable for structural use under cover. Structural timber adhesives are one of the important elements for the process of manufacturing glulam. Thus, adhesives are in demand both for bonding the laminations together and for the finger joints in the individual laminations. Large finger joints utilize in construction a comparable adhesive is necessary.

The delamination test is considered a stimulation of the exposure to weather where it imitates the stresses applied to the wood during its life but in much faster way and that result in the failure of the glue line between the wood sheets which is well known as delamination (Sanabria et al. 2013). If the water content of the wood rose up to the fiber saturated point, the wood expands. On the other hand, if its dried and the water content reduced the wood will shrink between the expanding and shrinking, there is a stress developed at the glue line, which led to a delamination so the main purpose of delamination test is to create a similar condition in a shorter time compared with what the environment exposure do (Mills 2013). Thus, this research was done to determine and compare the maximum and total delamination percentages for Mengkulang species.

If the water content of the wood rose up to the fiber saturated point, the wood expands. On the other hand, if it is dried and the water content is reduced the wood will shrink. The difference between the expanding and shrinking there is a stress developed at the glue line which led to a delamination, so the main purpose of delamination test is to create a similar condition in a shorter time compare with what the environment exposure do (Alfred Franklin and Christopher 2013).

The development of the first delamination test was reported from the forest product laboratory and it took a place in 1956 in the USA. The one cycle took 30 days of soaking and 30 days of drying at normal temperature, and it results in a low amount of failure comparing with a 1-year weather exposure.

The discovery of applying a vacuum and pressure under the water help increase the water content rapidly to the fiber saturated point and beyond, all that leads to a faster test method where the one cycle involves vacuum and pressure under the water and then drying for 6 h in controlled temperature and relative humidity and the cycle repeated for three times. The outcomes were more delamination comparing with 1-year weather exposure (Hoadley 2000).

3 Method

Test samples included 16 samples of glulam made of the Malaysian Mengkulang timber species. All the samples given are within the same sizes where 75 mm in length (L) and 130 mm in width (b) and 150 mm in depth (h) with four laminations as shown in Fig. 5. The adhesives used in all the glulam samples are PRF adhesive.

The test procedure and the test preparation were conducted according to the British Standard BS EN 391:1995/2002 by using Method A and Method C for eight specimens, respectively. This pressure vessel was made to withstand safely a pressure of at least 600 kPa (700 kPa absolute pressure) and a vacuum of at least 85 kPa (15 kPa absolute pressures). It was equipped with pumps that can give a pressure of at least 600 kPa (700 kPa absolute pressure) and of drawing a vacuum of at least 85 kPa (15 kPa absolute pressure). Figure 6 shows the pressure vessel that used to

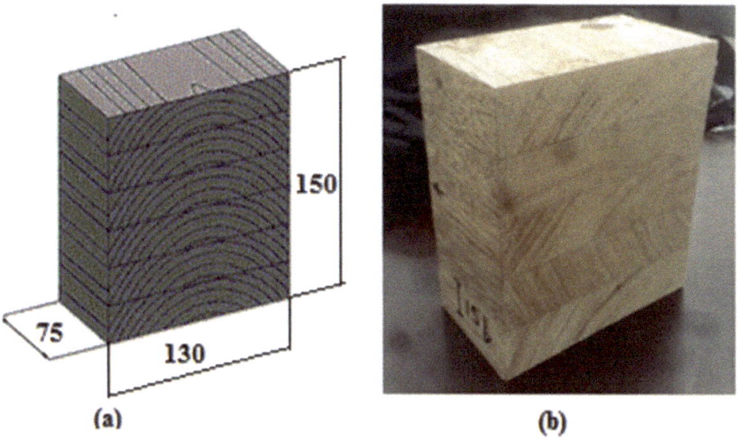

Fig. 5 Dimension of Mengkulang timber, **a** dimension of width, length, and height, **b** the real dimension

Fig. 6 Pressure vessel

apply pressure to the samples it can hold eight samples at one time. Drying duct is a machine where air is circulating at a velocity of 2–3 m/s and at a temperature and a relative humidity according to Table 2. The drying duct and function to dry the samples are displayed in Fig. 7. The units according to the British Standard BS EN 391:1995/2002 for vacuum pressure are between 70 and 85 kPa and for applied pressure are between 500 and 600 kPa; however, the reading gage of the pressure vessel was in Psi for the applied pressure and cmHg for the vacuum pressure, so the

Table 2 Temperature and relative humidity

Method	Temperature (°C)	Relative humidity (%)
A	60–70	<15
B	65–75	8–10
C	25–30	25–35

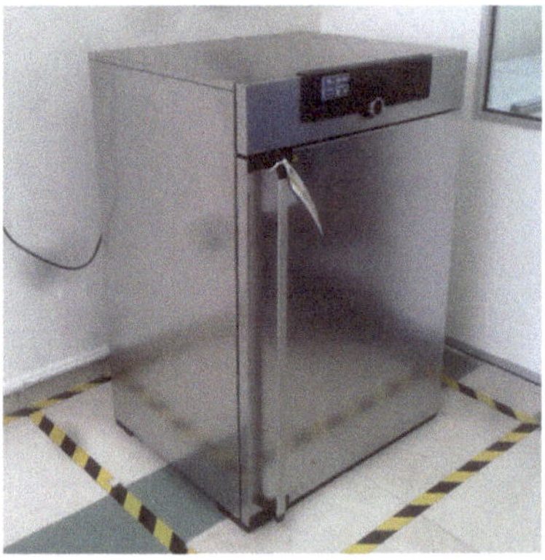

Fig. 7 Drying duct

unit was converted to get 80 Psi for the applied pressure and 24 cmHg for the vacuum pressure, and Fig. 8 presents the pressure vessel reading gages.

The total 16 specimens of all the two methods were weighed using a small balance and that will help to calculate the moisture content later. All the 16 specimens' mass was recorded in kg. The samples were dipped in water at a temperature between 10 and 20 °C completely about 8 specimen numbers at one time as shown in Fig. 9.

For Method A, the delamination specimens were dipped in water with a drawn vacuum of 70–85 kPa for 5 min. Then, a pressure was applied with 500–600 kPa for 1 h. After that, the vacuum pressure was repeated to complete two cycles impregnating process with a total time of 130 min. However, for Method C, the delamination specimens were dipped in water with a drawn vacuum of 70–85 kPa for 30 min. Then,

Fig. 8 Reading gages

Fig. 9 Submerged

a pressure was applied with 500–600 kPa for 2 h. After that, the vacuum pressure was repeated to complete two-cycle impregnating process with a total time of 5 h.

After all the specimens were taken out from the pressure vessel, they have been weighed while they are still wet. The test specimens were then dried at 60–70 °C in the drying duct with a relative humidity of less than 15% for 21–22 h for Method A. However, for Method C, the test specimens were then dried at 60–70 °C in the drying duct with a relative humidity of less than 15% for 90 h. Figure 10 shows the test specimen inside the oven before drying and Fig. 11 after drying. After the drying period, measurement was taken along the length of open glue lines on end grain surfaces of each test sample using a ruler all the delamination was recorded in mm as shown in Fig. 12. There was a range requirement of delamination length, which is from 3 to 5 mm, this range will only have considered as delamination.

Fig. 10 Before drying

(a) (b)

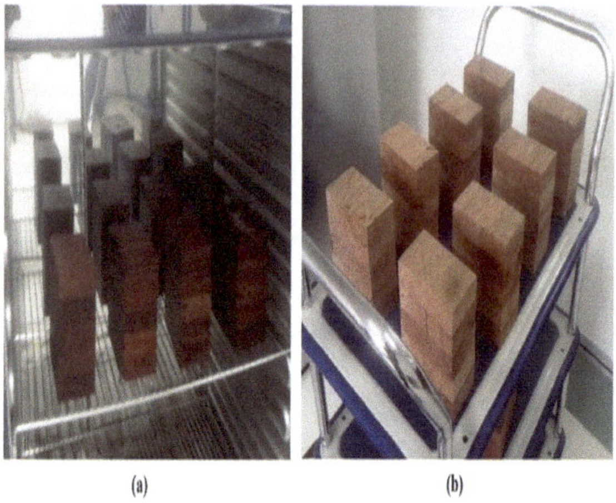

Fig. 11 After drying

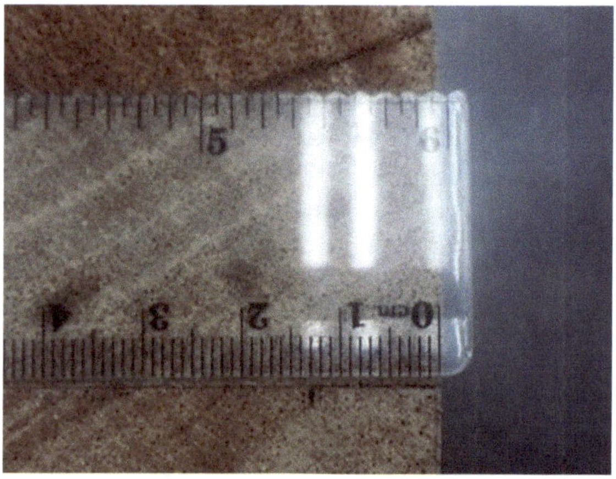

Fig. 12 Measuring delamination by ruler

The $l_{total,glueline}$ and it is found by adding measurement of all the blue and dashed blue line lengths together where the blue and dashed blue lines represent the glue lines as denoted in Fig. 13. Whilst, the $l_{glueline}$ was found by measuring the length of one blue line, and the $l_{total,delamination}$ was measured by adding the lengths of A, B, and C together where A, B, and C represent the delamination in the sample and $l_{max,delamination}$ was found by measuring A, which represent the maximum delamination length.

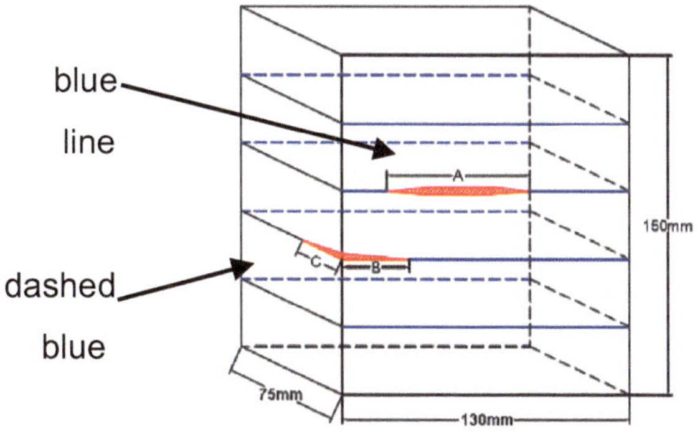

Fig. 13 Delamination measurement

4 Results and Discussion

Results and discussion are discussed about the outcomes of the delamination of Mengkulang glulam specimens. The analyses and the outcomes are sequenced according to the objectives. The first objective of this research is to determine the total delamination percentage and the maximum delamination percentage of glulam made of Mengkulang species, this goal is achieved in two stages starting with the manual measuring of $l_{tot,\mathrm{delam}}$, $l_{\mathrm{tot,Glueline}}$, $l_{\mathrm{max,Delam}}$, and l_{glueline} than substituted the measured data in the main formulas to calculate the total delamination percentage and the maximum delamination percentage as explained in the previous chapter. The second objective is to compare the percentage of total delamination and maximum delamination of Mengkulang glulam between the samples from Methods A and C. The last one is to determine the moisture content of all Mengkulang timber species before and after applying the delamination test.

There are eight samples with 75 mm in length (L) and 130 mm in width (b) and 150 mm in depth (h) with four laminations of Mengkulang glulam species used to study the delamination following Method A, which simulate the external effect especially in high temperatures. When the Method A of the delamination is done, and the outcome data were recorded as shown in Table 3.

Based on the results, the samples used in the delamination test Method A show the total delamination percentage 7.06% higher than the average is 1.01% in Method C as shown in Fig. 14. The maximum value in Method A is 25.67% in sample 5, on the other hand, the maximum value for Method C is 7.6% in sample number 1, and the average total delamination percentage for Method A is 7.06% for Method C, which represents the interior environment that means the glue lines in most of the samples did not affect due to the high pressure in the delamination test, and the Mengkulang glulam laminas did not separate and that because several factors such as temperature.

The temperature in the oven-drying process did not exceed 30 °C for almost 90 h compared with Method A where it samples dried at 70 °C for 21 h, it appears that the strength of the glue line bond becomes weak in the high temperatures despite the

Table 3 Data of delamination of Method A

Spec	$l_{tot,delam}$ (mm)	$l_{tot,gluline}$ (mm)	$l_{max,delam}$ (mm)	$l_{glueline}$ (mm)
1	137	1040	75	130
2	8	1040	3	130
3	15.5	1040	3	130
4	22	1040	10	130
5	267	1040	75	130
6	69	1040	19	130
7	20	1040	7	130
8	49	1040	9	130

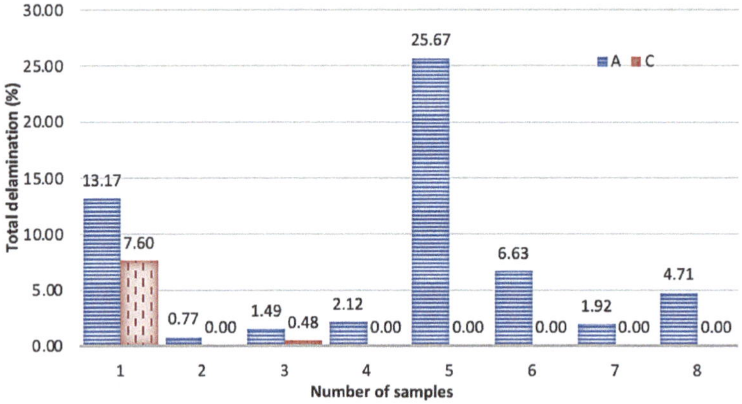

Fig. 14 Total delamination between methods A and C

duration of the drying. In the end, the effect generated from both methods was not enough to make the Mengkulang glulam fail inversely the result from both methods proved that the Mengkulang glulam timber was good enough to be used.

Table 4 shows individually to the total delamination percentage for Method A of the 8 samples of Mengkulang glulam found that differ in a range from 0.77% the lowest to 25.67% the highest and that because of the differences in the samples in terms of the characteristics and the composition between the glue and the timber laminas which form a bond. It is impossible to find two samples have the same characteristics and the composition 100% and that lead us with reference to the outcomes to understand that every sample varies in respect to the strength of the glue line, which plays the role of bonding the glulam laminas after been in an atmosphere with high pressure, high humidity, and high temperature.

Table 5 presents the total delamination percentage for Method C of the eight samples of Mengkulang glulam that delamination only occurs in two samples number 1 and number 3 with percentages of 7.6% and 0.48%, respectively, the rest have percentages of 0%. According to the standard requirement of delamination test

Table 4 Total delamination percentage and failure behavior for Method A

Specimens	Total delamination		Remarks
	<10%	>10%	
1	–	13.17	Fail
2	0.77	–	Pass
3	1.49	–	Pass
4	2.12	–	Pass
5	–	25.67	Fail
6	6.63	–	Pass
7	1.92	–	Pass
8	4.71	–	Pass
Average	7.06		Pass

Table 5 Total delamination percentage and failure behavior for method C

Specimens	Total delamination		Remarks
	<10%	>10%	
1	7.60	–	Pass
2	0.00	–	Pass
3	0.48	–	Pass
4	0.00	–	Pass
5	0.00	–	Pass
6	0.00	–	Pass
7	0.00	–	Pass
8	0.00	–	Pass
Average	1.01		Pass

Method C, the maximum value of the total delamination percentage for three cycles is equal or lower than 10% so any samples that go higher considered as failure comparing to our result, which all the eight samples did not cross the limit of maximum value thus is considered as acceptable according to the standard.

The samples used in the delamination test Method A give maximum delamination percentage higher than the ones used in Method C whereas most of the samples used in Method C give zero percentage as depicted in Fig. 15. The maximum value in Method A is 28.85% in the sample number 1 and 5, on the other hand, the maximum value for Method C is 7.31% in sample number 1 and the average total delamination percentage for Method A is 9.66% compared with 1.15% for Method C, which represents the interior environment. Eventually, the effect generated from both methods were not enough to make the Mengkulang glulam fail contrariwise the result from both methods proved that the Mengkulang glulam timber good enough to be used.

The percentage of moisture content of the samples before oven dried and after oven dried for Method A is presented in Fig. 16. The samples before oven-dried have higher percentage than the oven-dried samples almost two third where the before

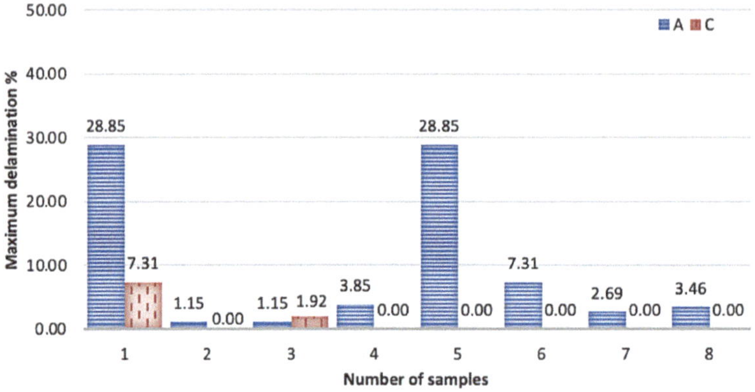

Fig. 15 Maximum delamination between methods A and C

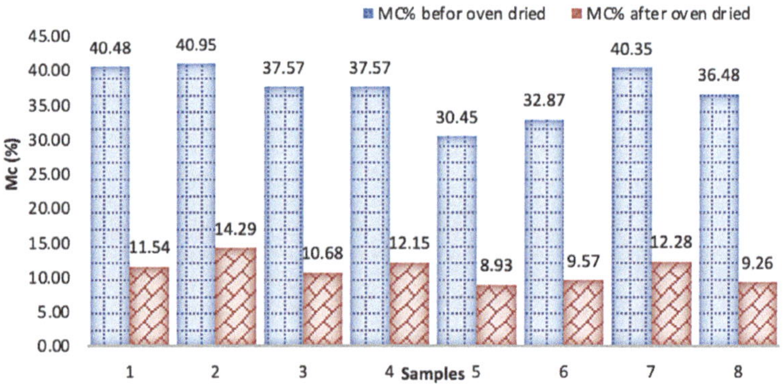

Fig. 16 Moisture content method A

oven-dried samples have an average of 37.09% of their weight as water comparing with 11.09% with the oven-dried samples and that due to the exposure to the high temperature in the oven, which help to reduce the water (Fig. 17).

The percentage of moisture content of the samples before oven dried and after oven dried for Method C is shown in Fig. 13. The samples before oven-dried have higher percentage than the oven-dried samples almost one third where the before oven-dried samples have an average of 46.55% of their weight as water compering with 29.28% with the oven-dried samples and that due to the exposure to the temperature in the oven, which help to reduce the water.

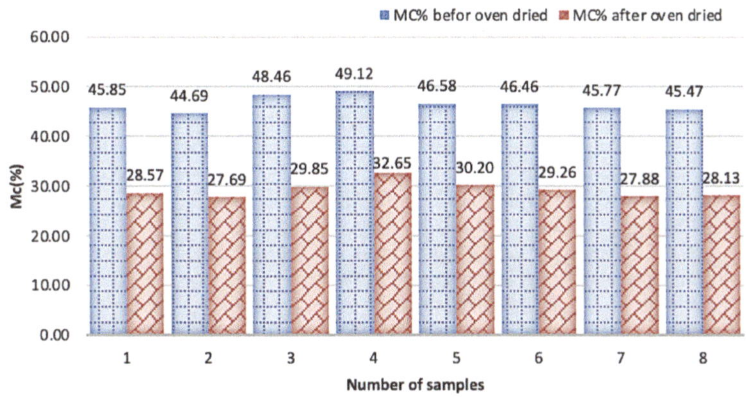

Fig. 17 Moisture content method C

5 Conclusion

Method A obtained a higher value compared with Method C due to the higher temperature that Method A experienced in the oven, which help weakening the glue bond yet both methods passed the maximum value for total and maximum delamination. This condition made glulam Mengkulang species bond with PRF suitable for use in exterior and interior conditions. The moisture content of glulam Mengkulang species contributes to bonding of PRF for Method A and C. The average moisture content percentage of eight samples for Method A was 37.09% before oven dried and 11.09% after oven dried where the high temperature of 70 °C and the low 12% of relative humidity played the major role in reducing the moisture content significantly within 21 h. On the other hand, the average moisture content percentage of eight samples for Method C was 46.55% before oven dried and 29.28% after oven dried where the temperature of 30 °C and the 25% of relative humidity help to reduce the moisture content but not as low as Method A, which represent the exterior condition. However, the samples in Method A had passed the requirement for moisture content according to MS 758: 2001 after the oven dried for treated glulam but the Method C after the oven dried failed.

Acknowledgements The authors gratefully acknowledge financial support from Universiti Teknologi MARA through MITRA PERDANA grant (600-IRMI/PERDANA 5/3/MITRA (001/2018)-2), SEGi University, Kota Damansara, and special assistant by Faculty of Civil Engineering laboratory technical staff that contributed to this research.

References

Alfred Franklin V, Christopher T (2013) Fracture energy estimation of DCB specimens made of glass/epoxy: an experimental study. Adv Mater Sci Eng 2013

Anonymous (1996) http://www.bcl-casa.kr/en/glulam_special_shaped_beams.php

Anonymous (2003) http://www.plywoodinspection.com/manufacturing-process-of-glulam/

Anonymous (2011) https://www.frim.gov.my/v1/fif/pdf_file/InFocusJUNE2011.pdf

Anonymous (2020) http://www.songsangplywood.com/timber.html

Bakar SA, Saleh AL, Mohamed ZB (2004) Factors affecting ultimate strength of solid and glulam timber beams. Jurnal Kejuruteraan Awam 16(1):38–47

Carlson A, Kim-Lee HJ, Wu J, Elvikis P, Cheng H, Kovalsky A, Turner KT (2011) Shear-enhanced adhesiveless transfer printing for use in deterministic materials assembly. Appl Phys Lett 98(26):264104

Cheng F, Hu Y (2011) Reliability analysis of timber structure design of poplar lumber with nondestructive testing methods. BioResources 6(3):3188–3198

Hilbers U, Neuenschwander J, Hasener J, Sanabria SJ, Niemz P, Thoemen H (2012) Observation of interference effects in air-coupled ultrasonic inspection of wood-based panels. Wood Sci Technol 46(5):979–990

Hoadley RB (2000) Understanding wood: a craftsman-s guide to wood technology. Taunton Press

Justo CF (2010) Intervención sobre los niveles de burnout y resiliencia en docentes de educación secundaria a través de un programa de conciencia plena (mindfulness)/Mindfulness program for increasing resilience and preventing burnouts in secondary school teachers. Revista Complutense De Educación 21(2):271–288

Mills C (2013) Determining the resistance to delamination of wood adhesives using a vacuum pressure impregnation method

Mohamad WHW, Ahmad Z, Jalil AA (2014) Bending strength of glulam from selected Malaysian hardwood timber. In: Advanced materials research, vol. 879, pp. 237–244. Trans Tech Publications

Ozelton EC, Baird JA (2006) Concentrated loadings. Timber designers' manual, Blackwell Science Ltd, pp 50–63

Sanabria SJ, Furrer R, Neuenschwander J, Niemz P, Sennhauser U (2013) Novel slanted incidence air-coupled ultrasound method for delamination assessment in individual bonding planes of structural multi-layered glued timber laminates. Ultrasonics 53(7):1309–1324

Sernek M, Resnik J, Kamke FA (2007) Penetration of liquid urea-formaldehyde adhesive into beech wood. Wood Fiber Sci 31(1):41–48

Wan Mohamad WH, Ahmad Z, Abd Jalil A (2014) Bending strength of glulam from selected Malaysian hardwood timber. Adv Mater Res 879:237–244. https://www.scientific.net/amr.879.237

Analysis of the Flexural Strength of Reinforced Beam with Bamboo by Empirical Modeling Using Statistical Model

W. A. Y. Banafea and T. A. R. Hussin

Abstract This research was done to increase the knowledge base of bamboo-reinforced concrete beam tested under the four-point load (one-third of support span) test. The objectives of the research were to investigate statistically the effects of various parameters of the flexural strength, determine the most influencing parameter, and develop empirical modeling to predict the flexural strength by using SPSS software. In addition, checking the efficiency of the developed model by calculating MAPE and MAD. Based on the results the parameters in this research have a certain amount of influence on the flexural strength, depending on the strength of the correlation between each parameter with the flexural strength and between them. The most influential parameter on the flexural strength was the modulus of elasticity of bamboo, which has the strongest correlation. Moreover, it was found that it had a certain effect on the accuracy of the empirical models (R^2) in some outputs when it is involved. In this research, 63 models were developed to predict flexural strength. It was found that model 1 from nonlinear regression power models had the highest accuracy. It had very close prediction results to the previous experimental results data, with less MAPE and MAD, which were 6.18% and 0.72, respectively.

Keywords Bamboo-reinforced concrete beam · Flexural strength · Empirical modeling · Prediction · Parameters

1 Introduction

The use of bamboo as an alternative to steel has become a sign of development construction methods. The use of bamboo composite materials in construction is more than steel in regions where bamboo is available (Brownell 2015). Since concrete is

W. A. Y. Banafea · T. A. R. Hussin (✉)
Faculty of Engineering, Built Environment, and Information Technology, SEGi University, Taman Sains Selangor, No. 9, Jalan Teknologi, PJU 5, 47810 Kota Damansara, Petaling Jaya, Selangor, Malaysia
e-mail: anitahussin@segi.edu.my

R. Hassan et al. (eds.), *Green Infrastructure*,
https://doi.org/10.1007/978-981-16-6383-3_9

only suitable for compression, steel that is good for tension is used to make concrete a more viable option for building complex projects and withstand more load efficiently. However, steel is expensive and limited. Alternatives such as bamboo are being investigated to cut down the cost and help use resources available efficiently. It is economically cheaper than steel and is fast growing. Also, it is widespread in eastern countries.

According to Rama and Rao et al. (2018), the use of steel has become very limited now, which made researchers direct search for cheaper natural alternatives than steel. Bamboo was a suitable alternative to strengthening concrete. Bamboo has been used for thousands of years to make homes, bridges, and other commodities, and it also has a fast-growing speed that makes it widely available. Studies conducted found that its mechanical properties are close to those of steel. The study of bamboo concrete reinforcement dates back several decades ago, after World War II (Hector Archila et al. 2018). These days there are ongoing studies to improve bamboo concrete by treating bamboo or increasing the percentage of reinforcement to bear more loads in the flexural test.

This research aims to investigate statistically various parameters that affect the flexural strength of beam reinforced with bamboo and determine the most influencing parameter of the flexural strength. Moreover, it aims to develop empirical models that can forecast the flexural strength and test the validation of the models to choose the suitable model. The empirical analysis was used to fulfill the objectives of this research by using SPSS software.

2 Literature Review

The development of empirical models that can predict the performance of concrete-reinforced steel or with different materials, whether in flexural and shear strength or other parameters, has greatly increased. Here are some of the researches that developed empirical models for reinforced concrete to predict its performance in addition to researches related to reinforced bamboo concrete beam, which is the focus of this research;

Otunyo and Odebiyi (2018) conducted a research study on the regression modeling of the strength properties of hybrid fiber-reinforced concrete made with polypropylene fiber (PPF) and alkali-resistant glass fiber (ARGF). The equations obtained from multiple regression analysis are as follows:

Compressive strength:

$$\mathrm{C} = 33.23 - 2.1104X + 0.0609Y + 0.0765Z \tag{1}$$

Split tensile strength:

$$\mathrm{T} = 2.26 - 0.089X + 0.0061Y + 0.0033Z \tag{2}$$

Flexural strength:

$$F = 4.52 - 0.4741X + 0.0203Y - 0.0227Z \quad (3)$$

where C, T, and F are the predicted compressive strength, split tensile strength, and flexural strength, respectively; X is the volume of hybrid fiber; Y is the volume of polypropylene fiber; and Z is the volume of alkali-resistant glass fiber in the concrete mix. The results showed that the regression equation for strength prediction and the estimated strength parameters are close to the experimental values.

Vijayan and Revathy (2016) conducted a research on multiple regression model for the prediction of flexural behavior of fiber-reinforced polymer (FRP) plated prestressed concrete (PSC) beams. The aim of the research was to develop a mathematical model by regression analysis for the prediction of load, deflection, and ductility for fiber-reinforced polymer plated prestressed concrete beams. The developed regression equations are:

Yield load:

$$18.17 - (0.000308\ f_{ck}) + (0.052\ t_f) + (0.047 f_{fu}) \quad (4)$$

Yield deflection:

$$3.72 - (0.00015\ f_{ck}) + (0.3232\ t_f) + (0.0081 f_{fu}) \quad (5)$$

Ultimate load:

$$48.51 - (0.00457\ f_{ck}) + (8.05\ t_f) + (0.22 f_{fu}) \quad (6)$$

Ultimate deflection:

$$37.07 - (0.00164\ f_{ck}) + (9.24\ t_f) + (0.06079\ f_{fu}) \quad (7)$$

Deflection ductility:

$$13.27 - (0.000833\ f_{ck}) + (1.38\ t_f) + (0.024 f_{fu}) \quad (8)$$

Energy ductility:

$$33.07 - (0.000125\ f_{ck}) + (2.546\ t_f) + (0.0101 f_{fu}) \quad (9)$$

Maximum crack width:

$$1.5 + (0.0000814\ f_{ck}) - (0.2123\ t_f) - (0.0027 f_{fu}) \quad (10)$$

The results showed that the predicted results were in considerable agreement with the experimental values. Hence, it was suggested that the developed models were best suited for the flexural performance of FRP plated prestressed concrete beams.

Another research was conducted by Shahnewaz and Shahria Alam (2014). This research aimed to develop the analytical equations to predict the shear strength of steel fiber-reinforced concrete (SFRC) beams by genetic algorithm (GA). The developed equations for this research are:

For SFRC deep beams:

$$\begin{aligned} V = {} & 0.2 + 0.034 f_{c'} + 19\rho^{0.087} - 5.8(a/d)^{1/2} + 3.4V_f^{0.4} \\ & - 800(l_f/d_f)^{-1.6} - 12((a/d)\ V_f)^{0.05} \\ & - 197((a/d)\ (l_f/d_f))^{-1.4} + 105(V_f(l_f/d_f))^{-2.12} \end{aligned} \tag{11}$$

For SFRC slender beams:

$$\begin{aligned} V = {} & 0.2 + 0.072(f_{c'})^{0.85} + 12.5\rho^{0.084} - 24(a/d)^{0.07} \\ & + 13.5V_f^{0.07} + 450(l_f/d_f)^{-2} - 0.0002((a/d)\ V_f)^{3.9} \\ & - 27.69((a/d)\ (l_f/d_f))^{-0.84} + 1181(V_f(l_f/d_f))^{-2.69} \\ & - 21.89((a/d)\ V_f)\ (l_f/d_f))^{-0.9} \end{aligned} \tag{12}$$

where b is the beam width, h is the height, d is the effective depth, a/d is shear span-to-effective depth ratio, f_c is the compressive strength of concrete, f_y is the yield strength of steel, ρ the longitudinal reinforcement ratio, l_f/d_f is the aspect ratio of fibers, and v_f is the fiber volume. The results showed that the interactions were significant. Therefore, nonlinear interaction terms were incorporated in the genetic algorithm to develop equations for deep and slender beams. The results showed that the proposed shear equations produced less scatter with a significant improvement in coefficient of variation, standard deviation, and average absolute error. A design example is presented for a reinforced concrete (RC) beam with steel fibers. It was observed that the steel fibers could reduce the stirrups requirements significantly in the RC beams by providing sufficient shear strength in fiber–concrete matrix.

A research was conducted by Muhtara et al. (2019). One of the objectives of the research was to increase bond stress, flexural capacity, and slip resistance by using a hose-clamp. Test results showed that the hose-clamp helped increase bond stress and flexural capacity and reduced the slip between the bamboo and the concrete. Due to different stress, strain, and elastic modulus properties of materials in steel and bamboo, the friction bond limit in steel reinforcement concrete occurs at 0.4 and 0.2 in bamboo reinforcement concrete.

Kavitha et al. (2018) conducted a research study on the flexural behavior of bamboo-reinforced concrete members. The flexural strength and load-deformation behavior of bamboo-reinforced beams were investigated. From the results, they found that bamboo had the potential to replace steel for reinforcements.

Siddika et al. (2017) conducted a research on evaluating bamboo reinforcements in a structural concrete member. The results indicated that the beams and columns reinforced with a good ratio of bamboo reinforcement showed the best strength-cost among plain and steel-reinforced concrete beams.

According to Atul Agarwal et al. (2014), another promising way of developing the reinforced beam is to increase the bond strength at the interface of the bamboo concrete composite. Based on the experiment results, the beam's load-carrying capacity increased up to 29.41% by using merely 1.49% by area of treated bamboo as reinforcement.

The study of development equations or empirical models for concrete reinforced with steel or different materials has been done to study their performance in several tests. However, there is a lack of research that investigates statistical parameters that affect the flexural strength of reinforced bamboo concrete beam tested under four-point load (one-third of support span) and determine the most influential parameter that affects flexural strength as well and develop an empirical equation to predict it. This research was adopted to do an empirical analysis from previous research to establish an empirical model to predict the flexural strength of the bamboo-reinforced concrete beam.

3 Research Methodology

In this study, Statistical Package for the Social Sciences (SPSS) software was used to develop a number of empirical models that can predict the flexural strength of beam reinforced with bamboo tested under four-point load (one-third of support span). The models established after went through several stages. The first stage was gathering data from previous research related to testing the flexural strength of a reinforced beam with bamboo under a four-point load (one-third of support span).

The second stage was about selecting and identifying the variables of data to be used in the study. To develop empirical modeling by using a statistical model, it must identify the variables. The dependent variable (DV) was the flexural strength which was the result of the test, and the independent variables (IVs) were the volumetric properties that influence the flexural strength or the DV.

In the third stage, plotting of the dependent variable versus independent variables in a graph was done. This stage was important to determine the relationships between each IVs to the DV in linear, power, and quadratic functions, and to predict which function can be suitable.

The fourth stage was about entering the DV and IVs to SPSS software for developing the models. Before starting to develop the models, the input needed to be checked for the outlier or the normality of distribution and that by frequency. After checking the outlier, the correlation and developing the models can be done properly. Determining the correlation for each IVs to the DV was done to identify the most influential parameter.

Then the developed models are analyzed using linear and nonlinear regression analysis, where the results of the regressions show the coefficients that can be taken to establish the model and the R^2 of models based on the inputs. The R^2 of the models is also used to identify the most influential parameter.

After developing all possible models, in the fifth stage, taking the models with the highest R^2 in each linear and nonlinear regression was done to check the validation by using the same data that was used for modeling, due to the limited amount of research that had same parameters. Therefore, the same data was used for validation.

The last stage was about selecting the suitable model that predicts the flexural strength, and that was based on two conditions. The first condition was R^2, where R^2 represents the model's accuracy. The second condition calculates the percentage of error of the results of the selected models to see how close they were to the actual flexural strength from previous research by the mean absolute percentage of error (MAPE) equation and the average difference between them by the mean absolute deviation (MAD) equation:

$$\text{MAPE} = \left(\frac{1}{n}\sum\frac{|A-P|}{|A|}\right) \times 100 \tag{13}$$

$$\text{MAD} = \left(\frac{1}{n}\sum|A-P|\right) \tag{14}$$

where A is the actual flexural strength from previous research, P is the predicted result from the models, and n is the number of models. Fig. 1 shows the flowchart of the methodology.

4 Results and Discussion

4.1 Collected Data and Selection Variables

Empirical models were established for flexural strength of the reinforced concrete fiber, unlike bamboo, where there was a lack of developing a model. Therefore, to develop a model a number of data were collected from several previous research, and the flexural strength of reinforced beam concrete with bamboo under four-point loads (one-third of support span) is tested by selecting and identifying the variable. The DV was flexural strength, and the IVs were reinforcement ratio, bamboo strength, bamboo modulus of elasticity, area of the reinforcement, and concrete strength, as shown in Table 1.

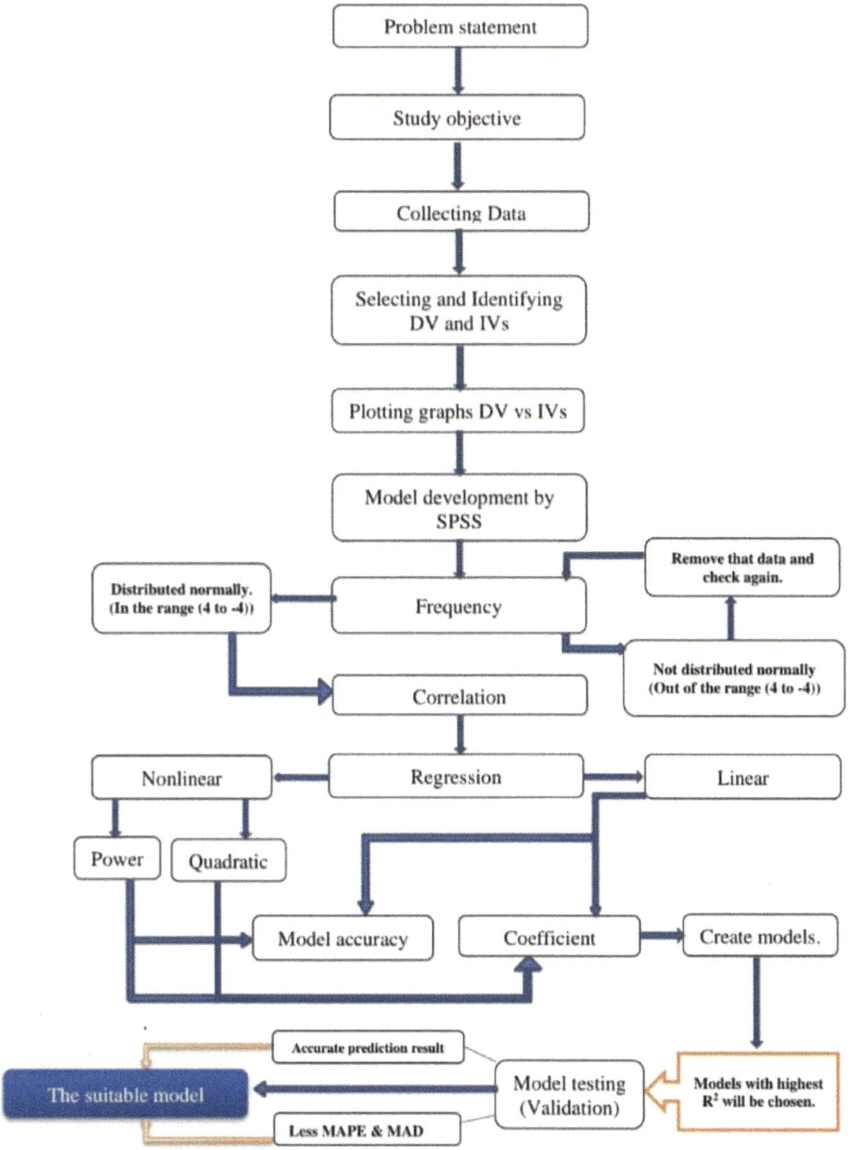

Fig. 1 The flowchart of the methodology

4.2 Plotting Graphs

After selecting and identifying the DV and IVs from Table 1, the graphs of the relationships between IVs and DV were plotted to determine the function of the empirical model approximately. Usually, this can be known from previous research; however,

Table 1 Experiment data from previous research

Authors	Flexural strength (MPa)	Load	Length	Width	Depth	Reinforcement ratio	Bamboo strength	Bamboo modulus of elasticity	Total area of reinforcement bamboo	Concrete strength
	$flt = PL / bd^2$	P (KN)	L (mm)	B (mm)	D (mm)	p (%)	σ = F/A (Mpa)	E = σ / ε (Mpa)	b x h or $\pi d^2/4$ (mm^2)	(MPa)
	DV					IV1	IV2	IV3	IV4	IV5
Atul Agarwal et al. (2014)	7.77	13.117	1000	75	150	1.49	185.93	24,460	140	20
	5.83	9.838	1000	75	150	1.49	185.93	24,460	140	20
Sevalia et al. (2013)	3.96	11.6	750	130	130	1.377	56.87	37,913.33	376	25.79
	4.27	12.5	750	130	130	2.168	56.87	37,913.33	593	25.79
Muhtara et al. (2019)	12.89	21.75	1000	75	150	1.24	126.68	17,235.74	140	31.31
	10.96	18.5	1000	75	150	1.24	126.68	17,235.74	140	31.31
	13.19	22.25	1000	75	150	1.24	126.68	17,235.74	140	31.31
	12.3	20.75	1000	75	150	1.24	126.68	17,235.74	140	31.31
	16.44	27.75	1000	75	150	1.78	126.68	17,235.74	200	31.31
	18.22	30.75	1000	75	150	1.78	126.68	17,235.74	200	31.31
	18.67	31.5	1000	75	150	1.78	126.68	17,235.74	200	31.31
	17.19	29	1000	75	150	1.78	126.68	17,235.74	200	31.31
	17.93	30.25	1000	75	150	4	126.68	17,235.74	450	31.31
	18.96	32	1000	75	150	4	126.68	17,235.74	450	31.31
	19.7	33.25	1000	75	150	4	126.68	17,235.74	450	31.31
	17.63	29.75	1000	75	150	4	126.68	17,235.74	450	31.31

since no empirical model for flexural strength of reinforced bamboo concrete beam has been done, therefore it can refer to the relationships between DV and IVs to predict function.

Figures 2, 3, 4, 5 and 6 show the relationships between the dependent variable (DV) which is the flexural strength of the reinforced bamboo beam versus each independent variable (IV) which are the ratio of reinforcement, bamboo strength, modulus of elasticity of bamboo, area of reinforcement, and concrete strength. The data was obtained from Table 1.

From the figures above, it can be observed that each IV has three R^2 from different functional relationships. These functional relationships are summarized in Table 2.

Table 2 indicates that the reinforcement ratio, bamboo strength, area of reinforcement, and concrete strength have a quadratic relationship based on R^2, while the

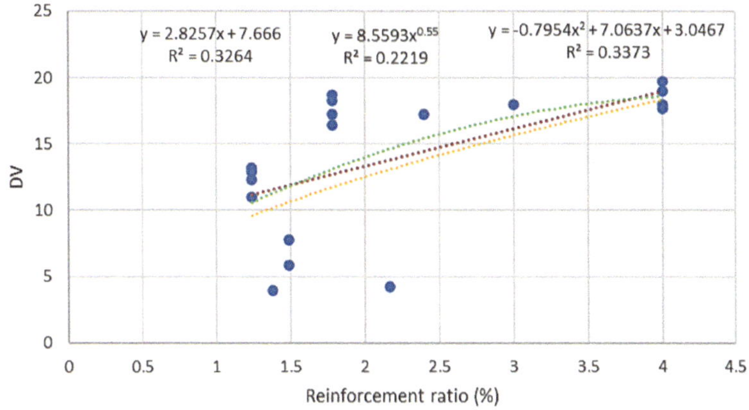

Fig. 2 Reinforcement ratio (%) relationship to flexural strength

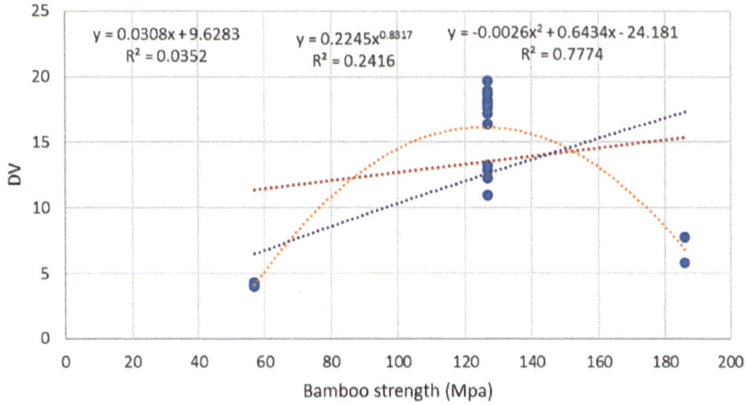

Fig. 3 Bamboo strength (MPa) relationship to flexural strength

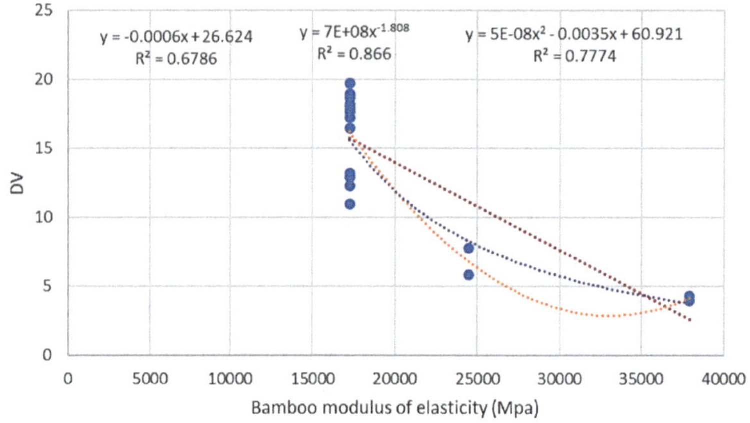

Fig. 4 Bamboo modulus of elasticity (MPa) relationship to flexural strength

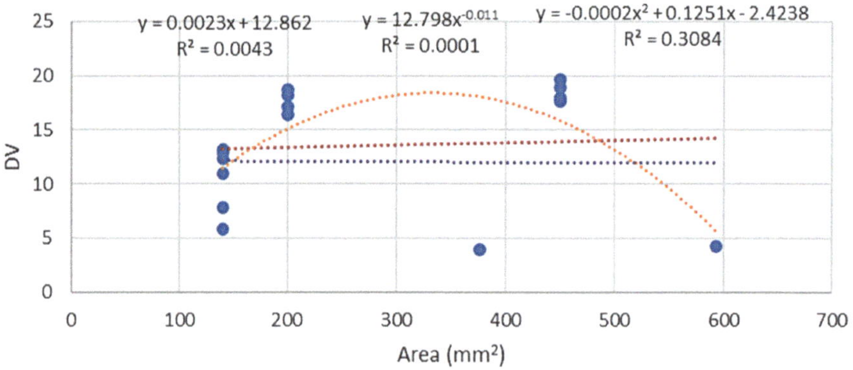

Fig. 5 Area of reinforcement (mm^2) relationship to flexural strength

bamboo modulus of elasticity has power. Therefore, it was concluded from Table 2 that the function of the accurate model might be quadratic where most of the IVs had a quadratic relationship.

4.3 *Developing Models*

4.3.1 Check the Outlier or Normality

After entering the data into the SPSS software to analyze, the data outlier or the normality of distribution was needed to be checked before modeling in frequency table in minimum and maximum rows, where it must be in the specific range which

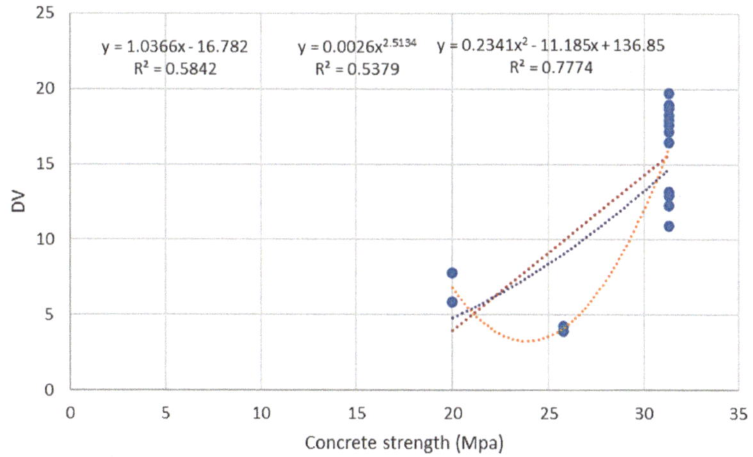

Fig. 6 Concrete strength (MPa) relationship to flexural strength

Table 2 Relationships of the IVs to DV

IV	Linear	Power	Quadratic
	R^2		
Reinforcement ratio	0.3093	0.1946	0.3094
Bamboo strength	0.0352	0.2416	0.7774
Bamboo modulus of elasticity	0.6786	0.866	0.7774
Area of reinforcement	0.0043	0.0001	0.3084
Concrete strength	0.5842	0.5379	0.7774

is from −4 to 4. Any data that is out of the range will be removed. As shown in Table 3, the data was in the range of a normal distribution which was the minimum −2.27350 and the maximum 2.03310; therefore, no need to remove any data.

4.3.2 Correlations

Data for the model was analyzed using Pearson's correlation. Table 4 indicates the correlation coefficients output from SPSS applying Pearson's correlation for flexural strength. The flexural strength was selected as a single dependent variable, and the influential independent variables were selected based on the high range of correlation coefficient for further model development purposes.

The independent variables were reinforcement ratio, bamboo strength, bamboo modulus of elasticity, area, and concrete strength. These variables were assessed for regression analysis to check the best-fitting regression models.

From Table 4, it was found that three variables have strong correlations with the flexural strength exceeding 50%, and they are the ratio of reinforcement, bamboo

Table 3 Frequency of the data

Statistics		Z-score (Flexural strength)	Z-score (Reinforcement ratio)	Z-score (Bamboo strength)	Z-score (Bamboo modulus of elasticity)	Z-score (Area of reinforcement)	Z-score (Concrete strength)
N	Valid	16	16	16	16	16	16
	Missing	0	0	0	0	0	0
Std. Deviation		1.00000000	1.00000000	1.00000000	1.00000000	1.00000000	1.00000000
Variance		1.000	1.000	1.000	1.000	1.000	1.000
Minimum		−1.73606	−0.81959	−2.05019	−0.48842	−0.86824	−2.27350
Maximum		1.12995	1.63168	1.81311	2.40726	2.03310	0.51953

Table 4 Correlation between dependent variable to the independent variables using Pearson's correlation

Correlations		Flexural strength	Reinforcement ratio	Bamboo strength	Bamboo modulus of elasticity	Area of reinforcement	Concrete strength
Flexural strength	Pearson's correlation	1	0.556[a]	0.188	−0.824[b]	0.065	0.764[b]
	Sig. (2-tailed)		0.025	0.487	0.000	0.810	0.001
	N	16	16	16	16	16	16
Reinforcement ratio	Pearson's correlation	0.556[*]	1	−0.045	−0.214	0.760[**]	0.286
	Sig. (2-tailed)	0.025		0.869	0.425	0.001	0.284
	N	16	16	16	16	16	16
Bamboo strength	Pearson's correlation	0.188	−0.045	1	−0.547[a]	−0.578[a]	−0.303
	Sig. (2-tailed)	0.487	0.869		0.028	0.019	0.255
	N	16	16	16	16	16	16
Bamboo modulus of elasticity	Pearson's correlation	−0.824[b]	−0.214	−0.547[a]	1	0.400	−0.632[b]
	Sig. (2-tailed)	0.000	0.425	0.028		0.125	0.009
	N	16	16	16	16	16	16
Area of reinforcement	Pearson's correlation	0.065	0.760[b]	−0.578[a]	0.400	1	0.080

(continued)

Table 4 (continued)

Correlations

		Flexural strength	Reinforcement ratio	Bamboo strength	Bamboo modulus of elasticity	Area of reinforcement	Concrete strength
	Sig. (2-tailed)	0.810	0.001	0.019	0.125		0.768
	N	16	16	16	16	16	16
Concrete strength	Pearson's correlation	0.764[b]	0.286	−0.303	−0.632[b]	0.080	1
	Sig. (2-tailed)	0.001	0.284	0.255	0.009	0.768	
	N	16	16	16	16	16	16

[a]Correlation is significant at the 0.05 level (2-tailed)
[b]Correlation is significant at the 0.01 level (2-tailed)

modulus of elasticity, and concrete strength. The modulus of elasticity and concrete strength show a higher correlation which is more significant. The ratio of the reinforcement has a less significant correlation than the bamboo modulus of elasticity and concrete strength, which was less considerable. However, for bamboo strength, it was found to have a weak correlation while the area has the weakest correlation between all IVs.

Table 4 also shows the correlation of the reinforcement ratio is strong with the area of reinforcement, and where the greater the area, the greater the ratio, which is a positive relationship. As for bamboo strength, it has good correlations with both bamboo modulus of elasticity and area, where bamboo strength naturally increases if its size and bending ability increase. The modulus of elasticity of bamboo has a strong negative correlation with concrete strength, where the modulus of elasticity of bamboo affects the strength of concrete. Since bamboo has less elasticity, it will fail inside the concrete, causing it to weaken and vice versa. Also, the strength of concrete affects the modulus of elasticity of bamboo, and that is, the higher the strength of the concrete, the less the percentage of bamboo exceeding the elastic limit, which reduces its failure.

4.4 Summary of Flexural Strength Models

Since all five independent variables were tested with the dependent variable, all possible regression model functions were chosen for this study. Flexural strength was modeled using single and multiple linear and nonlinear regressions to determine the largest number of possible models that might be of high R^2. The flexural strength model was regressed linearly with 21 models and nonlinearly (power and quadratic) with 42 models for a total of 63 models. Below a summary of the generated models has been gathered to be viewed and validated in order to enable the selection of the most dependable model to fit the prediction of DV.

Tables 5, 6 and 7 show the summary of the developed empirical models of all regressions. The symbols for each parameter are flexural strength (f_s), reinforcement ratio (ρ), bamboo strength (f_b), bamboo modulus of elasticity (E), area (A), and concrete strength (f_{cr}).

The R^2 of all the models in Table 5 did not exceed 90%. This is proportional to the relationships between the DV and IVs that were shown in Table 2, where IVs have a weaker relationship with DV in linear function than power and quadratic functions. In Tables 6 and 7, the power and quadratic regressions, respectively, showed a high percentage of R^2, where 9 models exceeded 90% in power regression and 13 models in quadratic regression. Among the 63 models, quadratic models 1, 2, and 9 have the highest percentage, 97.7% of R^2 as predicted by relationships graphs with the most robust relationship, and 97.6, 89% for the power model 1, linear model 1 and 2, respectively.

It was noticed from the outcomes of analyzing the IVs individually that the modulus of elasticity has the highest R^2 in linear and nonlinear power, which

Table 5 Summary of the developed empirical models of linear regression

Regression method	Model no	Model development	
		Model	R^2
Linear regression	1	$fs = -23.806 + 2.732(\rho) + 0.046(f_b) + 0(E) - 0.009(A) + 0.962(f_{cr})$	0.89
	2	$fs = -23.806 + 2.732(\rho) + 0.046(f_b) - 0.009(A) + 0.962(f_{cr})$	0.89
	3	$fs = -31.026 + 1.696(\rho) + 0.073(f_b) + 1.085(f_{cr})$	0.888
	4	$fs = -42.251 + 0.115(f_b) + 0.014(A) + 1.282(f_{cr})$	0.88
	5	$fs = 5.931 + 1.696(\rho) + 0(E) + 0.432(f_{cr})$	0.888
	6	$fs = 16.056 - 0.001(E) + 0.014(A) + 0.253(f_{cr})$	0.88
	7	$fs = 3.139 + 2.759(\rho) + 0.035(f_b)$	0.355
	8	$fs = 3.990 + 0.056(f_b) + 0.009(A)$	0.081
	9	$fs = 28.337 + 9.619(\rho) - 0.136(f_b) - 0.067(A)$	0.816
	10	$fs = -31.811 + 0.076(f_b) + 1.226(f_{cr})$	0.777
	11	$fs = -16.502 + 1.795(\rho) + 0.894(f_{cr})$	0.709
	12	$fs = -16.808 + 0(A) + 1.036(f_{cr})$	0.584
	13	$fs = 6.471 + 0(E) + 0.550(f_{cr})$	0.777
	14	$fs = 21.068 + 1.940(\rho) - 0.001(E)$	0.83
	15	$fs = 25.064 - 0.001(E) + 0.017(A)$	0.864
	16	$fs = -31.811 + 0.76(f_b) + 0(E) + 1.226(f_{cr})$	0.777
	17	$fs = 7.627 + 2.713(\rho)$	0.309
	18	$fs = 9.628 + 0.031(f_b)$	0.035
	19	$fs = 26.624 - 0.001(E)$	0.679
	20	$fs = 12.862 + 0.002(A)$	0.004
	21	$fs = -16.782 + 1.037(f_{cr})$	0.584

Table 6 Summary of the developed empirical models of nonlinear-power regression

Regression method	Model no	Model development	
		Model	R^2
Nonlinear power regression	1	$fs = (5.954 \times \rho^{61.995}) \times (5.954 \times f_b^{-35.721}) \times (5.950 \times E^{37.930}) \times (5.958 \times A^{-61.835}) \times (5.961 \times f_{cr}^{25.777})$	0.976
	2	$fs = (67.243 \times \rho^{2.798}) \times (67.244 \times f_b^{-1.508}) \times (67.244 \times A^{-2.539}) \times (67.244 \times f_{cr}^{1.465})$	0.911
	3	$fs = (0.023 \times \rho^{0.264}) \times (0.024 \times f_b^{1.016}) \times (0.024 \times f_{cr}^{2.592})$	0.906
	4	$fs = (0.009 \times f_b^{1.280}) \times (0.008 \times A^{0.264}) \times (0.009 \times f_{cr}^{2.711})$	0.905
	5	$fs = (199.958 \times \rho^{0.264}) \times (194.567 \times E^{-1.543}) \times (196.581 \times f_{cr}^{0.517})$	0.906
	6	$fs = (788.569 \times E^{-1.943}) \times (761.305 \times A^{0.264}) \times (771.437 \times f_{cr}^{0.098})$	0.905
	7	$fs = (1.252 \times \rho^{0.460}) \times (1.075 \times f_b^{0.412})$	0.42
	8	$fs = (0.183 \times f_b^{0.774}) \times (0.185 \times A^{0.413})$	0.315
	9	$fs = (35{,}890{,}344.04 \times \rho^{5.959}) \times (35{,}890{,}403.00 \times f_b^{-4.677}) \times (35{,}890{,}338.38 \times A^{-5.706})$	0.918
	10	$fs = (0.003 \times f_b^{1.028}) \times (0.002 \times f_{cr}^{2.813})$	0.777
	11	$fs = (0.008 \times \rho^{0.270}) \times (0.007 \times f_{cr}^{3.590})$	0.786
	12	$fs = (0 \times A^{0.231}) \times (0 \times f_{cr}^{5.488})$	0.74
	13	$fs = (2380.466 \times E^{-1.561}) \times (2369.476 \times f_{cr}^{0.714})$	0.777
	14	$fs = (38{,}963.981 \times \rho^{0.269}) \times (39{,}916.525 \times E^{-1.905})$	0.901
	15	$fs = (0.105 \times f_b^{0.837}) \times (0.105 \times E^{-0.290}) \times (0.104 \times f_{cr}^{2.422})$	0.777
	16	$fs = (34{,}164.216 \times E^{-2.003}) \times (33{,}750.119 \times A^{0.266})$	0.905
	17	$fs = (9.739 \times \rho^{0.459})$	0.309
	18	$fs = (2.637 \times f_b^{0.34})$	0.096
	19	$fs = (17{,}046{,}787.66 \times E^{-1.425})$	0.733
	20	$fs = (7.828 \times A^{0.099})$	0.02
	21	$fs = (0.00002525 \times f_{cr}^{3.819})$	0.654

Table 7 Summary of the developed empirical models of nonlinear quadratic regression

Regression method	Model no	Model development	
		Model	R^2
Nonlinear quadratic regression	1	$fs = ((4{,}361{,}348.903 \times \rho^2) + (-53{,}970{,}609.9 \times \rho) + (70{,}174{,}119.21)) + ((20{,}700.799 \times f_b{}^2) + (-3{,}033{,}385.011 \times f_b) + (38{,}787{,}661.47)) + ((0.699 \times E^2) + (-36{,}083.654 \times E) + (0.1)) + ((-360.605 \times A^2) + (489{,}799.686 \times A) + (-17{,}662{,}289.5)) + ((-215{,}490.689 \times f_{cr}{}^2) + (23{,}694{,}441 \times f_{cr}) + (-156{,}745{,}849))$	0.977
	2	$fs = ((-25{,}358.034 \times \rho^2) + (313{,}782.702 \times \rho) + (4{,}871{,}168.031)) + ((158.960 \times f_b{}^2) + (-40{,}023.625 \times f_b) + (-1{,}963{,}400.982)) + ((2.096 \times A^2) + (-2847.490 \times A) + (-3{,}969{,}191.707)) + ((2031.676 \times f_{cr}{}^2) + (-48{,}186.432 \times f_{cr}) + (3{,}105{,}158.467))$	0.977
	3	$fs = ((-2.430 \times \rho^2) + (14.729 \times \rho) + (-259{,}904.968)) + ((-8.478 \times f_b{}^2) + (1816.715 \times f_b) + (208{,}799.769)) + ((185.103 \times f_{cr}{}^2) + (-13{,}863.352 \times f_{cr}) + (209{,}612.352))$	0.946
	4	$fs = ((-0.623 \times f_b{}^2) + (-720.420 \times f_b) + (2{,}182{,}888.686)) + ((-0.000076 \times A^2) + (0.063 \times A) + (2{,}181{,}411.854)) + ((2651.792 \times f_{cr}{}^2) + (-140{,}857.396 \times f_{cr}) + (-2{,}452{,}381.024))$	0.922
	5	$fs = ((-2.428 \times \rho^2) + (14.718 \times \rho) + (-9{,}506{,}689.385)) + ((-0.021 \times E^2) + (1214.666 \times E) + (-477{,}783.348)) + ((10{,}227.652 \times f_{cr}{}^2) + (-298{,}735.093 \times f_{cr}) + (-5{,}491{,}311.515))$	0.946
	6	$fs = ((0.084 \times E^2) + (-4992.504 \times E) + (0.293)) + ((-0.00005527 \times A^2) + (0.049 \times A) + (-0.245)) + ((-66{,}899.103 \times f_{cr}{}^2) + (2{,}482{,}200.239 \times f_{cr}) + (48{,}944{,}716.10))$	0.908

(continued)

Table 7 (continued)

Regression method	Model no	Model development	
		Model	R^2
	7	$fs = ((-2.430 \times \rho^2) + (14.728 \times \rho) + (33{,}311.977)) + ((-0.002 \times f_b{}^2) + (0.630 \times f_b) + (-33{,}353.816))$	0.946
	8	$fs = ((-0.002 \times f_b{}^2) + (0.643 \times f_b) + (26{,}196.404)) + ((-0.000076 \times A^2) + (0.063 \times A) + (-26{,}232.721))$	0.922
	9	$fs = ((-48{,}593.803 \times \rho^2) + (601{,}319.483 \times \rho) + (300{,}764.049)) + ((66.105 \times f_b{}^2) + (-22{,}642.647 \times f_b) + (828{,}946.388)) + ((4.018 \times A^2) + (-5456.973 \times A) + (692{,}144.702))$	0.977
	10	$fs = ((-11.163 \times f_b{}^2) + (2462.859 \times f_b) + (123{,}252.282)) + ((24.972 \times f_{cr}{}^2) + (-6659.012 \times f_{cr}) + (-72{,}082.641))$	0.777
	11	$fs = ((-2.430 \times \rho^2) + (14.728 \times \rho) + (-6242.545)) + ((0.264 \times f_{cr}{}^2) + (-12.821 \times f_{cr}) + (6383.595))$	0.946
	12	$fs = ((-0.000076 \times A^2) + (0.063 \times A) + (-4190.749)) + ((0.330 \times f_{cr}{}^2) + (-16.320 \times f_{cr}) + (4384.741))$	0.922
	13	$fs = ((0.009 \times E^2) + (-200.207 \times E) + (114{,}499{,}496.6)) + ((175{,}505.464 \times f_{cr}{}^2) + (-8{,}890{,}602.527 \times f_{cr}) + (-7{,}439{,}316.264))$	0.79
	14	$fs = ((-2.426 \times \rho^2) + (14.708 \times \rho) + (-10{,}122{,}571.3)) + ((0.0000000399 \times E^2) + (-0.003 \times E) + (10{,}122{,}606.27))$	0.946
	15	$fs = ((0.0000000199 \times E^2) + (-0.002 \times E) + (-2{,}225{,}434.409)) + ((-0.000076 \times A^2) + (0.063 \times A) + (2{,}225{,}465.292))$	0.922

(continued)

confirmed the results of the correlation table, which means that modulus of elasticity has the most effect on the flexural strength, then concrete strength comes next, and after that are the ratio of the reinforcement, bamboo strength, and area of reinforcement.

In nonlinear quadratic regression, the results show that the modulus of elasticity, bamboo strength, and concrete strength in models 18, 19, and 21 have the same R^2 caused by limited data.

Table 7 (continued)

Regression method	Model no	Model development	
		Model	R^2
	16	$fs = ((-1137.751 \times f_b{}^2) + (161{,}196.482 \times f_b) + (-9{,}602{,}897.704)) + ((-0.078 \times E^2) + (4083.719 \times E) + (30{,}102{,}962.46)) + ((53{,}018.717 \times f_{cr}{}^2) + (-3{,}203{,}211.230 \times f_{cr}) + (-21{,}613{,}553.3))$	0.777
	17	$fs = ((-0.067 \times \rho^2) + (3.074 \times \rho) + (7.238))$	0.309
	18	$fs = ((-0.003 \times f_b{}^2) + (0.643 \times f_b) + (-24.181))$	0.777
	19	$fs = ((0.0000000531 \times E^2) + (-0.004 \times E) + (60.921))$	0.777
	20	$fs = ((0 \times A^2) + (0.125 \times A) + (-2.424))$	0.308
	21	$fs = ((0.234 \times f_{cr}{}^2) + (-11.185 \times f_{cr}) + (136.848))$	0.777

4.5 Validation

The results indicated that the DV has a better nonlinear relationship with IVs compared to linear. The highest R^2 in linear and nonlinear was chosen for model testing, as they are considered to have more accuracy.

The models below were chosen to check the validation in the model test:

- Linear from Table 5 = >Model 1
- Nonlinear power from Table 6 = >Model 1.
- Nonlinear quadratic from Table 7 = >Model 1, Model 2, and Model 9.

4.5.1 Linear

Models 1 and 2 are the same, but the difference is that the modulus of elasticity had been removed by the software, but here it has been displayed as multiplied by zero in model 1 to show the difference. They have the same result; therefore, they are added in the same Table 8. Through model testing for linear models, it becomes clear that models can predict the flexural strength almost accurately with MAPE 10.54% and MAD 1.42. As shown in Table 8, it is considered good. It was previously shown that the R2 for linear is 89%, which proves that there is a percentage of error that will appear. Based on the result given in Table 8, the models are valid.

Table 8 Validation for linear regression—models 1 and 2

Previous experimental result data	Model 1		
	$R^2 = 0.89$		
Flexural strength (*fs*) (MPa)	Predicted result of the model (MPa)	Error (%)	MAD
7.77	6.8	12.48	0.97
5.83	6.8	16.64	0.97
3.96	4	1.01	0.04
4.27	4.21	1.41	0.06
12.89	14.27	10.71	1.38
10.96	14.27	30.2	3.31
13.19	14.27	8.19	1.08
12.3	14.27	16.02	1.97
16.44	15.2	7.54	1.24
18.22	15.2	16.58	3.02
18.67	15.2	18.59	3.47
17.19	15.2	11.58	1.99
17.93	19.02	6.08	1.09
18.96	19.02	0.32	0.06
19.7	19.02	3.45	0.68
17.63	19.02	7.88	1.39
Average		10.54	1.42

4.5.2 Nonlinear Power

The results in Table 9 showed the predicted results for the model which was very close to the previous experimental results. The MAPE and MAD of the nonlinear power model were less than linear, as they reached 6.18% and 0.72, respectively, almost half of linear. The reason for that is the strength of the relationships between the DV and the IVs. The relationships in Table 2 showed that the IVs have a stronger power relationship than the linear with DV.

Moreover, the R2 of power was 97.6%, which is higher than the linear; thus, the error percentage was lower. Therefore, the model is valid.

4.5.3 Nonlinear Quadratic

The nonlinear quadratically developed models show unreliable results of prediction far from previous experiment results in Tables 10, 11, and 12, so it was easy to expect MAPE and MAD to be very high, based on those outputs. Although the nonlinear quadratic model has a strong relationship between the DV and IVs, with the highest R^2 of 97.7%, among the linear and nonlinear power models, the reason was due to

Table 9 Validation for nonlinear power regression—model 1

Previous experimental result data	Model 1		
	$R^2 = 0.976$		
Flexural strength (*fs*) (MPa)	Predicted result of the model (MPa)	Error (%)	MAD
7.77	6.8	12.48	0.97
5.83	6.8	16.64	0.97
3.96	4.18	5.56	0.22
4.27	4.05	5.15	0.22
12.89	12.35	4.19	0.54
10.96	12.35	12.68	1.39
13.19	12.35	6.37	0.84
12.3	12.35	0.41	0.05
16.44	17.63	7.24	1.19
18.22	17.63	3.24	0.59
18.67	17.63	5.57	1.04
17.19	17.63	2.56	0.44
17.93	18.58	3.63	0.65
18.96	18.58	2	0.38
19.7	18.58	5.69	1.12
17.63	18.58	5.39	0.95
Average		6.18	0.72

the limited amount of data. As the quadratic function needs an enormous amount of data to be able to give more accurate values, the models are invalid because of limited data.

Summary of the modeling and validation:

Through the analyzed data results, 6 models out of 63 models with the highest R^2 were extracted: models 1,2 from linear, model 1 from nonlinear power, and models 1, 2, and 9 from nonlinear quadratic.

The quadratic models had the highest R^2 among the rest, as it was 0.977, and this was expected by studying the relationships between DV and IVs in Table 2. Despite this, the model testing results of prediction were unreliable. The mean average percentage of error calculated for models 1, 2, and 9 was 2,834,633.1%, 437.54%, and 516.73%, respectively, and the mean absolute deviation was 195,388.94, 38.19, and 40.44, respectively, which were too large. Thus, they were considered invalid. The reason for that was the inadequate available data for this research.

The power model has a very close R^2 value to the quadratic where the difference between them was 0.001, meaning that R^2 of power was 0.976 and that the power relationship between DV and IVs was the second-best relationship after the quadratic in nonlinear. In addition, the results of the power model were very close to the

Table 10 Validation for nonlinear quadratic regression—model 1

Previous experimental result data	Model 1		
	$R^2 = 0.977$		
Flexural strength (*fs*) (MPa)	Predicted result of the model (MPa)	Error (%)	MAD
7.77	244,977.97	3,152,769.63	244,970.2
5.83	244,977.97	4,201,923.5	244,972.14
3.96	588,531.11	14,861,796.7	588,527.15
4.27	588,452.03	13,780,978	588,447.76
12.89	121,646.71	943,629.33	121,633.82
10.96	121,646.71	1,109,815.24	121,635.75
13.19	121,646.71	922,164.67	121,633.52
12.3	121,646.71	988,897.64	121,634.41
16.44	121,644.32	739,828.95	121,627.88
18.22	121,644.32	667,541.71	121,626.1
18.67	121,644.32	651,449.65	121,625.65
17.19	121,644.32	707,545.84	121,627.13
17.93	121,583.92	678,003.29	121,565.99
18.96	121,583.92	641,165.4	121,564.96
19.7	121,583.92	617,077.26	121,564.22
17.63	121,583.92	689,542.2	121,566.29
Average		2,834,633.1	195,388.94

previous experimental result; the mean average percentage of error and the mean absolute deviation were small, with 6.18% for mean absolute percentage of error and 0.72 for mean absolute deviation; thus, it was considered valid.

As for the linear models, they have the lowest R^2 among the rest of the models, where 0.89 was the highest R^2, but despite that, their results were reasonable compared to quadratic where the prediction results of the models were approximately close to the previous experimental result. The mean average percentage of error was 10.54% and 1.42 for mean absolute variation, which is less than double the power; thus, it was considered valid.

The output of developing the model also identified the most influential parameter in flexural strength. This parameter was a modulus of elasticity of bamboo, where it was found to have a strong correlation with flexural strength. Moreover, the R^2 for some models in linear and nonlinear power increased when the modulus of elasticity was involved. However, this increase was not shown in the nonlinear quadratic because of the limited data, and the R^2 for the modulus of elasticity was high as well when it was analyzed individually as linear and nonlinear.

Table 11 Validation for nonlinear quadratic regression—model 2

Previous experimental result data	Model 2		
	$R^2 = 0.977$		
Flexural strength *(fs)* (MPa)	Predicted result of the model (MPa)	Error (%)	MAD
7.77	−3.55	145.69	11.32
5.83	−3.55	160.89	9.38
3.96	−48.68	1329.29	52.64
4.27	−126.02	3051.29	130.29
12.89	3.39	73.7	9.5
10.96	3.39	69.07	7.57
13.19	3.39	74.3	9.8
12.3	3.39	72.44	8.91
16.44	1.16	92.94	15.28
18.22	1.16	93.63	17.06
18.67	1.16	93.79	17.51
17.19	1.16	93.25	16.03
17.93	−57.89	422.87	75.82
18.96	−57.89	405.33	76.85
19.7	−57.89	393.86	77.59
17.63	−57.89	428.36	75.52
Average		437.54	38.19

4.6 Selecting the Suitable Model

Table 13 shows the summarized suitable model that can be used in future studies to predict the flexural strength of the reinforced bamboo concrete beam.

The model was chosen based on model accuracy (R^2) and from the results of prediction of the flexural strength or the DV by calculating the MAPE and MAD. This model has a high R^2, as shown in Table 13, and the model's output was more accurate than other models with a minimum of MAPE and MAD.

5 Conclusion

Statistical analysis and mathematical models proposed by SPSS were used to investigate the flexural strength of a reinforced beam with bamboo tested under a four-point load (one-third of support span). The following are the conclusion of the objectives stated in this research:

I. The results showed that the parameters reinforcement ratio, bamboo strength, bamboo modulus of elasticity, area of reinforcement, and concrete strength

Table 12 Validation for nonlinear quadratic regression—model 9

Previous experimental result data	Model 9		
	$R^2 = 0.977$		
Flexural strength (*fs*) (MPa)	Predicted result of the model (MPa)	Error (%)	MAD
7.77	14.82	90.73	7.05
5.83	14.82	154.2	8.99
3.96	68.12	1620.2	64.16
4.27	163.96	3739.81	159.69
12.89	20.82	61.52	7.93
10.96	20.82	89.96	9.86
13.19	20.82	57.85	7.63
12.3	20.82	69.27	8.52
16.44	35.39	115.27	18.95
18.22	35.39	94.24	17.17
18.67	35.39	89.56	16.72
17.19	35.39	105.88	18.2
17.93	110.15	514.33	92.22
18.96	110.15	480.96	91.19
19.7	110.15	459.14	90.45
17.63	110.15	524.79	92.52
Average		516.73	40.44

Table 13 Results from the regression showing the best predicted flexural strength

Regression method	Model no	Empirical model development of flexural strength			
		Model	R^2	MAPE	MAD
Nonlinear regression (Power)	1	$fs = (5.954 \times \rho^{61.995}) \times (5.954 \times f_b{}^{-35.721}) \times (5.950 \times E^{37.930}) \times (5.958 \times A^{-61.835}) \times (5.961 \times f_{cr}{}^{25.777})$	0.976	6.18	0.72

have a certain amount of influence on the flexural strength, based on the strength of the correlation between each parameter with the flexural strength and between them. Reinforcement ratio, bamboo modulus of elasticity, and concrete strength have a significant correlation with flexural strength, and bamboo modulus of elasticity and concrete strength also have a significant correlation between them, which thus affect the flexural strength. If these parameters are taken into consideration, better flexural strength results can be obtained.

II. It was found that the parameter which has more influence on the flexural strength was the modulus of elasticity of bamboo; it has the strongest correlation compared to the rest of the parameters. Also by developing the models it was found that it had a certain effect on the accuracy of the empirical models (R^2) in some outputs in linear and nonlinear power multiple regressions when it is involved, where it increases their accuracy, and this reflects the impact and importance of this parameter on flexural strength.

    In addition, single regression also showed that the modulus of elasticity had the highest R^2 among the IVs. For nonlinear quadratic regression that influence did not appear due to limited data.

III. A model was developed to predict the flexural strength and that came after developing 63 models based on several independent variables that have some influence on the flexural strength. The developed model can be used for quick estimation of flexural strength. The developed empirical model is nonlinear regression power model 1.

$$fs = \left(5.954 \times \rho^{61.995}\right) \times \left(5.954 \text{ x} f_b^{-35.721}\right) \times \left(5.950 \times \text{E}^{37.930}\right) \times \left(5.958 \times \text{A}^{-61.835}\right) \times \left(5.961 \times f_{cr}^{25.777}\right). \quad (15)$$

where ρ is the reinforcement ratio, f_b is bamboo strength, E is the bamboo modulus of elasticity, A is the area, and f_{cr} is concrete strength. Based on the results of applying the models to predict the DV, it was found that model 1 from nonlinear regression power with R^2 of 0.976 has approached results to the actual value compared to others, as the MAPE was reached to 6.18% and 0.72 for the MAD, which is lesser than other models.

In the end, this research will help to determine the parameters by which the flexural strength of the reinforced concrete beam with bamboo can be improved. In addition, the empirical model that was constructed in this research will help to predict the flexural strength before the experiment is done by knowing several parameters.

Acknowledgements The authors would like to thank and acknowledge SEGi University, Kota Damansara for their support to this publication via research grant SEGiRF/2018-9/FoEBE-16/79.

References

Atul Agarwal BN (2014) Experimental investigation on chemically treated bamboo. Constr Build Mater 71:610–617

Brownell B (2015) Two natural rebar alternatives for concrete. The J Am Inst Architect

Hector A, Sebastian K et al (2018) Bamboo reinforced concrete: a critical review. Mater Struct 51:102

Kavitha S, Felix KT (2018) Experimental investigation on flexural behaviour of bamboo reinforced concrete members. Int J Adv Eng Technol 2:59–63

Muhtara, Dewi SM et al (2019). Enhancing bamboo reinforcement using a hose-clamp to increase bond stress and slip resistance. J Build Eng
Otunyo AW, Odebiyi AJ (2018) Regression modeling of the strength properties of concrete reinforced with polypropylene fiber and alkali resistant glass fiber. Niger J Technol 37:898–906
Rama PM, Rao SK (2018) Investigation on flexural behaviour of beam with bamboo as main rebars. Int J Recent Technol Eng 7:173–177
Sevalia JK, Siddhpura NB, Agrawal CS et al (2013) Study on bamboo as reinforcement in cement concrete. Int J Eng Res Appl 3:1181–1190
Siddika A, Al Mamun MA, Siddique MAB (2017) Evaluation of bamboo reinforcements in structural concrete member. J Construct Eng Project Manage
Shahnewaz M, Shahria AM (2014) Improved shear equations for steel fiber-reinforced concrete deep and slender beams. ACI Struct J 111:1–6
Vijayan DS, Revathy J (2016) Multiple regression model for the prediction of flexural behaviour of FRP plated pre-stressed concrete beams. Asian J Appl Sci 4:1299–1304

Application of Pavement Evaluation for Road Maintenance and Rehabilitation

Ahmad Kamil Arshad, Ekarizan Shaffie, Mohd Izzat A. Kamal, Mat Zin Hussain, and Nuryantizpura M. Rais

Abstract Timely maintenance and rehabilitation are required in order to maintain the road pavement in good condition. Proper evaluation of pavement is required prior to deciding the maintenance and rehabilitation strategies for the road. This paper discussed the process of pavement evaluation and the application of its results for use in the maintenance and rehabilitation of roads. The pavement life-cycle is first explained. The process of pavement evaluation is then explained and data collection for four major performance indicators, that is, visible distress, structural adequacy, surface friction and roughness are discussed, including the various equipment used for the evaluation. Finally, an explanation is made on how the results are analysed and decisions to determine the maintenance and rehabilitation strategies for the road network. It can be concluded that the systematic evaluation of pavement will lead to the optimization of resources used for pavement maintenance and rehabilitation.

Keywords Pavement evaluation · Pavement distress · Surface condition survey · Structural evaluation · Road maintenance · Road rehabilitation

A. K. Arshad (✉) · E. Shaffie · M. I. A. Kamal
Institute for Infrastructure Engineering and Sustainable Management (IIESM), Universiti Teknologi MARA, Shah Alam, Malaysia
e-mail: drahmadkamil@uitm.edu.my

E. Shaffie
e-mail: eka@uitm.edu.my

M. Z. Hussain
Kumpulan Ikram Sdn Bhd, Kajang, Malaysia
e-mail: matzin@ikram.com

N. M. Rais
Malaysia Institute of Transport (MITRANS), Universiti Teknologi MARA, Shah Alam, Malaysia
e-mail: yantiz@uitm.edu.my

R. Hassan et al. (eds.), *Green Infrastructure*,
https://doi.org/10.1007/978-981-16-6383-3_10

1 Introduction

As Malaysia relies heavily on road-based transportation, it is very important to maintain its road network. In a road network, pavements represent the largest capital investment; it is vital that the pavements are maintained to ensure that their condition is in a desirable state. Maintaining and operating pavements on a large road network typically involves complex decisions about how and when to resurface or apply other treatments to keep the road network performing and operating costs at a reasonable level. A road network needs to be regularly evaluated and maintained to meet the needs of the road user as a safe and efficient means of travel. If maintenance is neglected, the road condition will further deteriorate, causing not only the loss of life, injuries and damage due to accidents but also an increase in the vehicle operating costs.

2 Pavement Life-Cycle

A road pavement must fulfil two basic functions, that is structural and functional requirements (JKR Malaysia 1994; Asphalt Institute 2000). In terms of structural performance, the pavement must be of sufficient thickness and be composed of materials of sufficient quality, to be able to withstand the various loads that are applied to it by heavy vehicles. In terms of functional performance, the pavement must have a good riding quality to ensure comfortable travel for the road user and a surface having adequate drainage, skid resistance, reflectivity and line markings to ensure safe travel.

A pavement starts to deteriorate at an increasing rate once a road is opened to traffic. Initially, few distresses are present, and the pavement stays in relatively good condition. Over time as more traffic passes through it and as it ages due to the environment, more distresses develop, making it easier for subsequent damage to occur.

The timing of maintenance and rehabilitation actions can greatly influence their effectiveness and cost as well as overall pavement life. Once a pavement needs treatment, the sooner maintenance or rehabilitation activity is undertaken, the more cost-effective it will be.

Figure 1 demonstrates this concept. For the first 75% of pavement life, the pavement condition drops by about 40%. However, it only takes another 17% of pavement life for the pavement condition to drop another 40%. Additionally, in order to restore pavement condition to a predetermined level, it will cost 4–5 times as much if the pavement is allowed to deteriorate for even 2–3 years beyond the optimum rehabilitation point (Federal Highway Administration 1998).

This increase in cost is because of the following:

(1) The pavement condition must be improved by a greater amount (e.g., "very poor " to "very good" versus "fair" to "very good").

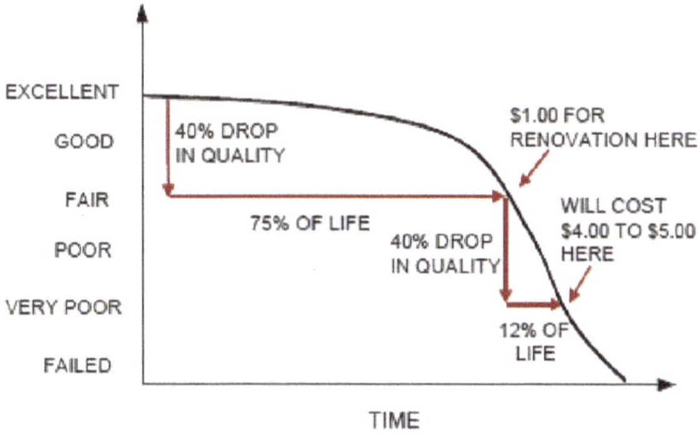

Fig. 1 Typical pavement condition life-cycle (Federal Highway Administration 1998)

(2) It costs more money per unit of pavement condition increase (e.g., it costs more to go from "very poor" to "poor" than it does from "fair" to "good").

Maintenance and rehabilitation can slow or reverse this deterioration.

3 Pavement Performance and Evaluation

Pavement performance can be defined as the measurable adequacy of a pavement's structural and functional service over a specified design period. A pavement provides functional service by giving road users a safe and comfortable ride for a specified design speed. Functional service comprised factors such as acceptable ride quality, adequate surface friction for safety, appropriate geometry for safety, the appearance of geometric adequacy and appearance of the condition.

A pavement provides structural service by supporting traffic loadings and withstanding environmental influences. The type and thickness of materials used to construct the pavement layers dictate how the pavement performs structurally.

Structural and functional adequacy is closely related but is not entirely interdependent. Structural deterioration of pavement is manifested to some extent in diminished functional adequacy, in the form of increased roughness, noise, and even hazard to vehicles and their occupants. However, some types of structural deterioration can occur and progress to fairly advanced stages without being noticeable to users. It is also possible for a pavement's functional adequacy to decrease without any significant change in structural adequacy (e.g., loss of skid resistance).

The evaluation of pavement condition is carried out at the network and project level (Transport Research Laboratory 1999). Pavement evaluation is carried out at the network level to determine the status of an entire pavement network as part of a pavement management system. The overall goal of the network-level pavement evaluation is to predict the current condition level of all the pavements within the road system. This will facilitate the prediction of the overall road network pavement performance and allow future budget forecasting (Ismail et al. 2013). This can be then used to help prioritize and select projects for pavement rehabilitation. Data obtained

Table 1 Test intervals for network and project level

Pavement test	Network level	Project level
Surface condition survey	100 m (Road scanner)	50 m (Manual survey)
Falling weight deflectometer	500 m	100 m
Coring and DCP	2000 m	250 m

from the network-level pavement evaluation is not sufficient for developing a proper rehabilitation alternative for a specific project. At the project level, a more detailed pavement evaluation is required to provide the information required for properly selecting a rehabilitation alternative and designing the project. The overall goal of the project level data collection is to obtain sufficient information to analyse and design rehabilitation options for specific pavement sections within the system (Austroads, 2018). This will facilitate the selection of the most desirable rehabilitation option, preparation of cost estimates, plans and specifications for the rehabilitation of the pavement section. Table 1 shows the intervals for the pavement evaluation tests carried out at both network and project levels. Table 2 shows the type of tests carried out for both network and project-level evaluation.

Pavement evaluation at the project level is a process in which systematic assessment of pavement condition is carried out to determine its modes of distress/deterioration before the appropriate treatment/rehabilitation design is proposed. The components of pavement evaluation include data collection, data analysis and rehabilitation design. Data collection involved surface condition survey, structural assessment and laboratory evaluation of pavement materials (Bennet et al. 2007). Data collected are then analysed and primary modes of distress for the pavement evaluated are identified. A further comprehensive analysis of data is carried out to identify the most suitable and evaluate the most economical rehabilitation technique before the implementation of the selected rehabilitation works for the pavement section. The process of project-level pavement evaluation and rehabilitation is shown in Fig. 2.

4 Performance Indicator

There are characteristics of pavements that can be measured quantitatively and can be correlated to the users' subjective assessments of performance. The four major performance indicators are visible distress, structural adequacy, surface friction and roughness (Garber and Hoel 2009). The following sections describe how these indicators are related to performance and how they can be measured.

Table 2 Type of tests for network and project level

Test	Purpose	Equipment and method
Falling Weight Deflectometer (FWD)	To determine pavement structural condition	Applying a load of 700 kPa and measuring the deflection bowl
Coring	To determine asphalt layer thickness and crack depth	Extract core samples using a rotary coring machine
Dynamic cone penetromete	To determine the thickness of the pavement layer	Releasing a standard weight onto the anvil and measures the rod penetration into the ground and then plot graph Penetration versus cumulative blows
Ground-penetrating radar	To determine the thickness of the pavement layer, change in construction and defects within the pavement	Sending electro-magnetic pulses down and registering the reflection time and dielectric discontinuities
Trial pit	To closely inspect pavement condition and to collect samples	Cutting and removing materials layer by layer
Road scanner	To capture pavement condition data, road geometry, survey mapping information and roadside assets	Using multi-laser profiler together with GPS and digital imaging system
Manual surface condition survey	To determine the category and extent of crack, surface defects, drainage condition, etc	Walking along the road and recording at 50 m block interval
Walking profiler	To determine road profiles and also the roughness index, IRI	Using measuring beam and sensor to measure distance and profile
Traffic count and axle load survey	To determine the damaging effect of commercial vehicles	Weigh vehicles using portable weigh-in-motion systems
Surface friction tester	To determine surface friction or skid resistance of the road surface	Measuring the friction force on a test wheel operated at a constant load and longitudinal reference slip
Grip tester	To measure surface friction or skid resistance	Directly measure the horizontal (drag) and vertical (load) force Friction = Load/Drag
Mini texture meter	To determine the macrotexture of the road surface	Using pulsing laser light projected onto the road surface and measuring surface displacement

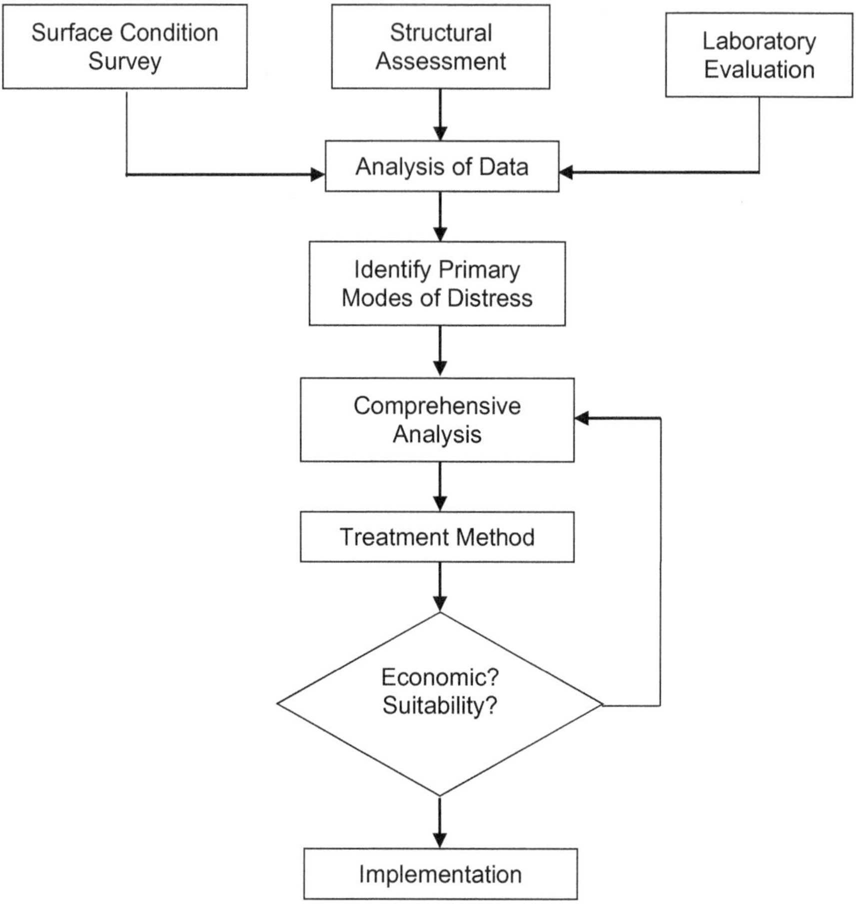

Fig. 2 Flowchart of pavement evaluation and rehabilitation process

4.1 Visible Distress

Surface condition survey is carried out to determine visible distress of the pavement surface. Distress occurs in pavements as a result of complex interactions of design, construction, materials, traffic, environment and maintenance procedures. Visible distress is quantified with respect to type, severity and quantity. Distress types, severities and quantities are determined during a distress survey of the pavement. The information can be manually observed (Fig. 3) or using high-speed survey equipment such as the road scanner (Fig. 4).

Pavement distress refers to the condition of a pavement surface in terms of its general surface appearance. Distress manifestations can be categorized according to the mechanism of distress, fractured, distorted or disintegrated (Brown et al. 2009). Fracture is the state of a pavement that is breaking. Example of fractures includes

Fig. 3 Manual measurement of rutting using a straight edge

Fig. 4 High-speed road scanner

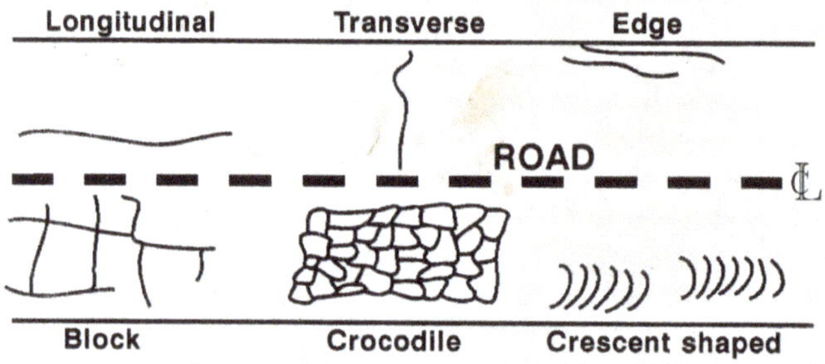

Fig. 5 Types of cracks (JKR Malaysia 1992)

cracks and spalling. Distortion is a permanent change in the shape of the pavement or pavement component. It can be manifested by ruts and corrugation of the surface. Disintegration is the decomposition or abrading of a pavement into its constitutive elements. Disintegration of the pavement includes ravelling, stripping and surface polishing.

Most road authorities, however, use the guidelines produced by FHWA for visual assessment of their pavements (Miller and Bellinger 2003). For example, Malaysia used a modified version of the guidelines to assess the surface condition of its flexible pavements (JKR Malaysia 1992). The distresses are categorized according to the following types: (a) cracks; (b) surface deformations; (c) surface defects; (d) patching defects; (e) potholes; and (f) edge defects. These distresses are shown in Figs. 5, 6, 7, 8 and 9.

4.2 *Structural Adequacy*

Structural performance is related to the ability of the pavement to sustain traffic loading. Deflection testing is usually performed to predict future structural performance. Nondestructive testing (NDT) devices, such as the falling weight deflectometer (FWD), are used to evaluate the structural capacity of the pavement (Khin Onn 2001). The FWD works by applying a load to the pavement that is similar in magnitude and duration to the load applied by a moving wheel of a vehicle (Fig. 10). The response of the pavement to that load is then measured by transducers placed at varying distances from that load, and the results can be analysed for both individual layers and the overall pavement structure (Lay 2009).

For the structural evaluation of pavements, the thickness of each pavement layer must be determined before the analysis can be carried out. For network evaluation, the ground-penetrating radar together with the dynamic cone penetrometer (Fig. 11)

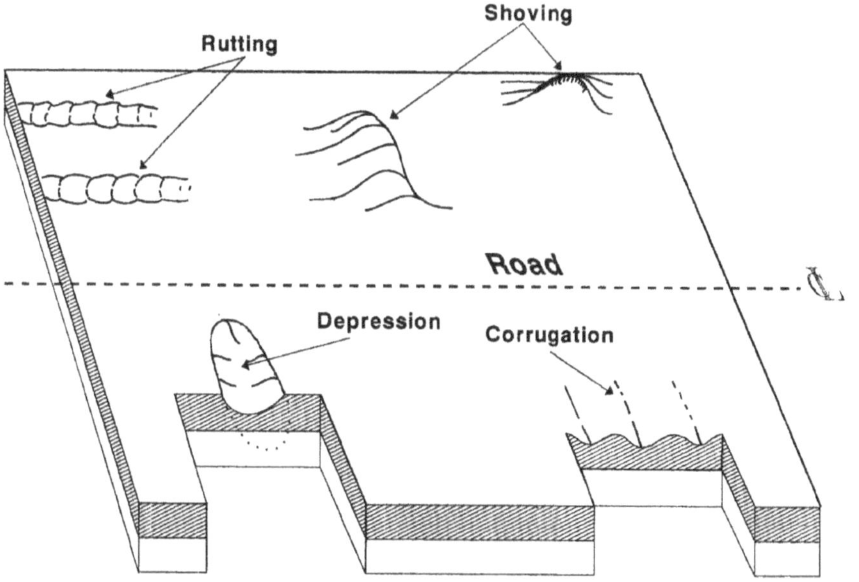

Fig. 6 Types of surface deformation (JKR Malaysia 1992)

Fig. 7 Types of surface defects (JKR Malaysia 1992)

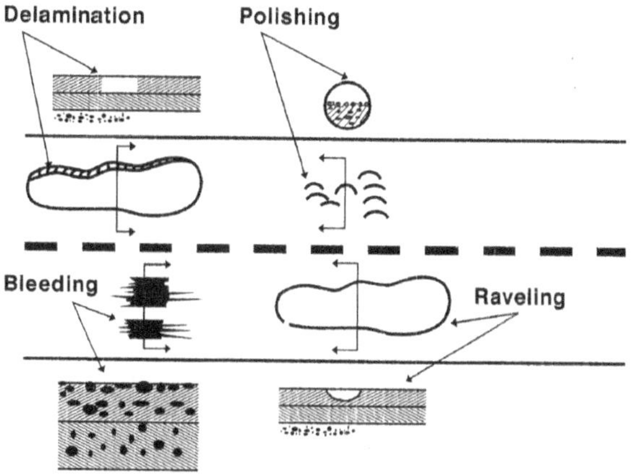

is used to determine the thickness of the pavement layers and the CBR value of the subgrade.

Traffic count and axle load study is carried out to determine the damaging effect of commercial vehicles on the existing pavement using portable weigh-in-motion systems such as the weighing pads (Transport Research Laboratory 2004). Data from the tests are then used to determine the existing pavement residual life and the required overlay thickness based on the projected future traffic.

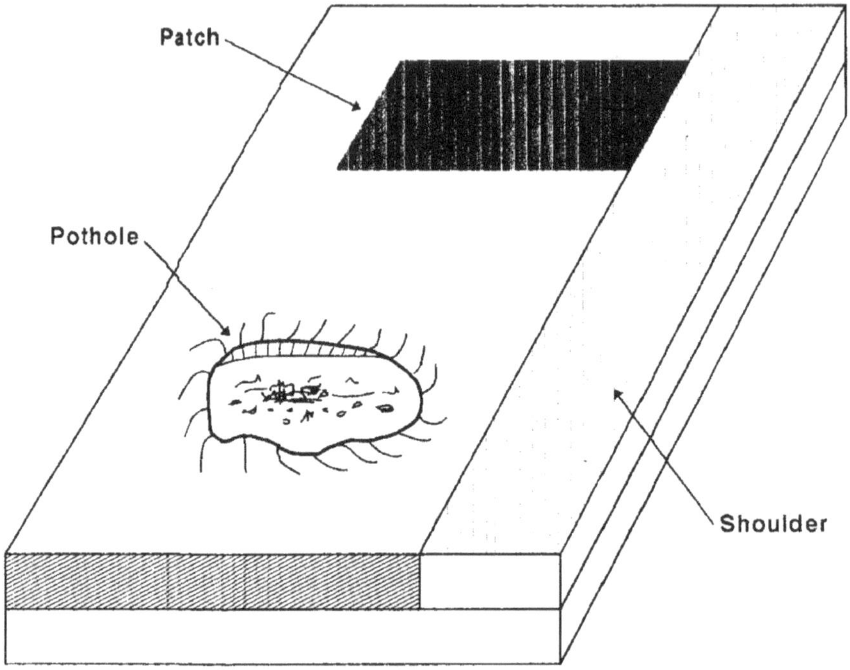

Fig. 8 Potholes and patching defects (JKR Malaysia 1992)

4.3 Surface Friction

Surface friction refers to the surface pavement characteristic that prevents the skidding of vehicles' tyres. The microtexture, macrotexture and transverse slope of the pavement influence a pavement's surface friction (Transport Research Laboratory 2004). Microtexture refers to the roughness of the aggregate particle surfaces. Microtexture contributes to the friction by adhesion with vehicles' tyres, contributing largely to skidding resistance at low speed (Shahid 2018). Macrotexture refers to the overall texture of the pavement, providing the surface texture needed to provide channels for water trapped between the road surface and the tyre to drain quickly, reinstating the contact between the tyres and the road surface (Shahid 2018). Adequate macrotexture is critical for vehicles' operations under wet condition. Transverse slope contributes to surface friction by removing water from the pavement surface. In Malaysia, a 2.5% transverse slope is provided for surface drainage. Aquaplaning is likely to occur if water is not removed immediately from the pavement's surface, particularly during rain.

Figure 12 shows the mini texture meter which is used to determine the macrotexture of the road surface. Using pulsing laser light projected onto the road surface and measuring surface displacement, the macrotexture of the road can be determined. Figure 13 shows the surface friction tester which is used to determine surface friction

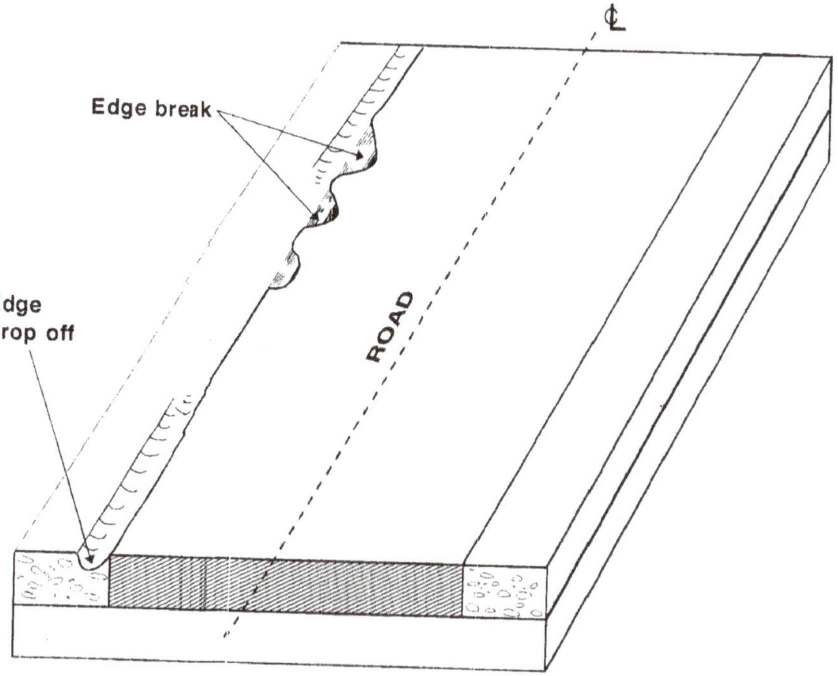

Fig. 9 Types of edge defects (JKR Malaysia 1992)

Fig. 10 Falling weight deflectometer

Fig. 11 Dynamic cone penetrometer

or skid resistance of the road surface. The skid resistance is determined by measuring the friction force on a test wheel operated at a constant load and longitudinal reference slip.

4.4 Roughness/Serviceability

Pavement roughness refers to the irregularities in the pavement surface that affect the smoothness of a ride. A road user assesses the condition of a pavement largely in terms of ride quality. Data collected by the high-speed road scanner with the attached multi-laser profiler (Fig. 14) is used to determine the ride quality in terms of international roughness index (IRI) (Shahid 2018). Figure 15 shows the walking profiler used to determine project-level road profile and also the international roughness index (IRI).

Fig.12 Mini texture meter

Fig. 13 Surface friction tester

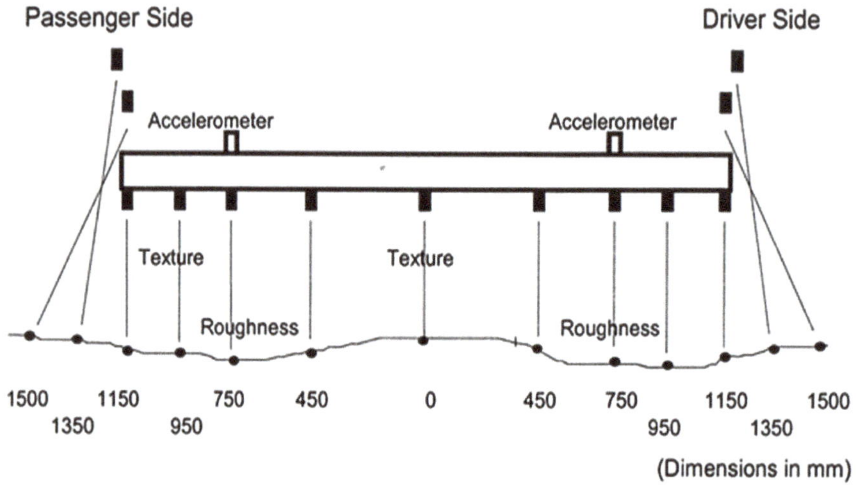

Fig. 14 Multi-laser profiler sensor configuration

Fig. 15 Walking profiler

5 Data Analysis and Determination of the Suitable Type of Rehabilitation

Generally, the road pavements will show some forms of distress and appropriate treatment was required to improve the functional and structural integrity of the roads. Various rehabilitation methods to accommodate the existing pavement condition are considered as shown in Table 3.

The collected data are analysed using suitable rehabilitation software to determine the condition of the existing road pavements. The pavements can be categorized as good, fair, bad or very bad based on the crack index (CI) and mean rut depth (RD) as summarized in Table 3.

The pavements are classified as good if there was no crack (C0) or RD less than 5 mm. The pavements are considered to be in fair condition if they exhibited single crack (C1) or multiple cracks but not interconnected (C2) or RD was between 5 and 15 mm. The pavements are categorized as bad when they exhibit interconnected cracks (C3) or crocodile cracks (C4) or when the RD is between 16 and 25 mm. Very bad condition is reached when severe crocodile cracks with spalling or block cracks (C5) are observed or when RD exceeds 25 mm. The overall pavement condition is determined by combining and weighting both CI and RD accordingly.

For structural evaluation of the pavement, the collected data is used in the analysis as shown in Fig. 16. First, the FWD data and the DCP results are analysed to determine the elastic moduli of the pavement layers and subgrade. The required overlay thickness is then determined, taking into account the elastic moduli, expected traffic loads, residual life and the recommended sectioning (Institute 1996). Together with the surface condition survey results, the proposed maintenance or rehabilitation method is then determined. The maintenance and rehabilitation strategies are discussed in the following section.

Table 3 Treatment matrix

Crack index		Rut depth (mm)			
		Good	Fair	Bad	Very bad
		<5	*5–15*	*16–25*	*>25*
Good	C0	T0	T1	T2	T3
Fair	C1 and C2	T1	T2	T3	T4
Bad	C3 and C4	T2	T3	T4	T5
Very bad	C5	T3	T4	T5	T5

Notes
Crack index: C0: No crack; C1: Single crack; C2: Multiple cracks, not connected; C3: Interconnected cracks; C4: Crocodile cracks; C5: Crocodile cracks and spalling or block cracks
Treatment types: T0: Do nothing; T1: Thin overlay; T2: Hot in-place recycling; T3: Structural overlay; T4: Cold in-place recycling; T5: Reconstruct

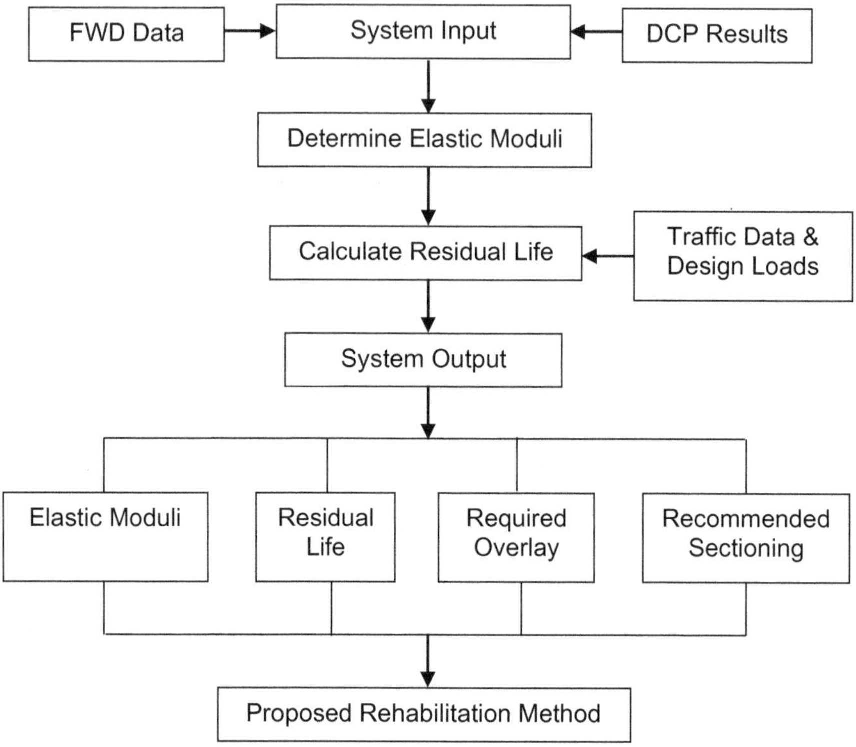

Fig. 16 Structural evaluation analysis

6 Maintenance and Rehabilitation Strategies

Maintenance and rehabilitation are the two principal treatments used to extend pavement life. In general, maintenance can slow the rate of deterioration by correcting small pavement defects before they worsen and contribute to further defects. Beyond a certain point, however, defects become too large for correction by maintenance. At this point, rehabilitation can be used to correct a large number of relatively severe defects, thus improving the pavement's condition.

In general, there are several levels of treatment to correct pavement deterioration as follows:

- Routine and periodic maintenance
- Rehabilitation
- Reconstruction.

Maintenance actions help slow down the rate of deterioration by identifying and addressing specific pavement deficiencies that contribute to overall deterioration. Rehabilitation is the act of repairing portions of an existing pavement to reset the deterioration process, while reconstruction is the total replacement of the deteriorated pavement.

6.1 Maintenance

Pavement maintenance describes all the methods and techniques used to prolong pavement life by slowing the deterioration rate. Thus, the performance of a pavement is directly tied to the timing, type and quality of the maintenance it receives. The following pavement maintenance operations are also known as preservation methods (Asphalt Institute 1996; Khin Onn 2003):

- Crack sealing
- Patching
- Fog seals
- Rejuvenation
- Sand seal
- Chip seal
- Slurry seal
- Cape seal
- Micro-surfacing
- Thin-lift overlay, and
- Ultrathin overlay.

6.2 Rehabilitation

While pavement maintenance can slow the rate of deterioration, it cannot stop the damage from occurring. Therefore, the effects of deterioration eventually need to be reversed by adding or replacing material in the existing pavement structure. This process is rehabilitation.

Rehabilitation options depend upon local conditions and pavement distress types but typically include (Asphalt Recycling and Reclaiming Association 2001; Kandhal and Mallick 1997):

- Hot mix asphalt (HMA) structural overlay
- Cold in-place recycling (CIPR)
- Hot in-place recycling (HIPR)
- Full-depth reclamation (FDR).

Rehabilitation essentially reverses the effects of deterioration by adding, replacing or recycling material in an existing pavement structure. The inclusion of recycling techniques may be appealing on many levels. First, it reuses in-place materials that are commonly some of the best materials available. Second, it is often a very cost-competitive option compared to reconstruction. Third, the disruption to the public tends to be less as the road can typically be used during the rehabilitation work. Especially in urban settings, this can greatly reduce user costs associated with a project.

Both recycling and reconstructing correct a myriad of distresses in a single effort. However, with recycling, the agency may choose to stage the construction steps into appropriate phases over an extended time. Staging improvements can reduce the funds needed to address a road today, while still correcting problems in the short term. Subsequent stages provide for a long-term solution, with the completion of all the planned stages.

6.3 Reconstruction

Reconstruction involves the removal and rebuilding of all or part of the road pavement using fresh material and new construction specifications (Nicholls 2017). Reconstruction can be either full or partial depending on which layer the new pavement is to be rebuilt. Full reconstruction is necessary when the subgrade layers, as well as the pavement layers, have deteriorated beyond repair. In this case the rebuilding will include the subgrade. However, partial reconstruction is required when the road base has been contaminated and has lost its inherent stability.

7 Conclusion

A road network needs to be regularly evaluated and maintained to meet the needs of the road user as a safe and efficient means of travel. The various methods of testing to evaluate the functional and structural performance of the pavement are discussed so that the condition of the pavement can be assessed. Data collected from the pavement evaluation process is used in the analysis of pavement condition, to determine the functional and structural condition of the pavement. From this, suitable maintenance or rehabilitation method can be applied to ensure that the structural and functional performance of the pavement is adequate. In conclusion, the evaluation of pavement is a systematic process that will lead to the optimization of resources used for pavement maintenance and rehabilitation, and for the longer use of the pavement.

Acknowledgements The authors would like to thank the management of the Institute for Infrastructure Engineering and Sustainable Management (IIESM) for the opportunity and support provided to the authors and Kumpulan IKRAM Sdn Bhd for the provision of supporting information in writing this article.

References

Asphalt Recycling and Reclaiming Association (2001) Basic asphalt recycling manual, ARRA, Annapolis
Austroads, (2018) Guide to pavement technology part 5—pavement evaluation and treatment design, 3rd edn. Austroads, Sydney
Bennet RB, Chamorro A, Chen C et al (2007) Data collection technologies for road management. The World Bank, Washington D.C.
Brown ER, Kandhal PS, Roberts FL et al (2009) Hot mix asphalt materials, mixture design, and construction, 3rd edn. NAPA Research and Education Foundation, Lanham
Federal Highway Administration (1998) Techniques for pavement rehabilitation: participants manual. FHWA, Washington DC
Garber NJ, Hoel LA (2009) Traffic & highway engineering, 4th edn. Cengage Learning, Toronto
Asphalt Institute (1996) Asphalt in pavement preservation & maintenance (MS-16), 3rd edn. Lexington, Asphalt Institute
Asphalt Institute (2000) Asphalt overlays for highway and street rehabilitation (MS-17). Lexington, Asphalt Institute
Ismail MP, Shahid MA, Abd Aziz S, Zahari MNF, Asanan A (2013) Pavement condition assessment and management of expressways based on key performance indicators. Paper presented at PIECE 2013 conference, Kuala Lumpur, Malaysia, 9–11 September 2013
Kandhal PS, Mallick RJ (1997) Pavement recycling guidelines for state and local governments-participants reference book. Federal Highway Administration, Washington, DC
Khin Onn L (2001) Pavement data collection strategy for JKR Malaysia. Paper presented at 20th ARRB conference, Melbourne, Australia, 19–21 March 2001
Khin Onn L (2003) Maintenance management of federal roads in the state of Selangor, Malaysia. Paper presented at the 21st ARRB Conference, Cairns, Australia, 18–23 May 2003
Lay MG (2009) Handbook of road technology, 4th edn. Spon Press, New York
JKR Malaysia (1992) A guide to the visual assessment of flexible pavement surface conditions. Ibu Pejabat JKR Malaysia, Kuala Lumpur
JKR Malaysia (1994) Interim guide to evaluation and rehabilitation of flexible road pavements. Ibu Pejabat JKR Malaysia, Kuala Lumpur
Miller JS, Bellinger WY (2003) Distress identification manual for the long-term pavement performance studies, 4th edn rev. Federal Highway Administration, McLean, VA
Nicholls JC (2017) Asphalt mixture specification and testing. CRC Press, Boca Raton
Shahid MA (2018) Maintenance management of pavements for expressways in Malaysia. Paper presented at the 10th Malaysian road conference & exhibition, Kuala Lumpur, Malaysia, 28–30 October, 2018
Transport Research Laboratory (1999) A guide to the pavement evaluation and maintenance of bitumen-surfaced roads in tropical and sub-tropical countries, Overseas Road Note 18. TRL Limited, Crawthorne
Transport Research Laboratory (2004) A guide to axle load surveys and traffic counts for determining traffic loading on pavements, TRL Limited, Crawthorne

Performance Evaluation of Stone Mastic Asphalt Containing Steel Fibre as Additive

Ekarizan Shaffie, Hanis Eizzati Ahmad, Ahmad Kamil Arshad, Wardati Hashim, Haryati Yaacob, and Fionna Shiong

Abstract This paper presents results and discussion regarding the addition of steel fibre in stone mastic asphalt (SMA) application. SMA is a gap-graded asphalt mixture that consists of a large proportion of coarse aggregate, a high percentage of asphalt binder and a substantial amount of filler that provides a durable surface course. However, SMA has its own problem which is asphalt binder drain-down due to the high asphalt binder content which leads to permanent deformation. There is a need to improve the performance of the HMA mix. Thus, the aim of this study is to utilize the steel fibre that has a high tensile strength to overcome the problem of SMA. Among the tests involve are marshall stability and flow, cantabro loss, resilient modulus and dynamic creep. From the properties and performance results of this study, the use of steel fibre improved the stability and stiffness of the mix in terms of dynamic creep and resilient modulus values. The addition of 0.3% steel fibre is the most optimum percentage to be used as a modifier in the SMA mixture. Therefore, it can be concluded that the addition of steel fibre in the SMA mixture significantly enhances the overall efficiency and performance.

E. Shaffie (✉) · A. K. Arshad
Institute for Infrastructure Engineering and Sustainability Management (IIESM), Universiti Teknologi MARA, 40450 Shah Alam, Selangor, Malaysia
e-mail: eka@uitm.edu.my

A. K. Arshad
e-mail: drahmadkamil@uitm.edu.my

E. Shaffie · A. K. Arshad · W. Hashim · F. Shiong
School of Civil Engineering, College of Engineering, Universiti Teknologi MARA, 40450 Shah Alam, Selangor, Malaysia
e-mail: wardati@uitm.edu.my

H. E. Ahmad
Geofrontier Sdn Bhd, No. 6-1, Pusat Perdagangan KLH, Menara KLH, Bandar Puchong Jaya, 47100 Puchong, Selangor Darul Ehsan, Malaysia

H. Yaacob
Faculty of Civil Engineering and Construction Research Alliance, Universiti Teknologi Malaysia, UTM Johor Bahru, 81310 Johor, Malaysia
e-mail: haryatiyaacob@utm.my

R. Hassan et al. (eds.), *Green Infrastructure*,
https://doi.org/10.1007/978-981-16-6383-3_11

Keywords Stone mastic asphalt · Marshall stability · Steel fibre · Dynamic creep · Polymer modified binder

1 Introduction

Malaysia is a country with a tropical monsoon climate all year round, with seasonal heavy rainfall followed by hot summer. The road pavement exposure to hot and wet conditions contributes to the occurrence of pavement distress in which will affect the performance of the road condition. According to the Road Transport Department, there are 31.2 million units of registered vehicles in Malaysia as of December 2019. In mid-2017, there are 28.13 million vehicle registrations. This means that the number of vehicles on the road increases per year, which will increase the road maintenance expenditure. Subsequently, the demand for better construction, long-lasting and more efficient road pavement will increase to prevent pavement distress. In addition, road pavements can also be subject to load stress due to a rise in traffic volume and tyre pressure. The weight of the vehicle would create more impact on the road surface, stripping it off and creating potholes due to the friction caused by the vehicles' tyre. This road distress can lower the quality and performance of the SMA. Consequently, the maintenance costs for the roads will rapidly increase. Moreover, an increase in traffic volume can cause the road pavement to permanent deformation as the road cannot resist the repetition load from the vehicle.

SMA has been used in this study. SMA is a gap-graded hot mix asphalt that contains a high percentage of course aggregate and binder filler. SMA has its characteristic that can sustain in a long time. It was originally developed in Germany as a treatment for heavy traffic roads and was very durable and rut-resistance because of the gap-graded aggregate structure (Roberts et al. 2009). SMA is commonly used due to its stable stone skeleton, good, bonded asphalt mixture and stabilizing agent. The stabilizing agent such as fibre is added to the asphalt mixture to improve the strength and to provide adequate asphalt mixture stability. SMA can be used to pave the high-volume highway and interstates as the road will receive a higher load that can be provided by SMA which is a strong asphalt mixture skeleton. Besides, SMA has a large proportion of course aggregate content that leads to having more void. The void is filled with a high percentage of asphalt binder, filler and any stabilizing agent. The coarse aggregate structures of SMA have stone-to-stone contact. SMA has its own engineering properties which are binder drainage during mixing, transportation and compaction (Arshad et al. 2016). This is due to the high percentage of asphalt binder in the mixture. The tensile strength of the mixture will become lower as well as the performance of the road pavement. A stabilizing agent (fibre) that is mixed in the asphalt mixture prevents the binder drainage. Many researchers have done their study in modifying the SMA. They have used many types of additives to improve the SMA performance. The use of a polymer as a stabilizing additive in SMA has been started in the mid-sixties (Ahmadinia et al. 2012).

Therefore, it is necessary to use an alternative material to replace as an additive in SMA. Many types of polymer have been introduced and are continuously being introduced as the innovation and improvement on the road. The example of the fibre is asbestos fibre, carbon fibre, cellulose fibre, glass fibre and many more. By using fibre in a modification of SMA, all the distress problems such as rutting, fatigue cracking, potholes, low-temperature cracking and others could be reduced, and the properties of the asphalt mixture will be improved (Liu 2015). Thus, the performance and the service life of the pavement can be enhanced well. A study by Shaffie et al. (2015) used nanopolyacrylate and nanocomposite as binder modifiers to evaluate the efficiency of SMA. PG76 binder was used to identify the optimum binder content for the SMA. The stripping efficiency was determined using the modified Lottman test (AASHTO T283). Six (6) SMA-modified mixes were made with nanopolyacrylate and nanocomposite-modified binder formulations at 2%, 4% and 6%, respectively. The major conclusions from study were: (1) tensile strength test showed that SMA-NC6 have the highest IDT; (2) the stripping test on both control and modified SMA mixtures showed that TSR values for all the mix pass the 80% limit which is least susceptible to stripping; and (3) the addition of nanopolyacrylate and nanocomposite-modified binders shows great potential in SMA mixtures, with substantially improved binder cohesion and adhesion properties. From the findings of this study, all SMA mixes are resistant to moisture damage.

To overcome problems of rutting, fatigue cracking, potholes, low-temperature cracking and others, the researcher had put their interest in SMA modification using steel fibre. The use of steel fibre in SMA can enhance the performance of road pavement and prevent the binder drain-down. Many research works on the use of steel fibre as a modifier in SMA were carried out for a few decades. However, there are still limited finding in terms of the optimum percentage of steel fibre used and the mechanical performance test. Hence, this study aims to evaluate the SMA mixture using different percentages of steel fibre and performance evaluation with respect to resilient modulus and dynamic creep testing. It is believed that steel fibre would provide better performance in SMA properties and provide long-lasting pavement that needs less surface maintenance and thus can reduce the cost (Hainin et al. 2012).

2 Materials and Methods

As shown in Fig. 1, this study was conducted using asphalt binder with a penetration grade of 60/70 and steel fibre. Gardner Global Enterprise provided steel fibre as an additive to the SMA mixture. Steel fibre was used at a percentage of 0, 0.3, 0.5 and 0.7% by weight of asphalt binder. The steel fibre has a diameter of 0.52 mm and a length of 35 mm. Steel fibres were dry mixed with aggregate from the Kajang Rocks Quarry. The aggregate was sieved into a size range with a nominal size of 20 mm, as required by the Malaysian Public Works Department. Ordinary Portland cement was used as a mineral filler.

Fig. 1 Steel fibre

One of the most important aspects of the design mixture is aggregate gradation. The asphalt mixture in this study was made with SMA 20 aggregate gradation. Each sample weighed in at 1200 g total aggregate. Table 1 shows the SMA 20 gradation.

The steel fibre was applied to the mixture before pouring the optimum binder content in this sample, which was carried out using the dry blending process. The optimum binder content by weight of mix was 6.2%, with the percentage of optimum binder content (OBC) based on previous studies (Jasni et al. 2020a, b). To achieve a uniform result, the steel fibre length was kept constant across all samples. The percentage of steel fibre used for each proportion was shown in Table 2. The mineral filler was used at a rate of 2% by the weight of the blend.

Table 1 SMA aggregate gradation

Sieve	Gradation		Percentage passing	Percentage retained	Mass passing	Mass retained
Size	Lower	Upper				
19.0	100	100	100	0.0	1200.0	0
12.5	85	95	90	10.0	1080.0	120
9.5	65	75	70	20.0	840.0	360
4.75	20	28	24	46.0	288.0	912
2.36	16	24	20	4.0	240.0	960
0.60	12	16	14	6.0	168.0	1032
0.30	12	15	13.5	0.5	162.0	1038
0.075	8	10	9	4.5	108.0	1092
Pan				9.0		108

Table 2 Weight of the steel fibre

Percentage of steel fibre used (%)	Weight of the steel fibre (g)
0 (Control sample)	0.0
0.3	3.7
0.5	6.2
0.7	8.7

2.1 Preparation of Marshall Samples

SMA mixture with 60/70 penetration grade asphalt binder, steel fibre and mineral filler was used to produce the asphalt mixture. A total of 12 Marshall samples with different percentages of steel fibre, which is 0, 0.3, 0.5, and 0.7% of steel fibre as shown in Fig. 2, were prepared according to ASTM D1559. A temperature range of 160–180 °C was used to heat the aggregates. This study used the dry blending process, in which the steel fibre was mixed with the aggregate before pouring the 6% asphalt binder. All samples were subjected to 50 blows for both sides up and down by using automatic Marshall compactor.

Fig. 2 Aggregate preparation for Marshall mix

Fig. 3 All samples immersed in the 60 °C water bath

2.2 Marshall Stability and Flow Test

Marshall stability and flow test were performed to control the quality of the asphalt mixture and the tensile strength capability of the modified SMA mixture. This method is widely used in paving projects to assess the effectiveness of asphalt mixtures. These stability and flow tests were conducted according to ASTM D1559. A cylindrical sample was submerged in a water bath for 30–40 min at 60 °C (Fig. 3). The sample was then loaded into the Marshall stability machine as shown in Fig. 4 and loaded at a constant rate of deformation of 5 mm per minute until failure was achieved. The sample was considered a loss when the load started to decrease rather than increase. The stability and flow test results were used to determine the asphalt mixture properties.

2.3 Cantabro Loss Test

The Cantabro loss test is used to assess the abrasion loss and aggregate toughness of a compacted asphalt mixture sample. The Los Angeles (LA) abrasion machine was used to perform this Cantabro test, which followed the ASTM D7064 standard procedure. Before testing, the samples as shown in Fig. 5 were held at room temperature for at least 6 h. This ensures that the sample is fully dry before being put into the LA machine. Before the test, the dry sample was weighted. Without the steel balls, one sample was put in the LA abrasion drum at a time. The machine was started and

Fig. 4 Marshall stability machine

set to 300 revolutions per minute with a velocity range of 188–208 rad/s. After 300 revolutions, the sample was removed and the loose material that had been broken off was discarded. The sample was then weighted to determine the final mass following abrasion.

2.4 Dynamic Creep Test

Dynamic creep was used to test the stiffness properties of modified SMA to assess permanent deformation. The test was performed according to the EN12697-25 standard method using the Universal Testing Machine, UTM-5 (Fig. 6). UTM-5 was set with the effect of different temperatures and loading conditions to suit the Malaysian climate condition. The dynamic creep test conditions used in this study are shown in Table 3. All the samples were placed in the UTM-5 cabinet for 2 h to ensure that they were all at the same temperature of 25 °C. After conditioning the sample at the desired test temperature, the sample was placed between two metal-platens and the two installed LVDTs were adjusted until the sample touched the metal-platen. A haversine load with 300 kPa was applied for 3600 cycles. If the test specimen failed in the tertiary stage, the period count was lower. To ensure a frictionless surface, a

Fig. 5 Sample condition after Cantabro loss test

Fig. 6 Universal Testing Machine (UTM-5)

Table 3 Creep test conditions

Stress level (kPa)	Termination cycle count	Termination Strain (% Fs)	Preload time (s)	Temperature (°C)
300	3600	5.0	60	25

thin layer of lubricant oil was added to the metal-platen surface. The test was run using software that was mounted on a PC monitor that was connected to the UTM-5 computer. The accumulated strains were recorded at each load cycle until they reached the cycle termination. The effects are measured and determined in terms of strain or final permanent deformation for various samples.

2.5 *Resilient Modulus Test*

The resilient modulus is a crucial component in mechanistic pavement design since it is used as an input to multilayer layer elastic analysis or finite element models to compute the pavement response under traffic loads. The test was carried out in accordance with ASTM D4123 using the Universal Testing Machine (UTM-5P) as shown in Fig. 7. This test is used to determine the elastic or resilient modulus. The stress–strain relationship of HMA, as measured by elastic or resilient modulus, is a key feature. Each sample was tested at 25 and 40 °C. The samples were conditioned at the selected test temperature for 4 h prior to testing in compliance with ASTM D4123. The resilient modulus test was performed using a cylindrical test sample and a repeated load of defined magnitude and cycle length. With a pulse width of 0.1 s and a rest time of 0.9 s, a haversine load of 1000 N peak force was applied. As a result,

Fig. 7 Universal Testing Machine (UTM-5P)

Table 4 Indirect tensile resilient modulus test parameter

Test condition	Parameter
Temperature (°C)	25 and 40
Load Pulse (ms)	100
Period (ms)	1000, 2000, 3000
Applied lpercentages of steel fibreoad	Induce 10% of the indirect tensile strength

the sample was subjected to one load cycle per second, with a Poisson's ratio of 0.35. Using the horizontal deformation and an expected Poisson's ratio, the resilient modulus can be determined after five pulses. The test was then replicated with the sample oriented at approximately 90°. Computer software was used to record all the data. The test parameters are shown in Table 4.

3 Results and Analysis

3.1 Marshall Stability and Flow Test

The effect of the Marshall stability with steel fibre content was shown in Fig. 8. In this study, four different percentages of steel fibre were used. A test sample containing 0% steel fibre was used to represent the control SMA mixture. The other three different percentages of steel fibre are 0.3, 0.5 and 0.7% by the weight of the sample mixtures. For each corresponding value of steel fibre material, the graph was plotted with values of Marshall stability. Three samples were checked for each percentage of steel fibre material, and the average result was used for the analysis. According

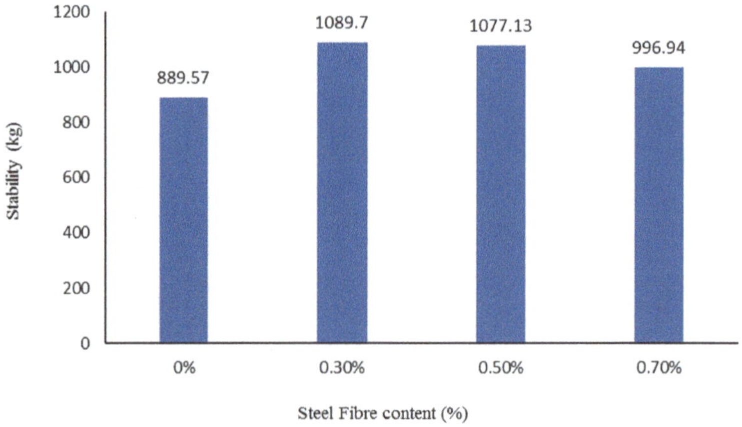

Fig. 8 Graph of Marshall stability with different steel fibre content

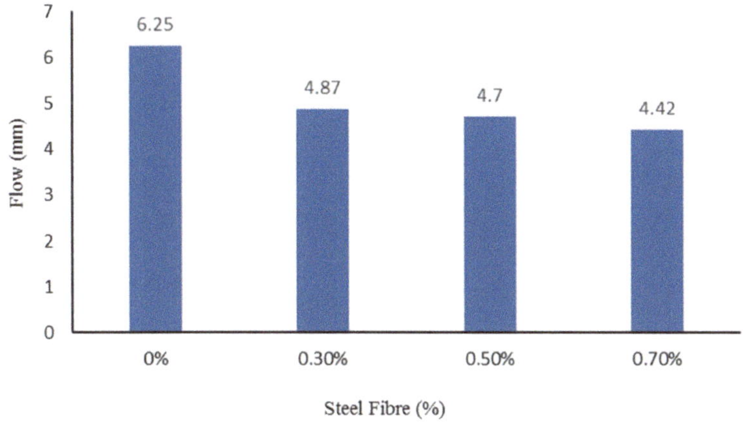

Fig. 9 Graph of Marshall flow with different steel fibre content

to the bar graph in Fig. 8, 0.3% steel fibre content has a maximum stability value of 1089.7 kg. This demonstrates that a higher imposed load can be sustained by 0.3%. Furthermore, the control sample, which had 0% steel fibre material, had the lowest stability value. This shows that the asphalt mixture without the steel fibre cannot sustain more load as compared to the asphalt mixture with steel fibre content. The stability test value increases as the steel fibre content increases, but it decreases once the asphalt mixture samples reach the maximum load, which is at 0.5% steel fibre content and above. This is due to the small amount of steel fibre added to the mixture, which allows the asphalt binder to fully bind all the sample mixture. Thus, 0.3% steel fibre content shows a good result of stability, indicating that the sample can withstand high loading.

Furthermore, the flow value represents the asphalt mixture's plasticity and flexibility properties. Figure 9 shows the difference in flow values for SMA mixtures with and without steel fibre material. The flow result for 0.7% steel fibre is 4.42 mm, which is the lowest among the others. The highest flow value, 6.25 mm, is found in the control sample. The trend for the flow value of the SMA mixture decreases after adding steel fibre. This is due to the stiffness of the steel fibre in the SMA mixtures. The more the steel fibre added into the mixtures, the less will be the flexibility of the sample. Thus, the probability of the sample to deform will become less.

3.2 Cantabro Loss Test

The Cantabro loss test is the test that measures the resistance to abrasion loss of compacted asphalt mixture. The term "resistance to loss abrasion" refers to the ability of a sample to withstand a high load without disintegrating. It is also an indicator of aggregate hardness and abrasion resistance. The Cantabro test is useful for predicting

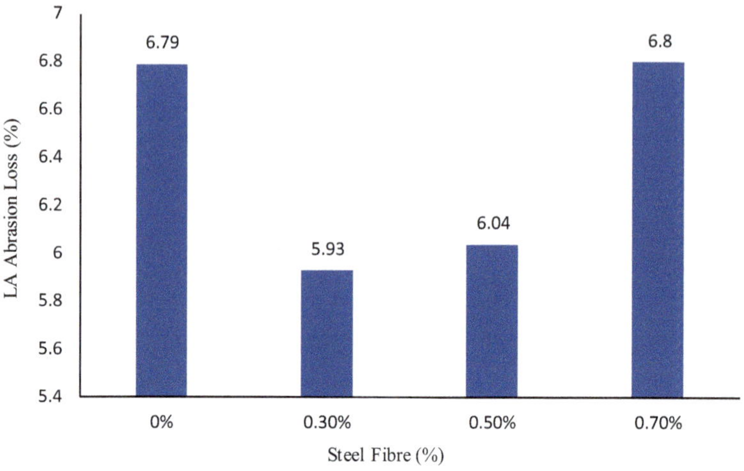

Fig. 10 Graph of abrasion loss with different steel fibre content

the lifetime of an SMA mixture (Jasni et al. 2020a, b). Figure 10 showed the abrasion loss result between the percentage differences of steel fibre. The SMA mixture containing 0.3% steel fibre has the lowest abrasion loss value of 5.9%. The abrasion value was then followed by 0.5%, 0% and 0.7% steel fibre, respectively. As steel fibre was added, the graph pattern showed a fluctuation in abrasion loss. The highest abrasion loss value is 0.7% of steel fibre, which is higher than the controlled sample (0% steel fibre). This indicates that using too much steel fibre in SMA asphalt would result in poor performance and abrasion resistance because the SMA mixture will disintegrate between each other. In this analysis, 0.3% steel fibre was found to be the most appropriate percentage to use in SMA mixtures. As compared to the other proportions, this proportion will minimize abrasion loss by up to 12%. A good proportion of steel fibre can improve the abrasion resistance and hardness of the asphalt mixture (Jasni et al. 2020a, b).

3.3 Dynamic Creep Test

Dynamic creep is the accumulation of deformation with the loading cycle and time. Dynamic creep curve can be obtained from the dynamic creep result.

The dynamic creep test results on the SMA mixture with different percentages of steel fibre content were shown in Fig. 11. As seen in the dynamic creep curve, the vertical strain of the mixture increases with increasing steel fibre material. This means that using steel fibre in an SMA mixture reduces the resistance to permanent deformation. However, as compared to the 0.3% steel fibre content, the control sample has a high value of strain. This demonstrates that the control SMA mixture is not resistant to permanent deformation. Based on the dynamic creep curve, the addition

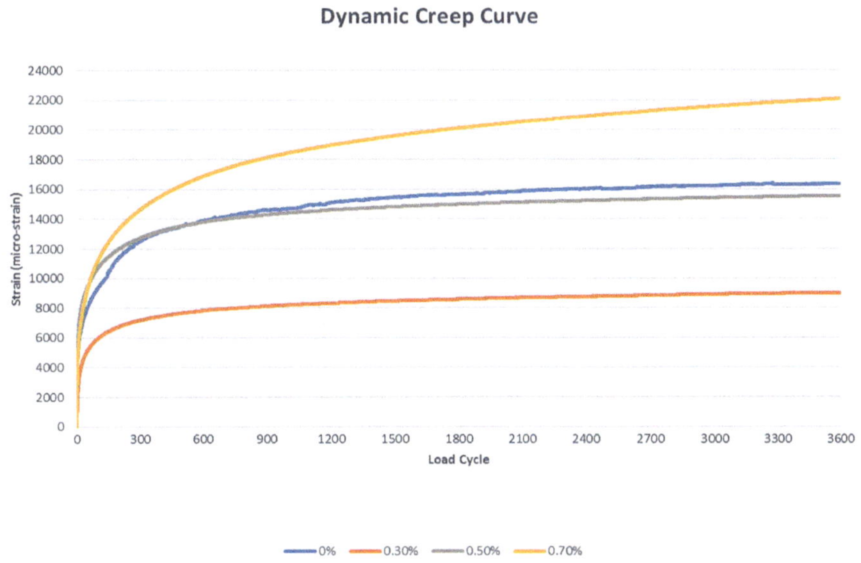

Fig. 11 Dynamic creep curve of SMA mixture

of 0.3% steel fibre to the mix decreases the strain value considerably as compared to the control mix. This means that the SMA mixture containing 0.3% steel fibre is less prone to permanent deformation. Thus, 0.3% of steel fibre is the optimum value for the SMA mixture in terms of permanent deformation.

Creep strain slope is the slope at the second stage which is determined from the regression line that excludes the primary stage. During the secondary creep phase, the creep strain rate is nearly constant under constant stress loading conditions (Katman et al. 2015). As a result, CSS is a term that describes the permanent deformation characteristics that are influenced by load cycling. Hence, CSS is another term to reflect the permanent deformation characteristics affected by the load cycling. Equation 1 has been used to calculate the CSS value.

$$CSS = \frac{\log\varepsilon_{3600} - \log\varepsilon_{1200}}{\log 3600 - \log 1200} \tag{1}$$

The CSS results for SMA mixtures with and without steel fibre are shown in Fig. 12. The graph shows a decrease in steel fibre content from 0% to 0.3%, with CSS values of 0.09 and 0.07, respectively. Furthermore, the CSS value graph is rising at 0.5 and 0.7% steel fibre. The same finding was observed as in the dynamic creep curve. An increase in the value of steel fibre content leads to an increase in CSS value for the SMA mixture. The larger the CSS value means that the SMA mixture is less resistant to permanent deformation (Napiah et al. 2010). However, as compared to the control sample, 0.3% of steel fibre has a lower CSS value (0% steel fibre). This indicates that the control sample has low resistance to permanent deformation. When

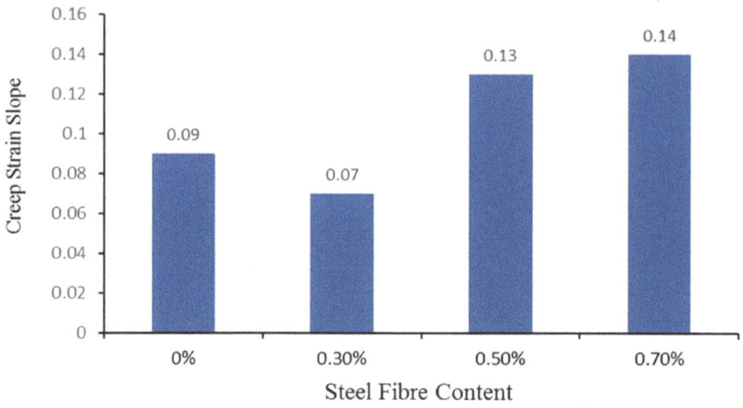

Fig. 12 Creep strain slope (CSS) of SMA mixture

0.3% steel fibre is added to the SMA mixture, the maximum resistance to permanent deformation is achieved.

3.4 Resilient Modulus Test

Resilient modulus testing is one of the tests used to assess the performance of steel fibre in the modified asphalt mixture. It was performed at two different temperatures, 25 and 40 °C. A data of resilient modulus against the form of mixtures at both temperatures (25 and 40 °C) is shown in Table 5.

Figure 13 shows the resilient modulus values at 25 °C. The resilient modulus of the control mixture (0% steel fibre) increased by up to 27% after adding 0.3%. The same pattern was observed for 0.5% and 0.7%, which showed an increase of 22% and 21%, respectively, after steel fibre was added. Steel fibre mixtures have slightly greater resilient modulus than control mixtures in all three samples. The increase in

Table 5 Indirect tensile resilient modulus test parameter

Temperature (°C)	Period (ms)	Resilient modulus (MPA)			
		Fibre content			
		0%	0.3%	0.5%	0.7%
25	1000	1101	1648	1466	1453
	2000	1047	1456	1340	1326
	3000	1066	1332	1309	1295
40	1000	289	726	593	586
	2000	295	649	684	657
	3000	297	705	514	498

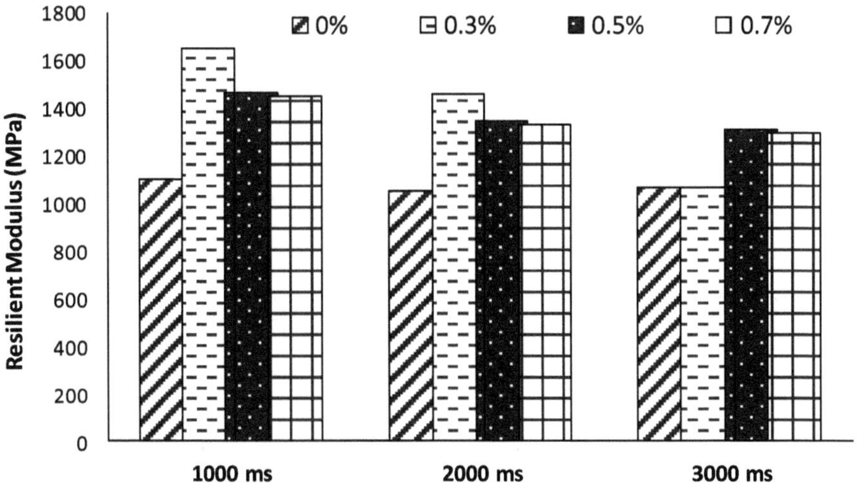

Fig. 13 Resilient modulus at 25 °C

resilient modulus might be due to the higher modulus of elasticity and low ability of extension of the fibre and randomly oriented in a different direction (Arshad et al. 2019).

Figure 14 shows the resilient modulus values at 40 °C. When the temperature increases from 25 to 40 °C, the resilient modulus values are reduced by about 58%. When measured under the same conditions, the resilient modulus of 0.5% and 0.7% decreased by as much as 50% and 49%, respectively. From this finding, the sample containing steel fibre had significantly higher stiffness than the control sample. When

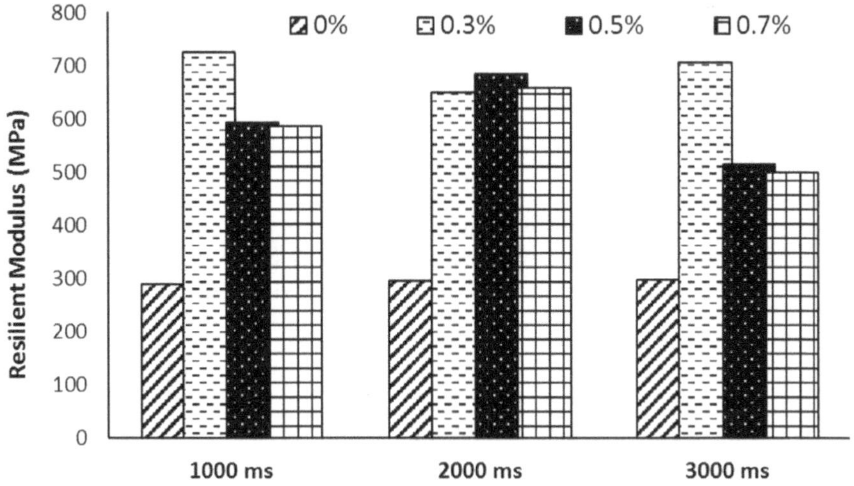

Fig. 14 Resilient modulus at 40 °C

the mixture was subjected to a lower temperature, the viscosity of the asphalt binder increased, reducing the strain and capacity of the mixture to flow. As a result, the resilient modulus is greater. When the sample was subjected to a higher temperature, the pressure and capacity of the mixture to flow improved as the asphalt binder became less viscous and the resilient modulus decreased. As a result, the resilient modulus decreased as the temperature increased.

4 Conclusion

Based on the results obtained from the laboratory works and the detailed analysis, the following conclusions are drawn:

1. The properties of the asphalt mixture have been improved by adding steel fibre. The increase in stability shows that the strength of the asphalt mixture to resist the traffic load is improved. It shows that steel fibre can prolong the deterioration of asphalt pavement. Furthermore, as the steel fibre content increases, the flow test value decreases.
2. Steel fibre's unique characteristics and properties have led to a major improvement in the efficiency of SMA mixtures. SMA-improved steel fibre is found to improve abrasion resistance. A Cantabro test revealed that 0.3% steel fibre would significantly reduce the high amount of abrasion loss compared to other percentages of steel fibre.
3. From the dynamic creep curve, the vertical strain increases as the steel fibre content increases. The strain value for the control sample is higher than the 0.3% steel fibre material. The SMA mixture is less susceptible to irreversible deformation as the strain value decreases. The creep strain slope also revealed that steel fibre with a content of 0.3% has the lowest value. The lower the CSS value, the more resistant the SMA mixture is to permanent deformation.
4. The resilient modulus performance determined for the mix using the steel fibre mix additive is higher than the control sample at both temperatures of 25 and 40 °C and at all the repetition periods tested.
5. Based on the performance results of this study, the use of steel fibre can improve the stiffness of the mix in terms of dynamic creep and resilient modulus values. 0.3% steel fibre by weight of the mixture is the most optimum steel fibre material to be used as a modifier in the SMA mixture. These proved that the steel fibre can enhance the performance of the hot mix asphalt (HMA).

As a recommendation, the morphology study using scanning electron microscope images and performance evaluation based on rutting, stripping, fatigue and other mechanical efficiency have been proposed for future research to gain a better understanding of this substance in asphalt mixture.

Acknowledgements A special thanks go to the Research Management Institute (RMI) of Universiti Teknologi MARA for providing the financial support under the FRGS (600-IRMI/FRGS 5/3

(379/2019). The authors would like to thank the School of Civil Engineering, College of Engineering, Universiti Teknologi Mara, 40450 Shah Alam, Selangor for providing the experimental facilities and to all technicians at Highway and Traffic Engineering Laboratory.

References

Ahmadinia E, Zargar M, Karim MR, Abdelaziz M, Ahmadinia E (2012) Performance evaluation of utilization of waste Polyethylene Terephthalate (PET) in stone mastic asphalt. Constr Build Mater 36:984–989

Arshad AK, Mansor S, Shafie E, Hashim W (2016) Performance of stone mastic asphalt mix using selected fibres. Jurnal Teknologi 78(7–2):99–103

Arshad AK, Shaffie E, Hashim W, Syahirah N, Mustapa F (2019) Resilient modulus performance of warm mix asphaltic concrete using cecabase RT additive. Int J Recent Technol Eng 8(3S3):399–403

Hainin R, Reshi WF, Niroumand H (2012) The importance of stone mastic asphalt in construction. Electron J Geotech Eng 17:49–74

Jasni NE, Masri KA, Ramadhansyah PJ, Arshad AK, Shaffie E, Ahmad J, Norhidayah AH (2020a) Mechanical performance of stone mastic asphalt incorporating steel fiber. IOP Conf Ser Mater Sci Eng

Jasni NE, Masri KA, Ramadhansyah PJ, Arshad AK, Shaffie E, Ahmad J, Norhidayah AH, Fauzi NANM, Masri KA, Ramadhansyah PJ, Samsudin MS, Ismail A, Arshad AK, Shaffie E, Norhidayah AH, Hainin MR (2020b) Mechanical performance of stone mastic asphalt incorporating steel fiber. IOP Conf Ser Mater Sci Eng 712(1).

Katman HY, Ibrahim MR, Karim MR, Salim Mashaan N, Koting S (2015) Evaluation of permanent deformation of unmodified and rubber-reinforced SMA asphalt mixtures using dynamic creep test. Ad Mater Sci Eng 2015

Liu Y (2015) Performance of modified asphalt binder with tire rubber powder. Jurnal Teknologi 4:55–60

Napiah, Madzlan B, Kamaruddin IG (2010) The creep performance of base and polymer modified bitumen mixes containing different types of sand as fine aggregate. J East Asia Soc Transp Stud 8

Roberts FL, Kandhal PS Brown ER Lee D-Y, Kennedy TW (2009) Hot mix asphalt materials, mixture design and construction, 3rd edn. NCAT, Alabama

Shaffie E, Ahmad J, Arshad AK, Kamarun D (2015) Using nanopolyacrylate and nanocomposite as binder modifiers for Stone Mastic Asphalt (SMA). Appl Mech Mater 802:381–386

Post-construction Complexity Factors Impacting Infrastructure Project Performance in Malaysia

Akhtarul Norfaiza Che Nen, Che Maznah Mat Isa, and Che Khairil Izam Bin Che Ibrahim

Abstract Poor project performance has been imperative to both private and public construction stakeholders to ensure organizational sustainability. Most previous studies generally focused on project delays, cost overruns, and lack of quality during the construction phase, but still lacking literature on infrastructure project performance during the post-construction phase. This chapter focuses on the identification of post-construction complexity factors for infrastructure projects impacting the three dimensions of project performance, namely time, cost, and quality. A questionnaire survey was administered to 106 construction professionals involved in infrastructure projects in Malaysia. A Rasch model was used to validate the instrument item and person in the questionnaires. A confirmatory factor analysis based on a structural equation modelling (SEM) using partial least square (SEM-PLS) revealed that the top three post-construction cost factors that have significantly impacted the project performance are inadequate warranty and maintenance manuals, insufficient documents certifying all sub-contractors bills after collecting all documents and attending snags, and insufficient documents to prepare completion certificate based on all documents. Thus, these findings have empirically contributed to integrating complexity assessment for infrastructure projects during the post-construction life cycle in line with the Twelfth Malaysia Plan for the construction industry.

Keywords Infrastructure · Project performance · Cost · SEM-PLS

1 Introduction

The construction industry is considered one of the most exclusive, unpredictable, and tough industries and has a poor reputation for managing risk, with many significant

A. N. Che Nen · C. M. Mat Isa (✉) · C. K. I. B. Che Ibrahim
School of Civil Engineering, College of Engineering, Universiti Teknologi MARA, Shah Alam, Selangor, Malaysia
e-mail: chema982@uitm.edu.my

C. K. I. B. Che Ibrahim
e-mail: chekhairil449@uitm.edu.my

R. Hassan et al. (eds.), *Green Infrastructure*,
https://doi.org/10.1007/978-981-16-6383-3_12

efforts neglected to meet project performance requirements. A construction process is viewed as the most unpredictable endeavour where of late the development business is having incredible trouble in adapting to expanding complexity at significant construction projects (Baccarini 1996).

Project complexity can be cited as an important issue as it commits uncertainties that are highly related to the task management process which is very much for the construction industry (Yong and Othman 2017).

As a result of the difficulties encountered during the development period, these are the primary factors that contribute to the project's complexity. The construction industry is unpredictable in its inclination because it includes a large number of gatherings as owners (clients), contractors, consultants, stakeholders, and regulators. Many local construction projects report poor performance as a result of numerous project and explicit causes (Enshassi et al. 2009). Project performance can be estimated and assessed using numerous execution markers that can be identified with various measurement groups, for example, time, cost, quality, customer fulfilment, customer changes, business execution, health, and safety (Cheung et al. 2004). Furthermore, due to the complexity of the plan and construction process, project completion in terms of cost, time overrun, quality, performance, and safety is truly poor in the construction industry (Wood and Gidado 2008). The effectiveness of performance estimation is an issue of growing importance to both industrialists and academics. In construction projects, the standard performance zones are cost, time, and quality. Thus, an effective project is one that meets the requirements of time, cost, quality, performance, and safety (Rahman et al. 2013; Masrom et al. 2015).

In general, construction projects, including road infrastructure projects, are made up of many interconnected parts, which meets the criteria of complexity. Nonetheless, previous research on complexity in construction projects has been limited, with most studies focusing solely on conceptual factors that contribute to complexity in the construction industry (Baccarini 1996; Williams 1999; Sinha et al. 2006; Brockmann and Girmscheid 2007; Wood and Gidado 2008; Pitsis et al. 2004). The literature reviews revealed an existing or deficiency of understanding of complexity factors in infrastructure projects. As a result, the purpose of this paper is to identify the complexity factors that affect project time, cost, quality, and performance, specifically during the post-construction stage.

2 Literature Review

A construction project is getting bigger and more complex, so it is important to understand the concept and management of the complexity of the project. The construction industry has faced great difficulties and has to overcome the problem of increasing complexity of major construction projects as well as declining project management performance, especially infrastructure projects (Williams 2002; Bosch-Rekveldt 2011). Although a comprehensive study of the concept of complexity has been made, there is no consensus on the complexity of the project. According to a study by Sinha et al. (2006), in fact, the whole intuitive concept of complexity is unlikely to be accurately understood by a single concept or definition. Thus, the meaning and term "complexity" can be felt to have various connotations not only in different fields but also in the same field. Project complexity needs to be understood and measured efficiently to help project management become better (Vidal and Marle 2008; Bryde 2008). Various factors of complexity occur during project construction, especially megaprojects, and each factor is interrelated, resulting in various levels of complexity with specific characteristics (Lucas 2005; Chan et al. 2004). In general, difficulties associated with decision-making and goal achievement seem to stem from complexity. Therefore, an understanding of project complexity is very important for project management (Remington et al. 2009). Thus, the success of a project ultimately depends on the performance of the project and the complexity of the related project (Chan et al. 2004). Although the importance of understanding project complexity in project management has been widely acknowledged (Vidal and Marle 2008; Parsons-Hann and Liu 2005), however, preventive and remedial measures are insufficient for project complexity level assessment, mainly because project complexity is interrelated with observer subjectivity (Corning 1998; Pich et al. 2002).

Brockmann and Girmscheid (2007) revealed that the complexity level of the project changes through the life cycle of the project. Established on six case studies and 35 interviews, the authors argue that the complexity level of the project is the highest at the start of the project and it decreases as the project is executed. Also, some issues such as change orders or legal disputes might change the complexity level o f the project by a significant amount.

3 Conceptual Framework and Research Hypothesis

Based on the literature review, the research proposes the following hypothesis and conceptual model as shown in Fig. 1.

Based on that, we are interested in testing the following hypotheses:

H1: Post-construction phases complexity factors related to time positively influence and affect the project performance (time, cost, and quality).

H2: Post-construction phases complexity factors related to cost positively influence and affect the project performance (time, cost, and quality).

H3: Post-construction phases complexity factors related to quality positively influence and affect the project performance (time, cost, and quality).

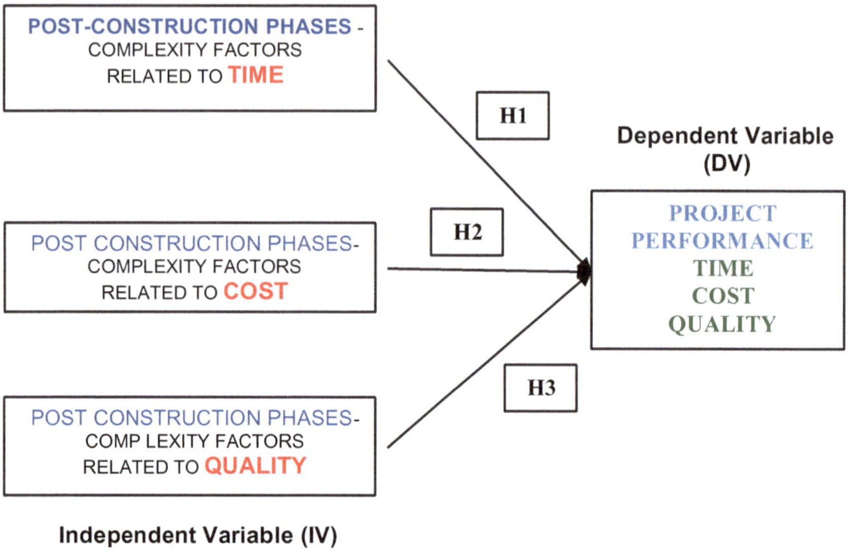

Fig. 1 Conceptual model

4 Methodology

4.1 Research Approach

Survey questions were designed to measure the complexity factors that significantly affect project performance during the post-construction stage in infrastructure projects by adapting a quantitative method.

4.2 Target Respondents and Sampling Design

A random sampling method was used to target the respondents who were clients, consultants, contractors, sub-contractors, and other parties involved in infrastructure projects based on a sampling frame provided by the Construction Industry Development Board (CIDB), Malaysia and Master Builders Association Malaysia (MBAM). The targeted number of respondents was around 384 based on Krecjie and Morgan (1970) table (Table 1).

Table 1 Table for determining sample size from a given population. *Source* from Krecjie and Morgan (1970)

N	S	N	S	N	S
10	10	220	140	1200	291
15	14	230	144	1300	297
20	19	240	148	1400	302
25	24	250	152	1500	306
30	28	260	155	1600	310
35	32	270	159	1700	313
40	36	280	162	1800	317
45	40	290	165	1900	320
50	44	300	169	2000	322
55	48	320	175	2200	327
60	52	340	181	2400	331
65	56	360	186	2600	335
70	59	380	191	2800	338
75	63	400	196	3000	341
80	66	420	201	3500	346
85	70	440	205	4000	351
90	73	460	210	4500	354
95	76	480	214	5000	357
100	80	500	217	6000	361
110	86	550	226	7000	364
120	92	600	234	8000	367
130	97	650	242	9000	368
140	103	700	248	10,000	370
150	108	750	254	15,000	375
160	113	800	260	20,000	377
170	118	850	265	30,000	379
180	123	900	269	40,000	380
190	127	950	274	50,000	381
200	132	1000	278	75,000	382
210	136	1100	285	1,000,000	384

[a]*Note* N is population size
[b] S is sample size

4.3 Questionnaire Design

There are three sections in the survey questionnaire: Part A studies the demographics of the respondent; Part B focuses on the factors of complexity during pre-construction, construction, and post-construction stages that affected project performance (time, cost, and quality) in infrastructure projects. Therefore, respondents are required to answer based on their work experience in infrastructure projects. A Likert five-point scale was used (1: Strongly Disagree; 2: Disagree; 3: Agree; 4: Strongly Agree; 5: Strongly Agree). Meanwhile, Part C is related to the impact mitigation plan on critical complexity impact to project performance in infrastructure projects. This paper only discusses Part A and Part B during the post-construction phases.

4.4 Data Analysis

In this study, the Rasch model was used to validate the items in the survey questionnaire and the respondents who answered the survey questionnaire. Subsequently, SmartPLS3 (Partial Least Squares-Structural Equation Modeling (PLS-SEM)) was used for statistical analysis. Measurement models using PLS-SEM are convergent validity, discriminant validity (heterotrait–monotrait (HTMT)), and in terms of the structural model, the path coefficients, and the coefficient of determination (R^2) will be measured.

5 Results and Discussion

5.1 Respondent's Demographic Characteristics

In total, about 106 respondents answered the survey questionnaires resulted in a response rate of 27.6%. The response rate is reasonable as most of the studies conducted in Malaysia resulted in the rate falling between 10 and 20% (Ramayah et al. 2005; Ainin et al. 2010). The demographic characteristics of the respondents are given in Table 2.

It shows that 67% of the respondents are involved in the private sectors and 33% of the respondents are in the public sectors. Table 2 shows the percentage distribution of the respondent's years of working experience where more than 70% have more than 5 years of working experience in the construction industry. Table 2 presents the percentage distribution of the respondent's working background, which are around 26.4% working as consultants and clients, 33% as contractors, and 3.8% as sub-contractors.

Table 2 also presents the percentage distribution of respondent's position in a construction organization.

Table 2 Demographic of respondents

	Frequency	Percentage
Respondent background		
Client	28	26.4
Consultant	28	26.4
Contractor	35	33.0
Sub-contractor	4	3.8
Others	11	10.4
Respondent's position		
Project director	5	4.7
Project manager	22	20.8
Contract manager	4	3.8
Design manager	1	0.9
Project engineer	32	30.2
Construction manager	2	1.9
Risk manager	0	0
Site supervisor	2	1.9
Quantity surveyor	5	4.7
Others	33	31.1
Respondent working experience		
<5 years	24	22.6
6–10 years	29	27.4
11–15 years	27	25.5
16–20 years	6	5.7
>20 years	20	18.9
Type of sector		
Public sector	35	33.0
Private sector	71	67.0

5.2 Reliability and Validity (Rasch Model)

Table 3 shows the acceptable reliability values for the person (respondents) and item reliability. This table is part of the rating scale instrument developed by Fisher (2007) based on Rasch literature and his extensive experience in conducting Rasch analysis in different settings.

Table 4 shows the result of reliability for person and item by using Rasch model. Referring to Table 3, the rating scale for person reliability is acceptable for time, cost, and quality performance during the post-construction stage. Meanwhile, item reliability for cost performance is poor in rating scale measurement before the person is deleted (N = 106). During measurement after the person is deleted (N = 87), it

Table 3 Rating scale

Person and item measurement reliability	
Poor	<0.67
Fair	0.67–0.80
Good	0.81–0.90
Very good	0.91–0.94
Excellent	>0.94

Table 4 Person and item reliability measurement before and after person deleted

Stage	Measurement before person deleted (N = 106)		Measurement after person deleted (N = 87)	
	Person reliability	Item reliability	Person reliability	Item reliability
Post-construction affecting time performance	0.91	0.74	0.90	0.82
Post-construction affecting cost performance	0.90	0.56	0.89	0.90
Post-construction affecting quality performance	0.88	0.76	0.89	0.90

shows that all item reliability is acceptable for time, cost, and quality performance in the post-construction stage.

5.3 *Measurement Model*

(1) *Convergent Validity*

In order to measure the reliability of each item, the factor loading should be measured. According to Hair et al. (2017), a threshold value of equal or greater than 0.7 for each item's loading is considered reliable. In addition, the Cronbach's alpha and composite reliability values should be equal to or greater than 0.7.

Based on Table 5, we can observe that all the items are reliable and satisfactory. Besides, the average variance extracted (AVE) is defined as the grand mean value of the squared loadings of the items related to the construct and is the common measure for establishing the convergent validity. A value of 0.5 or greater for the AVE specifies that the construct elucidates more than half of the variance of its items (Hair et al. 2017).

Table 5 Convergent validity

Construct	Item	Details	Loadings	CR	AVE
Post-construction quality	B3_10_Quality	Lack of workers to rectify all defects work	0.879	0.922	0.797
	B3_7_Quality	No proper defect liability inspection	0.895		
	B3_8_Quality	Insufficient time to rectify all defects works without a proper schedule of defects	0.905		
Post-construction time	B3_1_Time	Insufficient testing and commissioning work	0.872	0.908	0.767
	B3_2_Time	Inadequate of all building drawings from sub-contractor and consultant and approval by an architect	0.92		
	B3_8_Time	Insufficient time to rectify all defects works without a proper schedule of defects	0.833		
Post-construction cost	B3_3_Cost	Inadequate warranty and maintenance manuals	0.91	0.918	0.79
	B3_4_Cost	Insufficient document to certify all sub-contractors bills after collecting all documents and attending snags	0.838		
	B3_6_Cost	Insufficient document to prepare completion certificate based on all documents	0.917		
Performance	COST	[redacted]	**0.908**	0.806	0.598
	QUALITY		**0.453**		
	TIME		**0.875**		

Table 5 indicates that the Cronbach's alpha and composite reliability values are greater than 0.7, and the AVE values are greater than 0.5. Thus, the constructs' convergent validity is established.

(2) *Discriminant Validity*

In order to establish the discriminant validity, the heterotrait–monotrait ratio should be examined. Regarding the heterotrait–monotrait ratio (HTMT), a value of less than 0.85 for HTMT should be confirmed. According to Table 6, it can be deduced that the HTMT criterion is met, thus indicating that the discriminant validity is established.

Table 6 Discriminant validity (heterotrait–monotrait)

	1	2	3	4
1. Performance	[redacted]			
2. Post-construction cost	0.168	[redacted]		
3. Post-construction quality	0.166	0.638	[redacted]	
4. Post-construction time	0.145	0.776	0.839	[redacted]

5.4 *Structural Model Assessment*

The explanatory power of the model is evaluated by measuring the discrepancy amount in the dependent variables of the model. According to Hair et al. (2017), the R2 and path coefficients are the essential measures for assessing the structural model. As shown in Fig. 2, the model has R2 value of 12.6% for project performance.

In terms of path analysis, Fig. 2 and Table 7 demonstrate the path coefficients and p-values for each hypothesis. There are three hypotheses from the model, which are:

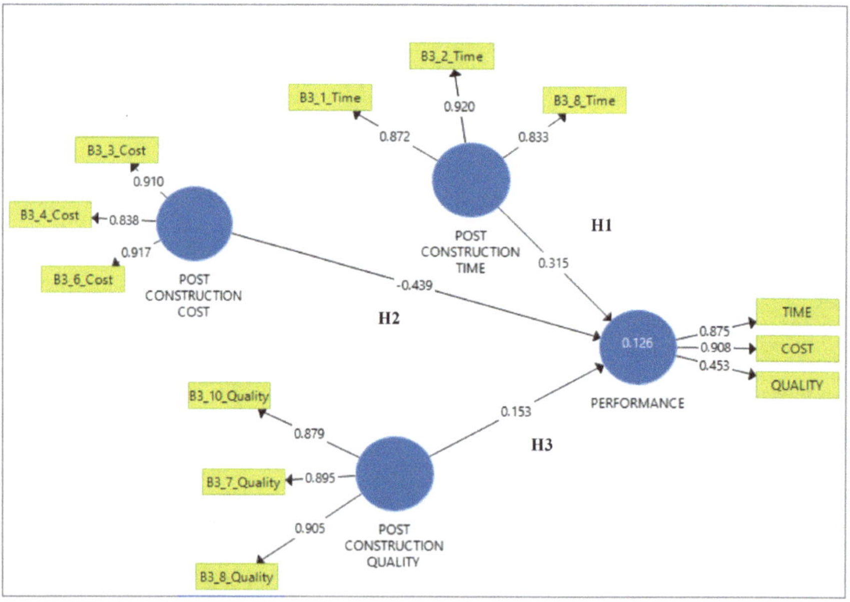

Fig. 2 Path analysis results (R^2)

Table 7 Hypothesis test results

Hypothesis	Relationship	Path coefficient	p-value	Remarks
H2	POST-CONSTRUCTION COST -> PERFORMANCE	−0.439	0.015	Significance
H3	POST-CONSTRUCTION QUALITY -> PERFORMANCE	0.153	0.155	Not significance
H1	POST-CONSTRUCTION TIME -> PERFORMANCE	0.315	0.06	Not significance

Hypothesis 1(H1): Post-construction time (complexity factors) influences and affects project performance (time, cost, and quality).

Hypothesis 2 (H2): Post-construction cost (complexity factors) influences and affects project performance (time, cost, and quality).

Hypothesis 3 (H3): Post-construction quality (complexity factors) influences and affects project performance (time, cost, and quality).

From the results, it shows that only one hypothesis is supported, which in turn indicates that one of the paths is significant between the independent and dependent variables. H2 ($B = -0.439$, $p < 0.05$) describes the path between post-construction cost influence and affects project performance (time, cost, and quality). From these results, it shows that during post-construction stage, complexity factors affected more the cost performance.

6 Discussion

The demographic profiles of the respondents show that most of them are involved in private sectors as compared to public sectors. The majority of them have acquired more than 5 years of working experience in the construction industry. There was a fair distribution of respondents from consultants, clients, and contractors. Thus, the descriptive analysis indicates the various respondents have good experience and background to give valid and relevant feedbacks to this study. The research instrument used is the survey questionnaires which were validated using Rasch model analysis based on items and persons. Thus, the validated instrument contributed to reliable and valid data, and the results can be used further in PLS-SEM to develop a path model. The results from Rasch model analysis indicate that all item reliability is acceptable for time, cost, and quality performance during post-construction phases with validated persons giving their opinions. Based on the PLS-SEM path analysis one path is significant between the independent (post-construction complexity factors) and dependent variables (project performance—time, cost, and quality). The significant values describe the path between post-construction cost and project performance. From these results, it shows that during post-construction phases, complexity factors significantly affected the cost performance, due to inadequate warranty and maintenance manuals, insufficient document certifying all sub-contractors bills after collecting all documents and attending snags, and insufficient document to prepare completion certificate based on all documents.

7 Conclusion

This chapter focuses on complexity factors during the post-construction phase in relation to the three dimensions of performance, namely time, cost, and quality, specifically for infrastructure projects. A questionnaire survey was administered to 106 construction professionals involved in infrastructure projects in Malaysia. This study determines the complexity factors affecting project performance during the post-construction stage. The PLS-SEM approach is used to assess the measurement and structural models. The outcomes reveal that the complexity factors are significantly effective in terms of cost performance during the post-construction stage. This chapter is limited to complexity factors during the post-construction phase affecting project performance. However, these findings have empirically contributed to integrating complexity assessment for infrastructure projects during the post-construction life cycle in line with the Twelfth Malaysia Plan for the construction industry. The Twelfth Malaysia Plan (12MP) will be aligned with the shared prosperity initiative encompassing three dimensions, namely economic empowerment (4.0 Industrial Revolution), environmental sustainability (green technology, risk management), and social re-engineering (cost of living). Further research can be carried out during pre and construction phases to determine a comprehensive scenario in infrastructure projects.

Acknowledgements The authors would like to thank the Faculty of Civil Engineering, UiTM, for giving support. They are also grateful to the professionals and managers from Malaysian construction firms, CIDB Malaysia, and other institutions for their participation in this research.

References

Ainin S, Kamarulzaman Y, Farinda AG, Che Azmi A (2010) Business and entrepreneur characteristics influence on business performance of professional small medium enterprises. In: Proceedings of the 5th European conference on entrepreneurship and innovation, University of Athens, Greece, pp 31–41

Baccarini D (1996) The concept of project complexity—a review. Int J Project Manage 14(4):201–204. https://doi.org/10.1016/0263-7863(95)00093-3

Bosch-Rekveldt M (2011) Managing project complexity. Delft Centre for Project management, The Netherlands. ISBN 978–94-91005-00-8

Brockmann C, Girmscheid G (2007) Complexity of megaprojects. CIB World Building Congress, pp 219–230

Bryde D (2008) Perceptions of the impact of project sponsorship practices on project success. Int J Project Manage 26(8):800–809. https://doi.org/10.1016/j.ijproman.2007.12.001

Chan APC, Scott D, Chan APL (2004) Factors affecting the success of a construction project. J Constr Eng Manag 130(1):153–155. https://doi.org/10.1061/(ASCE)0733-9364(2004)130:1(153)
Cheung SO, Suen HCH, Cheung KKW (2004) PPMS: a web-based construction project performance monitoring system. Autom Constr 13(3):361–376. https://doi.org/10.1016/j.autcon.2003.12.001
Corning PA (1998) Complexity is just a word! Technol Forecast Soc Change 59(2):197–200
Enshassi A, Mohamed S, Abushaban S (2009) Factors affecting the performance of construction projects in the Gaza strip. J Civ Eng Manag 15(3):269–280. https://doi.org/10.3846/1392-3730.2009.15.269-280
Fisher WJ (2007) Rating scale instrument quality criteria. Rasch Measur Trans 21(1):1095
Hair JF, Hult GTM, Ringle CM, Sarsdedt M (2017) A primer on partial least squares structural equation modeling (PLS-SEM). SAGE Publication, Inc. https://doi.org/10.1080/1743727x.2015.1005806
Krejcie RV, Morgan DW (1970) Determining sample size for research activities. Educ Psychol Measur 30:607–610. https://doi.org/10.1891/9780826138446.0006
Lucas C (2005) The philosophy of complexity. On-line at www.calresco.org. Accessed 8 Aug 2005
Masrom MAN, Rahim MHIA, Mohamed S, Chen GK, Yunus R (2015) Successful criteria for large infrastructure projects in Malaysia. Procedia Eng 125:143–149. https://doi.org/10.1016/j.proeng.2015.11.021
Parsons-Hann H, Liu K (2005) Measuring requirements complexity to increase the probability of project success. In: ICEIS 2005—proceedings of the 7th international conference on enterprise information systems, pp 434–438. https://doi.org/10.5220/0002548104340438
Pich MT, Loch CH, De Meyer A (2002) On uncertainty, ambiguity, and complexity in project management. Manage Sci 48(8):1008–1023. https://doi.org/10.1287/mnsc.48.8.1008.163
Pitsis TS, Komberger M, Clegg S (2004) The art of managing relationships in interorganizational collaboration. Project Manage J 7(3):47–67. https://doi.org/10.1002/pmj.20019
Rahman IA, Memon AH, Tarmizi A, Karim A (2013) Significant factors causing cost overruns in large construction projects in Malaysia. J Appl Sci 13(2):286–293
Ramayah T, Lim YC, Mohamed S (2005) SME e-readiness in Malaysia : implications for planning and implementation. Sasin J Manag 11(1):103–120
Remington K, Zolin R, Turner R (2009) A model of project complexity: distinguishing dimensions of complexity from severity. In: International research network of project management conference (IRNOP)
Sinha S, Kumar B, Thomson A (2006) Measuring project complexity: a project manager's tool. Archit Eng Design Manag 2(3):187–202. https://doi.org/10.1080/17452007.2006.9684615
Vidal LA, Marle F (2008) Understanding project complexity: implications on project management. Kybernetes 37(8):1094–1110. https://doi.org/10.1108/03684920810884928
Williams TM (1999) The need for new paradigms for complex projects. Int J Project Manage 17(5):269–273. https://doi.org/10.1016/S0263-7863(98)00047-7
Williams T (2002) Modelling complex projects
Wood H, Gidado K (2008) Project complexity in construction. Cobra 2008:1–13
Yong K, Othman M (2017) Categorization of project complexity factors in Malaysian construction industry. Int J Prop Sci 7(1):1–15. https://doi.org/10.22452/ijps.vol7no1.3

SWOT Analysis of Green Technology Application for the Development of Low Carbon Cities

Nur Kamaliah Mustaffa, Sakhiah Abdul Kudus, Muhamad Fuad Shukor, Ridza'uddin Mohd Radzi, and Verona Ramas Anak Joseph

Abstract The integration of green technology in urban development has become an imperative measure towards effective implementation of low carbon city policy. Therefore, this study aims to identify the strengths-weaknesses-opportunities-threats (SWOT) of green technologies adoption in low carbon city development. Responses from fifty-seven (57) respondents and five (5) experts of local city players were collected through survey questionnaires and interviews. The results showed that the role of government is recognized as an effective catalyst for green development while the lack of awareness and technical capacity identified as the weaknesses in adopting green technology in low carbon city development. Findings from the study also indicate that green technology transfer provides opportunity to promote economic development while green investment risk as a threat towards successful adoption of green technology in low carbon cities. The findings of this study could help policy makers to identify specific SWOT in green technologies adoption and develop suitable strategies towards conceptualization of low carbon city enhancement in Malaysia. This study contributes to the current body of knowledge concerning the insights of current status of green technology application for the development of low carbon city in developing countries especially Malaysia.

N. K. Mustaffa (✉) · S. A. Kudus · V. R. A. Joseph
School of Civil Engineering, College of Engineering, Universiti Teknologi MARA Shah Alam, 40450 Shah Alam, Selangor, Malaysia
e-mail: nurkamaliah@uitm.edu.my

S. A. Kudus
e-mail: sakhiah@uitm.edu.my

M. F. Shukor
School of Civil Engineering, College of Engineering, Universiti Teknologi MARA Pasir Gudang, 81750 Masai, Johor, Malaysia
e-mail: fuad9845@uitm.edu.my

R. M. Radzi
Department of Engineering, Ipoh City Council, Jalan Sultan Abdul Jalil, Greentown, 30450 Ipoh, Perak, Malaysia
e-mail: ridzauddin@mbi.gov.my

R. Hassan et al. (eds.), *Green Infrastructure*,
https://doi.org/10.1007/978-981-16-6383-3_13

Keywords SWOT analysis · Green technology · Low carbon cities · Climate change · And urban development

1 Introduction

The last two decades have seen growing concern worldwide about the effects of climate change. The detrimental effect of climate change has changed the physical and biological properties of the earth's surface and eventually lead to natural disasters. It is common knowledge that carbon dioxide emissions are mainly responsible for global warming. Carbon dioxide from the use of fossil fuels is the largest source of greenhouse gas (GHG) emissions due to human activities (Alkhayyat and Pehlivan 2017). According to Masson-Delmotte (2018), human activities are estimated to have caused approximately 1.0 °C of global warming above pre-industrial levels, with a likely range between 0.8 and 1.2 °C. If the increment continues at the current rate, global warming is predicted to reach 1.5 °C between 2030 and 2052. Increases in carbon dioxide (CO_2) emissions would result in an increase in atmospheric CO_2 concentration, and consequently rising global temperatures and sea level.

The rapid urbanization through various human activities is the foremost driving factor towards climate changes. Urban development coupled with population growth is posing significant risks on living conditions, the environment, and causing degradation of the ecosystems. According to UN DESA (2014), more than 3.6 billion people out of 7.1 billion world's population live in cities. It is also further forecasted that the urban population will achieve 6.7 billion by 2050. The rapidly growing significance of cities in the developing world, and the sustained importance in the developed world, has contributed to the rising awareness on the mitigation of climate change. Cities currently account for more than 67% of energy use and produce 71–76% of greenhouse gas (GHG) emissions (Edenhofer et al. 2014). The detrimental effect of climate change has managed to attract significant interest from a society looking for sustainability in urban development decisions. Developing a low carbon city is a prerequisite to develop low-emission climate-resilient plans and strategies, and to ensure that societies are resilient to cope with the impacts of climate change.

Malaysia is one of the developing countries that has faced urbanization and economic modernization process. According to Department of Statistic Malaysia (2020), Malaysia's population in 2021 is estimated at 32.7 million with the growth rate at 1.27 percent as compared to 32.3 million in 2020. The rising populations in urban areas increase pressure on the resources, apparently causing degradation of the ecosystems. Incorporating low carbon city concept in urban development is an opportunity to reduce carbon emissions while offering tremendous socio-economic benefits. Low carbon city embraces and integrates the fundamental principles of sustainability through adopting a high level of energy efficiency and utilization of low carbon technology and sources in cities (Ho et al. 2013a, b). Despite the vast implementation of low carbon concept due to it being a promising approach to reduce carbon as well as mitigate climate change (Gouldson et al. 2016), there are still limitations and challenges that continually emerge in various directions for the adoption of low carbon technology. The challenge incorporates with low carbon investment, collaboration capacity development and finance and investment which limit its implementation (UNDP 2019). Intergovernmental Panel on Climate Change (2014) reports that lack the political will and the institutional with financial capacities needed to shift to more energy and carbon-efficient development paths in many cities, and particularly those in the developing countries. Thus, this study aims to explore the current state of green technology adoption for low carbon city development from the perspective of local stakeholders. Subsequently, the research objectives are: (1) to determine the current understanding on green technology and low carbon city concept, and (2) to identify the key factors of strengths-weaknesses-opportunities-threats in adopting green technology for low carbon city development. The exploration on the internal and external factors of different concepts and parametrises like strengths-weaknesses-opportunities-threats to strategize green technology would provide the foundation and platform that further enables the implementation of low carbon city which in line with SDG 11: Sustainable Cities and Communities and SDG 13: Climate Action.

2 Literature Review

The environmental problem that the society faces today, due to the rapidly growing number of contaminants, is not regional but worldwide. The climate change caused by greenhouse gases (GHGs) is not a new phenomenon and is a factor for several other issues like increase in temperature, rising sea level and frequent extreme climate events with human activities in cities contributing 80% of GHG emissions (Satterthwaite 2008). It is an indispensable prerequisite to control the emission of GHGs at the global level in cities because it was projected that by 2050, 6.3 billion people will be living in cities with the increment of 70% from current value (UNDESA 2015). The biggest proportion of GHGs is from carbon dioxide (CO_2), and CO_2 is considered as the major gas prominent to climate change (Shi et al. 2017). This event increases the vulnerability to a population who live in low-elevation coastal areas especially in urban areas.

Malaysia is one of the vulnerable countries to the effects of climate change. Malaysian Energy Commission (2016) reported that the amount of CO_2 emissions in 2014 was equivalent to 220.5 Mtonne, meanwhile the amount of GHG in 2012 was 279,098 ktonne. Additionally, the report from The United Nations Framework Convention on Climate Change (UNFCCC) mentioned that Malaysia's recent climate change in terms of the country's temperature, rainfall and sea levels have increased significantly in the last 40 years and are expected to remain increasing until 2050. In this respect, carbon emission reduction has become a top agenda for all countries throughout the world in alignment with Sustainable Development Goal (SDG).

Low carbon city (LCC) is a sustainable urbanization approach to reduce carbon emissions while offering great economic prospects. The city that comprises societies that consume sustainable green technology, green practices and emit relatively low carbon or GHG are the criteria of LCC (KeTTHA 2017a, b). Malaysia has established the Low Carbon Cities Framework (LCCF) in 2011 to provide guidance for local authorities, agencies and communities on guidelines to shift their cities and institution into low carbon cities (KeTTHA 2011). The LCCF looks at tackling carbon emissions reduction in four main areas which are urban environment, urban infrastructure, urban transportation, and buildings consist of four key elements, thirteen performance criteria and thirty-five sub-criteria. This framework is one of the Malaysia's government initiatives to encourage local authorities to adopt LCCF with more dependence on clean sources of energy and fewer CO_2 emissions (Rugg 2014). In 2020, the Ministry of Energy, Science, Technology, Environment, and Climate Change (MESTECC) Malaysia and GreenTech Malaysia has come out with a new initiative called as the Low Carbon Cities 2030 Challenge (LCC2030C) aimed at promoting the transformation of cities into low carbon cities.

Another approach to reducing GHGs emission by educating the society with the concept of low carbon society (LCS). LCS is defined as a society that takes necessary actions to actively contribute towards principles of sustainable development including reversing the current trend of climate change (Skea and Nishioka 2008). The Malaysian government has recognized the importance of LCS through establishing the LCS framework called Iskandar Malaysia 2025 (Iskandar Malaysia 2025, 2013). This blueprint is to assist Iskandar Regional Development Authority (IRDA) with carbon mitigation strategies to reduce 40% of GHG emission (Iskandar Malaysia 2025, 2013).

Many cities in Malaysia have initiated projects and programs based on LCCF and undertake several strategies and actions towards achieving low carbon vision (Iskandar Malaysia 2025, 2013). The Malaysia's government represented by GreenTech Malaysia is working with local authorities to create low-carbon zones in the state capital and major urban areas across the country to ensure better coordination of the implementation of low carbon actions. In 2018, 52 of the 154 have been trained by GreenTech for low carbon cities and 19 local authorities have measured their emissions baseline and developed a low carbon action plan (Khoo 2019). The assessment system of LCCF implemented in Malaysia comprises a carbon calculator which assists local authorities in determining their current baseline of carbon emission (KeTTHA 2017a, b).

The adoption of green technologies in urban areas is one of the approaches to highly achieve energy efficiency and promote sustainable development (Ho et al. 2013a, b). Increasing the awareness and knowledge on the adoption of green technologies and its importance in urban areas through discussions, seminars, training, and workshops by the government with the support of stakeholders can be adopted to promote green technologies in building (Ametepey et al. 2015). Chel and Kaushik (2018) describe four aspects for energy conservation in mitigating emissions of CO_2 in building which are (i) comfort passive building design and its orientation for harnessing solar energy, (ii) low embodied energy materials for building construction, (iii) energy-efficient domestic appliance to conserve the building operational energy, (iv) building integrated renewable energy technologies. Therefore, this study is undertaken to explore the current state of green technology adoption for low carbon city development and consequently to determine the key factors of strengths-weaknesses-opportunities-threats in adopting green technology for low carbon city development.

3 Research Methodology

This study was performed in 2020 through adopting mixed methods. Mix of data collection methods in terms of in-depth interviews and questionnaire surveys were adopted with a cross sectional review on past research regarding green technology and low carbon city to identify the key factors of strengths, weaknesses, opportunities and threats that would address the research objectives. The details of the interviews and questionnaire survey are presented in the following section.

The main research instrument used was a well-structured questionnaire containing both closed and open-ended questions. A face-to-face survey method was conducted with individual respondents during the site visit to allow clarifications and provide high-quality data. Fifty-seven (57) respondents including town planners, engineers, green consultants, city operators, government agencies and academia that were involved with LCC development participated in this study. These participants were identified as the key players who were involved in planning and implementing the low carbon city concept in the local city.

3.1 Questionnaire Survey

The objectives of the survey were to determine the current understanding on green technology and low carbon concept gain insight on the key factors of strengths, weaknesses, opportunities and threats towards the implementation of green technology and to strategize the smart city enabling factors. The data gathered were measured on a Likert Scale and analysed quantitatively. The questionnaire had three (3) sections as follows:

a. Part A: Respondent profile background.
b. Part B: Current understanding on Green Technology and Low Carbon City concept
c. Part C: Identifying the key factors of the strengths, weaknesses, opportunities and threats of green technology adoption in development of low carbon city.

A set of questionnaires was designed to solicit the respondents' opinion on the degree of understanding and significance of green technology adoption in low carbon city (LCC). A five-point Likert scale (1: not significant/very poor understanding to 5: extremely relevant) was used for this purpose.

3.2 Interviews with Experts

Interviews were also conducted to seek expert opinions on the key external and internal factors and drivers towards successful adoption of green technology in low carbon city. This interview is to support the findings from the questionnaire survey. There were five (5) experts who have engaged with low carbon city development and have shared their insights on green technology and low carbon city development in Malaysia. The interviewees' designations are as follows: two (2) government officers; two (2) town planners and one (1) engineer. The questions were designed to allow interviewees to provide their expert knowledge on green practices as follows:

a. the key external and internal factors towards adoption of green technology in low carbon city.
b. the motivating factors to improve green practices implementation in Malaysia
c. an open question for any other issues related to the topic.

3.3 Data Analysis

Descriptive analysis and descriptive statistics were carried out, including the percentage frequency distribution, mean and standard deviation, reliability, test and content analysis.

3.3.1 Reliability Test

Reliability analysis with Cronbach's alpha was calculated for the variables with 24 items on the collected information in the questionnaire, for a Cronbach Alpha value of 0.808. This provides the evidence that a value greater than 0.70 was considered acceptable (Nunnally and Bernstein 1994).

3.3.2 Respondents Profile

Fifty-seven (57) respondents including town planners, engineers, consultants, facility managers, administration, academicians and government officers involved with local city development participated in this study. The differences in the range of working experience and roles of position would enable more comprehensive analysis on the SWOT analysis as it covers opinion from several stakeholders' perspective. There were good mixtures of designations of respondents with various professions involved

Table 1 Demographic profile of respondents

	Respondent characteristics	No. of respondents (total = 57)
Years of experience	Less than 10 years	11
	Between 11 and 20 years	36
	Between 21 and 30 years	10
Expertise	Engineering and design	6
	Building	4
	Transportation	8
	Landscaping	4
	Town planning	13
	Education	9
	Administration	10
	Developer	3

in LCC projects. From the 57 responses, the respondents have described their expertise in engaging from various departments including town planning (13), administration (10), education (9), transportation (8), engineering and design (6), both building and landscaping (4), respectively and developer (3). On average, respondents had working experience between 11 and 20 years while the majority age ranging from 25 to 44 years. Table 1 shows the demographic profile of the respondents.

3.3.3 Content Analysis

The outcomes from the respondents have been analysed through converting raw narrative data (notes, audiotapes) into partially processed data (transcripts), and were then coded manually. The keywords most related to the quotes have been extracted and discussed in the following section.

4 Results and Discussion

4.1 Determination of the Current Understanding on Green Technology and Low Carbon City Concept

Table 2 shows that forty-one respondents claimed that they are involved directly in LCC projects while 16 of them are indirectly involved in any of LCC projects.

Table 2 Cross tabulation on the involvement in LCC project and the level of familiarity with green technology

		Involvement in LCC project	
		Directly	Indirectly
Understand on LCC concept	Level of understanding ≤ Quite well	12	13
	Level of understanding ≥ Well	29	3
	Total	41	16
Familiar with green technology	Level of familiarity ≤ Quite well	28	10
	Level of familiarity ≥ Well	13	6
	Total	41	16
Green technology adoption important in LCC	Level of important ≤ Neutral	3	0
	Level of important ≥ Significant	38	16
	Total	41	16

From the 41-respondents who had experience in LCC projects, 12 of them stated that they understood the LCC concept "quite well" or less while the other 29 respondents felt that they understood the concept of LCC "well" or "very well". For the respondent who is indirectly involved in any LCC projects, 13 of them claimed that their understanding on LCC concept is "quite well" or less and the remaining 3 have understood the LCC concept "well" or "very well". Although the respondents were involved directly in LCC projects, 28 of them indicated that they are "quite well" or less familiar with green technology that can be adopted in LCC projects while 11 of them stated "well" or "very well" familiar in LCC projects. The similar trend of familiarity can be observed for those who are not involved directly in LCC projects. Majority of respondents felt that it is important to adopt green technology in LCC projects. Overall, it can be deduced that although the majority of respondents have been involved directly in the LCC project as well as understand the LCC concept, however, the majority of them are not familiar with green technology.

4.2 *Strengths, Weaknesses, Opportunities, and Threats of Green Technology Adoption in Development of Low Carbon City*

The respondents were required to indicate the perspectives on the key factors related to strengths, weaknesses, opportunities and threats of green technology adoption in development of low carbon cities. Table 3 shows the descriptive statistics of the mean (M) and standard deviation (SD) for all factors identified from previous studies.

Table 3 Mean and standard deviation (N = 57) on key factors of strengths, weaknesses, opportunities and threats of green technology adoption in low carbon city

Strength	Mean	Std Dvd	Rank
Participation from financial institution	3.984	1.102	6
Financial incentives from government	4.702	0.462	1
Government support and initiatives in green technology adoption	4.544	0.629	2
Competent and proactive green technology promotion team and local authority	4.474	0.684	3
Extensive promotion from government to promote green technology adoption	4.316	0.783	5
Green technology policy framework	4.404	0.728	4
Weakness	Mean	Std Dvd	Rank
Lack of motivation and enthusiasm of agencies to participate in LCC development	4.596	0.593	3
Unfamiliarity and lack of understanding on green technologies	4.632	0.555	1
Lack of knowledge and information on availability of green technology and supplier	4.456	0.867	4
Lack of technical skills and expert on green technology	4.386	0.881	5
Lack of interest from client and market demand	4.614	0.559	2
Lack of local institutes and facilities for research and development (R&D) of green technology	4.123	0.881	6
Opportunities	Mean	Std Dvd	Rank
Create new eco-friendly product and services	4.456	0.709	2
Technology transfer	4.544	0.629	1
Job creation opportunity	4.281	0.726	4
Optimization cost and productivity	4.193	0.789	6
Better environmental management	4.228	0.756	5
Strong business market	4.404	0.704	3
Threat	Mean	Std Dvd	Rank
Economic uncertainty	4.140	0.934	4
Complexity of green technology	4.333	0.893	2
Political context	3.965	0.906	6
Drive from industry associations (non-green conscious)	4.070	0.821	5
Insecure green technology market	4.246	0.931	3
High initial investment risk with long payback period	4.404	0.884	1

The overall results from the descriptive statistics show that mean scores of almost all items related to strengths, weaknesses, opportunities and threats of incorporating green technology in low carbon city are above 3.00, meaning that the respondents provided good responses on the agreements (Aghili et al. 2017).

4.2.1 Key Factors of Strength in Incorporating Green Technology in Low Carbon City

The first rank factor is "financial incentives from government" (M = 4.702; SD = 0.462; N = 57), followed by the second factor, "government support and initiatives in green technology adoption" (M = 4.544; SD = 0.629; N = 57). The third rank factor is "competent and proactive green technology promotion team and local authority" (M = 4.474; SD = 0.684; N = 57), followed by "green technology policy framework" (M = 4.404; SD = 0.728; N = 57) and the fifth rank factor is "extensive promotion from government to promote green technology adoption" (M = 4.316; SD = 0.783; N = 57). The lowest rank factor is "participation from financial institution" (M = 3.984; SD = 1.102; N = 57).

4.2.2 Key Factors of Weakness in Incorporating Green Technology in Low Carbon City

The first rank factor is "unfamiliarity and lack of understanding on green technologies" (M = 4.632; SD = 0.555; N = 57), followed by the second factor, "lack of interest from client and market demand" (M = 4.614; SD = 0.559; N = 57). The third rank factor is "lack of motivation and enthusiasm of agencies to participate in LCC development" (M = 4.596; SD = 0.593; N = 57) followed by "lack of knowledge and information on availability of green technology and supplier" (M = 4.456; SD = 0.867; N = 57). The fifth rank factor is "lack of technical skills and expert on green technology" (M = 4.386; SD = 0.881; N = 57).

4.2.3 Key Factors of Opportunities in Incorporating Green Technology in Low Carbon City

The first rank factor is "technology transfer" (M = 4.544; SD = 0.629; N = 57) which is part of promoting economic development through commercializing innovative technology. The second factor, "create new eco-friendly product and services" (M = 4.456; SD = 0.709; N = 57). The third rank factor is "strong business market" (M = 4.404; SD = 0.704; N = 57) followed by "job creation opportunity" (M = 4.281; SD = 0.726; N = 57). While the tangible benefits of green technology such as improving productivity performance are recognizable, intangible benefits on environmental and social offer the lucrative solutions on the reduction of carbon emissions, energy, and waste, conserve natural resources and improve occupant health and comfort.

4.2.4 Key Factors of Threat in Incorporating Green Technology in Low Carbon City

For threat in incorporating green technology in low carbon city, the first rank factor is "high initial investment risk with long payback period" (M = 4.404; SD = 0.884; N = 57), followed by the second factors, "complexity of green technology" (M = 4.333; SD = 0.893; N = 57). The third rank factor is "insecure green technology market" (M = 4.246; SD = 0.931; N = 57), followed by "economic uncertainty" (M = 4.140; SD = 0.934; N = 57) and the fifth rank factor is "drive from industry associations (non-green conscious)" (M = 4.070; SD = 0.821; N = 57).

4.3 Interviewee's Opinion on the Strengths, Weaknesses, Opportunities, and Threats of Green Technology Adoption in Development of Low Carbon City

Interviews were conducted to seek expert opinions on the strength-weakness-opportunities-threat in adopting green technology towards successful low carbon city. Overall, most of the interviewees have recognized the support from the government in terms of introducing initiatives, provide financial assistance, promotional strategies and develop strategic framework as the strengths towards successful adoption of green technology in low carbon city. Broekhoff et al. (2019) discussed the key roles of government in promoting green technologies including develop and implement urban policies, provide incentives structures and building local capacities and resources. Chan et al. (2018) and Azeem et al (2017) claim that mandatory government policy is the most substantial measure in promoting green practices. The application of the green technology is influenced by the realization of the institutional agencies responsibilities on the implementation of low carbon policies in the country (Ho et al. 2013a, b). Furthermore, imposing policies is part of top-down approach by the government to promote green initiatives to change the industry's attitudes towards efficient development and implementation of policies and regulations (Ruparathna and Hewage 2015).

Despite the strengths of green technological adoption, this study identifies key weakness that could affect the successful implementation of green technology. Unfamiliarity, lack of interest from clients and market demand, lack of motivation and lack of knowledge and technical skills on green technologies are the weakness of the adoption of green technology in low carbon city development. According to Persson and Grönkvist (2015), the lack of common understanding on the sustainability is a major barrier towards successful uptake of green practices. Interviewees believed that green technology is newer to developing countries and that the experts and professionals who have experienced and technical knowledge of it are limited in number. Unfamiliarity and lack of understanding with green technologies are the main barriers towards its slow commercialization, application and innovation process in green technologies.

Technological advancement in the green sector also provides vast opportunities towards reducing carbon emissions, while creating better urban centres for the nation's citizen. The first ranked opportunity is that technology transfer to provide knowledge spillovers on environment-friendly technologies. Green technology transfer promotes economic development through commercializing innovative technology through disseminating the benefits of green operations from developed to developing countries. The use of technology to create products and services that are environmentally friendly also potentially leads to reduced cost, energy consumption and improved productivity and work performance. Interviewees emphasized that the green technology provides opportunities to green investors to help the environment as well as take advantage of inherent cost savings. Green start-ups company could exploit business market opportunities on technology innovation through coordination between national and international business markets to build sustainable solutions (Halepoto et al. 2015).

Green technology investments can be considered as good investment although there is risk associated with investing in any new technology as well as an emerging market. Majority of interviewees highlighted that the extra financial cost that is needed to invest in green technology becomes a hurdle to green investment. The risk in investing green technology is often twofold; the requirement of highly trained workers to deal with new green technology and additional expenses for maintenance and repair (Azeem et al. 2017). The complexity, insecure market and economic uncertainty would hinder the adoption of green technology in developing countries.

Based on validation with the experts during the interview session, the top key factors of strengths-weaknesses-opportunities-threats in adoption of green technology towards successful development of low carbon city were tabulated in the SWOT analysis matrix in Table 4.

Table 4 SWOT analysis matrix for green technology adoption in development of low carbon city

Strengths 1. Financial incentives from government 2. Government support and initiatives in green technology adoption 3. Competent and proactive green technology promotion team and local authority 4. Green technology policy framework	**Weaknesses** 1. Unfamiliarity and lack of understanding on green technologies 2. Lack of knowledge and information on availability of green technology and supplier 3. Lack of interest from client and market demand 4. Lack of technical skills and expert on green technology
Opportunities 1. Technology transfer 2. Create new eco-friendly product and services 3. Job creation opportunity 4. Strong business market	**Threats** • High initial investment risk with long payback period • Complexity of green technology • Insecure green technology market • Economic uncertainty

5 Recommendation and Way Forward

With the insights drawn from the questionnaires and interviews results, the solutions on the way forward to improve the effectiveness green technology adoption in development of low carbon city in Malaysia are as follows:

- *Government Support*
  The role of government is recognized as an effective mechanism for green development and calls for the efficient development of policies and regulations (Ruparathna and Hewage 2015). Introducing green policies, regulation and standards are seen as effective agents to change industry behaviours and promote the development of new technologies (Carmichael et al. 2018). The government of Malaysia has started initiatives in regard to green development (Bohari et al. 2017; Chua and Oh 2011). The Malaysian government continues to prioritize green adoption to spur economic multiplier effects by introducing the National Green Technology Policy (NGTP), which was launched in 2009 as a driver to recognize green technology and promote sustainable development. Various green technology initiatives have been promoted as part of green growth strategy including incentives, subsidy programs, tax exemption, investment tax allowance to encourage investments in green equipment production and the adoption of green technology.
- *Building Knowledge and Institutional Capacity Development*
  Increasing the awareness and motivation through building the knowledge and skills, technical capacity, information and good practices will enhance current knowledge management systems related to the green technologies and low carbon cities in Malaysia. Strengthening the capabilities and culture of the institutional towards low carbon climate-resilient development and integrated urban planning would expedite the process of nations in meeting sustainability goals.

6 Conclusion

This study explores the internal and external factors of different concepts; strengths-weaknesses-opportunities-threats of green technologies adoption in low carbon city development to strategize green technology from several stakeholders' perspectives. Responses from fifty-seven (57) respondents and five (5) experts of local city players who were involved in planning and implementing the low carbon city concept in the local city were collected through survey questionnaires and interviews. The results were analysed using descriptive analysis and content analysis. The results showed that the role of government is recognized as an effective catalyst for green development while the lack of awareness and technical capacity identified as the weaknesses in adopting green technology in low carbon city development. Finding from the study also indicates that green technology transfer provides opportunity to promote economic development while green investment risk as a threat towards successful adoption of green technology in low carbon cities. The findings of this study could help policy makers identify specific problems in green technologies adoption and develop suitable strategies towards conceptualization of low carbon city enhancement in Malaysia. This study contributes to the current body of knowledge concerning the insights of current status of green technology application for the development of low carbon cities in developing countries especially Malaysia.

Acknowledgements The authors would like to express their gratitude to the Ministry of Higher Education Malaysia and Universiti Teknologi MARA, Shah Alam for supporting this publication via research grant 600-IRMI/FRGS 5/3 (333/2019).

References

Aghili N, Mohammed AHB, Sheau-Ting L (2017) Management key practices for improving green building performance. Adv Sci Lett 23(9):8874–8876
Alkhayyat A, Pehlivan S (2017) Energy consumption analysis in conventional buildings: case study of Sweden. Energy 1(1)
Ametepey O, Aigbavboa C, Ansah K (2015) Barriers to successful implementation of sustainable construction in the Ghanaian construction industry. Procedia Manuf 3:1682–1689
Azeem S, Naeem MA, Waheed A, Thaheem MJ (2017) Examining barriers and measures to promote the adoption of green building practices in Pakistan. Smart Sustain Built Environ
Bohari AAM, Skitmore M, Xia B, Teo M (2017) Green oriented procurement for building projects: preliminary findings from Malaysia. J Clean Prod 148:690–700
Broekhoff D, Gillenwater M, Colbert-Sangree T, Cage P (2019) Securing climate benefit: a guide to using carbon offsets. Stockh Environ Inst Greenh Gas Manag Inst 60
Carmichael DG, Mustaffa NK, Shen X (2018) A utility measure of attitudes to lower-emissions production in construction. J Clean Prod 202(23–32)
Chan APC, Darko A, Olanipekun AO, Ameyaw EE (2018) Critical barriers to green building technologies adoption in developing countries: the case of Ghana. J Clean Prod 172:1067–1079
Chel A, Kaushik G (2018) Renewable energy technologies for sustainable development of energy efficient building. Alex Eng J 57(2):655–669
Chua SC, Oh TH (2011) Green progress and prospect in Malaysia. Renew Sustain Energy Rev 15(6):2850–2861
Department of Statistic Malaysia (DOSM) (2020) Malaysia population and demography. https://www.dosm.gov.my/. Accessed 17 Feb 2021
Edenhofer O, Pichs-Madruga R, Sokona Y, Agrawala, S, Bashmakov IA, Blanco G, Zwickel T (2014) Summary for policymakers
Gouldson A, Colenbrander S, Sudmant A, Papargyropoulou E, Kerr N, McAnulla F, Hall S (2016) Cities and climate change mitigation: economic opportunities and governance challenges in Asia. Cities 54:11–19
Halepoto IA, Sahito AA, Uqaili MA, Chowdhry BS, Riaz T (2015) Multi-criteria assessment of smart city transformation based on SWOT analysis. In 2015 5th national symposium on information technology: towards new smart world (NSITNSW). IEEE, p 1
Ho CS, Matsuoka Y, Simson J, Gomi K (2013a) Low carbon urban development strategy in Malaysia. The case of Iskandar Malaysia development corridor. Habitat Int 37:43–51
Ho CS, Matsuoka Y, Chau LW, Teh BT, Simson JJ, Gomi K (2013b) Blueprint for the development of low carbon society scenarios for Asian regions-case study of Iskandar Malaysia. In: IOP conference series: Earth and environmental science. IOP Publishing 16(1), p 012125

KeTTHA (2011) Low Carbon Cities Framework and assessment system. Ministry of Energy Green Technology and Water (KeTTHA) Putrajaya, Malaysia
KeTTHA (2017a) Low carbon cities framework. Ministry of Energy, Green Technology and Water (KeTTHA) Putrajaya, Malaysia
KeTTHA (2017b) Green technology master plan Malaysia 2017–2030 (GTMP), Ministry of Energy Green Technology and Water (KeTTHA) Putrajaya, Malaysia
Khoo E (2019) Malaysia continues efforts to reduce carbon footprint. City & Country, The Edge Malaysia
Malaysian Energy Commission (2016) Malaysian energy statistics handbook 2016. https://meih.st.gov.my/documents/10620/bcce78a2-5d54-49ae-b0dc-549dcacf93ae. Accessed Apr 2021
Masson-Delmotte V, Zhai P, Pörtner HO, Roberts D, Skea J, Shukla PR, Waterfield T (2018) Global warming of 1.5 C. IPCC Spec Rep Impacts Glob Warm 1:1–9
Nunnally JC, Bernstein IH (1994) The assessment of reliability. Psychom Theory 3:248–292
Persson J, Grönkvist S (2015) Drivers for and barriers to low-energy buildings in Sweden. J Clean Prod 109:296–304
Ruparathna R, Hewage K (2015) Sustainable procurement in the Canadian construction industry: current practices, drivers and opportunities. J Clean Prod 109:305–314
Rugg, R (2014) Low carbon cities are the core of Malaysia's green future. Carbon Trust. www.carbontrust.com/news/2014/08/lowcarbon-cities-malaysia-green-future/. Accessed Apr 2021
Satterthwaite D (2008) Cities' contribution to global warming: notes on the allocation of greenhouse gas emissions. Environ Urban 20(2):539–549
Shi Q, Chen J, Shen L (2017) Driving factors of the changes in the carbon emissions in the Chinese construction industry. J Clean Prod 166:615–627
Skea JIM, Nishioka S (2008) Policies and practices for a low-carbon society UNODRR (2015) United National Office for disaster risk reduction, extreme weather tied to over 600,000 deaths over 2 decades
UTM, IRDA, Kyoto University, Okayama University, National Institute for Environmental Studies (2012) Low carbon society blueprint for Iskandar Malaysia 2025. UTM-Low Carbon Asia Research Center, Johor Bahru, Malaysia
UNDESA (2015) World urbanization prospects—the 2015 revision Department of economic and social affairs, New York
UNDP (2019) Low carbon cities—Malaysia's response to global climate emergency. https://www.my.undp.org/content/malaysia/en/home/newscentre/articles/2019/lowcarboncity.html. Accessed Dec 2019

Governance Practices in Poverty Alleviation Projects: Case Study from Stewardship-Driven Perspective and Sustainability Context

Abdul Muhaimin Abdul Latiff, Che Maznah Mat Isa, and Aini Jaapar

Abstract The successful planning and implementation of poverty alleviation projects are crucial for establishing optimum benefits to its intended target groups in line with the Sustainable Development Goals charted by the United Nations. This chapter focuses on exploring the project governance practices in poverty alleviation projects from public officials' perspective through the lens of stewardship theory. Project governance is a framework and functions to guide project management activities to meet organisational strategic and operational goals, leading to project success. By investigating three (3) public poverty alleviation projects in Malaysia, it is discovered that project governance is practised and positively interplayed with several project governance elements throughout the project planning and implementation. Several insights and rationales of the decision-making process from the sustainability contexts are also explored where the target groups should be allowed to participate in decision making in the front-end of the project, while lessons learnt from previous projects are invaluable input that should be compiled and documented for future references. Although the fragmented nature of public projects is considered a point of weakness, project delivery success could be achieved by utilising project governance practices applicable in public poverty alleviation projects.

Keywords Project governance · Stewardship · Poverty alleviation · Sustainability · Malaysia

1 Introduction

Good governance practices are crucial in alleviating poverty (Sarker et al. 2018). In Malaysia, poverty incident stood at 3.6% in 2007, 1.7% in 2012, 0.6% in 2014

A. M. A. Latiff · C. M. M. Isa (✉) · A. Jaapar
Faculty of Architecture, Planning and Surveying, School of Civil Engineering, College of Engineering, Universiti Teknologi MARA, Shah Alam, Malaysia
e-mail: chema982@uitm.edu.my

A. Jaapar
e-mail: ainijaapar@uitm.edu.my

R. Hassan et al. (eds.), *Green Infrastructure*,
https://doi.org/10.1007/978-981-16-6383-3_14

and further reduced to 0.4% in 2016 (DOSM 2017). However, the review of the national poverty line income (PLI) definition and methodology in 2020 has significantly increased the PLI value to RM 2208 from RM 980 in 2016, thus contribute to 5.6% of poverty incidents in 2019 (DOSM 2020). The government has spent more than RM 2 billion for the past ten years in addressing the poverty issue in Malaysia (CSDU 2017) which various ministries and departments have executed a number of programmes and projects. However, existing social protection programmes and projects in Malaysia are poorly coordinated, heavily siloed, which cross over many ministries and agencies and often ineffective (Nixon et al. 2017; UNHR 2020). This reality shows that even with the low poverty incident recorded, there are many weaknesses in managing poverty alleviation projects, leading to the question of its sustainability in helping the poor people.

Concerning that, Poverty Task Force (2002) found that public sector inefficiency, lack of accountability in the institutional structures, and widespread corruption contribute to the weak governance in the poverty alleviation programme. Besides, Nawi et al. (2018) found that poor planning at the project initiation stage, lack of awareness in project monitoring, and late decision making by the authority contribute to the delay in public projects in Malaysia. These circumstances indicate that the governance practices in poverty alleviation projects are poorly understood by the project actors, contributing to the project cost and time overrun and the failure for the project to operate as planned (National Audit Department 2017).

The poverty alleviation project stands as a temporary project-based organisation in which the main goal is to deliver the right projects to the right people effectively. The successful planning and execution of poverty alleviation projects are vital to ensure the poor people received maximum benefits from these projects and lift their quality of life. Hence, this is where the project governance term is utilised to improve project delivery. Project governance is a framework and functions to guide project management activities to meet organisational strategic and operational goals (Project Management Institute 2016). There is evidence that project governance could lead to project success (Young et al. 2019) and at the same time minimise the risk of project failure (Zhang et al. 2016).

While there is no research available in the literature that could determine the project governance practices in the context of public poverty alleviation projects, there are pressing needs to address the better understanding of project governance at both strategic and operational level of project organisation because of the complex structure of development projects and the unsustainable spending of public money (Renz 2007). Hence, the current chapter focuses on the practices of project governance elements in poverty alleviation projects from the stewardship perspectives towards understanding how project governance elements are practised in poverty alleviation projects. Hence, by positioning the study's theoretical notion of stewardship towards project governance practices, this chapter offers significant contribution in demonstrating how stewardship theory could describe how project governance is practised in the context of public officials involved in the planning and implementation of poverty alleviation projects. In light of the problem statement as discussed above, the study question is derived as follow:

"How project governance elements are practised in the planning and implementation of the poverty alleviation projects?

The following sections present the literature review, the methodology, the findings, discussion, and conclusions, recommendations, and limitations.

2 Literature Review

The literature review is leaning towards stewardship-driven governance practices in the context of public actors involved in the planning and implementing poverty alleviation programmes.

2.1 Poverty Alleviation

Poverty alleviation is an effort implemented by any organisation or civil society to comb poverty through various initiatives and programmes, which will ultimately enhance the well-being of poor people and get them out of poverty. Poverty alleviation involves heavily from drafting the policy, engagement with the stakeholders, drawing the arrangement of initiatives and programmes, and ideating and selecting the best projects that can be implemented within the overall budget provided and ensuring the projects are executed accordingly.

The government needs to spend money in order to alleviate poverty. Fan and Chan-Kang (2008) reported that public infrastructure development, particularly in rural areas, significantly contributes to poverty reduction and economic growth. However, Celikay and Gumus (2017) discovered a negative relationship between social spending and poverty in the short run, while the relationship is positive in the long run, suggesting that social spending is fostering in nature rather than alleviating poverty. Much of the previous research has touched on the importance of effective policies and delivery system to support poverty alleviation initiatives. Sagaran and Nair (2015) suggested a paradigm shift and systemic overhaul in the implementation and delivery mechanisms of poverty alleviation initiatives to ensure its sustainability, while Thang and Baharuddin (2011) recommended good governance institutional change to be given priority in order for poverty alleviation to effectively executed. Further, Abdalla and Elhadary (2012) revealed that bureaucracy and tedious procedures in directing funding contribute to the problem of mismanagement and corruption, while Li et al. (2016) proposed a comprehensive assessment system in identifying the poor households through establishing poverty monitoring system and dynamic modification mechanism which will offer better accuracy and efficiency in the poverty alleviation programme. However, the previous studies have not been able to determine how the delivery system of poverty alleviation programmes and projects could be further improved from the perspective of public officials that execute the projects.

2.2 Governance in Poverty Alleviation

Good governance is generally viewed as an essential component of poverty alleviation (World Bank 2000), and the quality of governance at each level of government machinery is essential in ensuring effective poverty alleviation programmes (ADB 1999). Grindle (2004) explained that good governance in poverty alleviation programme should touch all elements of the public sector, including decision-making practices, project delivery to the target groups and the engagement with target groups in multiple channels, while good governance is linked with less multidimensional poverty especially in middle-income countries (Jindra and Vaz 2019). Besides, Kwon and Kim (2014) identified incompetent officials and capacity issues as primary factors that affected the deliverables of poverty alleviation efforts in the least developed countries, while lack of transparency and accountability as well as excessive control and bureaucracy are among factors that contribute to failure to address rural poverty (Khan 2000).

On the other hand, Siwar (2006) stated that proper planning and implementation capability, efficient use of resources, consistent monitoring and practical impact assessment are vital for an effective delivery system that could ensure the attainment of poverty alleviation programmes. Also, the application of governance in the poverty alleviation programme has to be comprehensive, where all stakeholders should be involved in the whole process of the programme. Although good governance is generally essential in alleviating poverty (Grindle 2004), more details and transparent indicators are needed to address the specific contexts and challenges faced by each country (Jomo and Chowdhury 2016) which suggests that more studies are needed to understand the governance based on its specific contexts.

Much literature on governance in poverty alleviation seems to have been based on its association with policymaking and delivery system, while how the project actors perceive the governance at organisational and project level are not fully understood. As such, there is a gap, particularly in the practices of project governance in poverty alleviation projects.

2.3 The Gap of Project Governance in Poverty Alleviation

Project governance could be identified as a system involving any elements pertaining to the governance of a particular project (Musawir et al. 2020). From the literature, it is found that key challenges in delivering poverty alleviation projects to the target groups revolve around the governance at the policy and programme level and the extent at the project level. While the works of Renz (2007) has contributed to the project governance knowledge, particularly in understanding the linkages between corporate governance, project management and development (nonprofit NGO) where several elements such as system management, mission management, integrity management, extended stakeholder management, and audit management have been concluded as a basis for project governance model, it did not focus directly to the poverty alleviation programme from the public sector perspective. A recent study by Khan et al. (2019) proposed adopting a project governance framework in public infrastructure development where more studies are needed to understand the project governance application in public sector organisations.

On the other hand, it is discovered that improved coordination mechanisms at the local level could balance between the poverty alleviation efforts and sustainability (Barrett 1996), while Klakegg and Haavaldsen (2011) discovered a lack of commitment from key stakeholders and conflict with regards to the objectives and strategies at the front-end of projects are among the reasons for lack of major public project sustainability. However, a recent study by Irfan and Hassan (2019) proved that project governance and sustainability positively contribute to project success. It is also in line with Khan et al. (2020) idea that precise and responsive project governance with a systematic approach to manage projects will enhance project performance in public infrastructure projects.

An organisation that enjoy shared values with stakeholder orientation signals that stewardship relationships are in place (Joslin and Müller 2016). Stewardship stressed the involvement-oriented philosophy where employees will develop a behaviour of self-control when given responsibilities and challenges and have a collectivist culture and positive attitude towards group harmony to avoid any conflict and confrontation (Davis et al. 1997). Trust and collaboration are two central intrinsic values related to each other, where trust level influences the level of collaboration, which in turn contributes to the project's success (Bond-Barnard et al. 2018). Although Imam and Zaheer (2021) concurred that trust and strong ties contribute to the project success through shared leadership, it could not explain the real motivation of project actors and its connection with other elements that could influence the governance of the projects. Together these studies provide important insights on how the factoring of trust and collaboration together with other project governance element could be the basis for better understanding in terms of project governance practices in the poverty alleviation programme.

To conclude this section, the literature review has touched on the governance practices in poverty alleviation projects through stewardship lens governance. It has been shown from this review that project governance could be practised in the framework of sustainability, but still, they are poorly unstated. Hence, a better understanding could be achieved by exploring the practices and how they are interplayed with each other.

3 Methodology

This section explains how the study was conducted, as well as strategies to address validity and reliability.

3.1 *Study Approach*

A case study research strategy is adopted in this study. The case study focuses on detailed descriptions, interpretations and explanations as well as involves actors who participate in the social process (Swanborn 2010), and it is suitable to be utilised in this study since it involved several project actors in conducting activities to ensure the deliverables of the project benefits to the main project's target groups.

3.2 *Selection of Cases*

Multiple case studies were designed and used where each case was selected to predict similar results (a literal replication) (Yin 2018). Apart from the case study selection justification based on research design, the projects for the case study are selected to showcase the varieties of the poverty alleviation projects and the impact on the beneficiaries. As a result, three projects related to housing schemes for the poor people under the national poverty alleviation programmes have been selected for the study, as displayed in Table 1.

Table 1 Basic information of the projects selected as case studies

Code	Case study (Project)	Cost per project (RM)	Total beneficiaries (Unit/individual/km) *For projects constructed between 2011 and 2019*
CS1	Housing (rural area)	46,000 (per unit)	27,013 units 27,013 participants
CS2	Area development, including housing and basic amenities (rural area)	7,800,000	965 units 965 participants
CS3	Housing and basic amenities (urban area)	82,100,000	46,223 units 46,223 participants

3.3 *Selection of Informants*

A total of 19 interviews were conducted throughout the data collection process from May 2019 until January 2020. Purposive sampling was utilised during the selection of informants since this study seeks to select the informants who best represent the case studies to answer the research questions. Thus, a list of potential informants is obtained and subsequently selected based on their ability to deliberately offer insight regarding the research problem and central phenomenon in this study.

3.4 *Data Collection*

A case study protocol was developed as a guiding document during data collection, and it is a substantial way of enhancing the case study reliability (Yin 2018). Multiple sources of evidence were used for data collection, such as semi-structured interviews, document analysis, direct observations, and field visits, as suggested by Yin (2018). On top of that, memo and reflection notes are utilised throughout the research process, and it helps during data analysis and interpretation (Creswell and Poth 2018). Data collection for each case study is considered complete once there is an emergence of regularities in the data along with the exhausted sources and having a point of saturation.

3.5 *Validity and Reliability*

In terms of validity and reliability, operational measures of the theoretical constructs were based on the theoretical lens used in this study. Hence, the current study used multiple sources of evidence to enhance the construct validity of the study (Yin 2018). Besides, early findings of each case study were sent through email to three key informants of the case, and the feedbacks and critical comments by them functioned as a validation process to increase the credibility of the study (Merriam and Tisdell 2016). Also, the application of case study protocol is consistent with Yin (2018) to enhance the reliability of the case study while the researcher keeps all the evidence following the sequence of the data collected to boost the confidence and reliability of the study (Merriam and Tisdell 2016; Yin 2018).

4 Findings

This section provides the case study background and findings of the analysis conducted through the cross-case analysis.

4.1 *Background of Case Study*

4.1.1 Case Study 1 (CS1)

CS1 is a poor people housing project developed to enhance the quality of life of low-income families by constructing a new house to provide a comfortable and safe home for them as a basis for continuous family development. The implementing agency of this project is IACS1 (an acronym), responsible for conducting the verification process for the participants, project procurement, project monitoring, and delivering the completed project to the participants. CS1 was implemented by the IACS1 in a rural area of the southern state in peninsular Malaysia.

4.1.2 Case Study 2 (CS2)

CS2 is a regional development project that is an integrated project developed to create a new and well-planned housing area for rural people exposed to natural disaster and safety and health threats, and poor people categorised under the *e-Kasih* system. The PPK project typically involves constructing several houses and necessary infrastructure and amenities such as access road, Muslim worship and playground. The implementing agency for CS2 is IACS2 (an acronym), a regional development authority located in the northern state of peninsular Malaysia and responsible for selecting the participants and implementing the project on behalf of the ministry.

4.1.3 Case Study 3 (CS3)

For CS3, a newly completed public housing project for low-income earners, including the urban poor and the squatters in Kuala Lumpur was selected, and the implementation agency for this project is the development division in the ministry or IACS3. The main project scopes are the construction of multi-stories 500 units of a house where each house size is measuring at 700 s.f. and it includes the construction of a multi-storey car park, main hall, kindergarten, shop lots, utility substation, playground and trash house. As this project cost is more than RM 50 million, it went through the value management (VM) process, which was coordinated by the IACS3, and the process is followed by the negotiation of project price between the government and the contractor as this is a direct negotiation project.

4.2 *The Findings*

Initially, nine themes emerged as the outcome of the within-case analysis of all five case studies. Subsequently, these themes are further consolidated to only five themes during cross-case analysis to reflect the applicability to all cases, as shown in Table 2.

Table 2 Project governance elements found in cross case analysis

	Project governance element	Project governance sub-element
1	Altruistic empathy	– Empathy to poor people life – Spirit to uplift the poor
2	Intrinsic motivation	– Trust – Collaboration
3	Effective leadership	– Empowerment – Ability to motivate and inspire – Knowledgeable
4	Decision making	– Ethical decision making – Collective decision making
5	Shared vision	– Alignment with government objectives – Shared objectives

4.2.1 Altruistic Empathy

Based on the analysis of each case study, it is discovered that public officials' behaviour is driven by altruistic empathy, and it makes a strong presence in all cases. The informants believe that motivation that drives them to conduct the project effectively is related to empathy and altruism beliefs, which influence them in executing any task related to the project. Further analysis found that altruistic motivation is a compilation of the motivations of why government officials that hold power in delivering poverty alleviation projects to the poor people conduct their work. The empathetic feelings towards the poor people's lives sparkled the officials' hearts and motivated them to rekindle the spirit to effectively organise and execute works to uplift the lives of the poor people.

Meanwhile, having experiences growing up in a difficult life enables several public officials to better understand and empathise with the target groups. Therefore, although facing challenges and pressure from various parties, it is discovered that public officials have a strong determination to improve poor people living with ethical values and integrity in the presence of altruistic empathy. The spirit to uplift the poor is high, and the officials are motivated to execute it. CS2_3, an informant attached to the ministry affirms:

> So, people said the land value is expensive, but we do not account that; that is why we spent hundreds of millions. We do not take into consideration on the ROI; yes, we do not account for it since if we consider that, we have failed in development ... the only reason is that we consider they are poor...

4.2.2 Intrinsic Motivation

Trust and collaboration are the backbones of intrinsic motivation, where trust is needed to foster collaboration among public officials and project actors. However, in CS1, the contractor failed to complete the project on time although trusted by the IACS1, while there is a high level of trust between the ministry as the main organisation and the implementing agencies in CS1, CS2 and CS3 since both parties have long working relationship and the implementing agencies are under the ministry's purview. In CS2, the project is unique because continuous engagement between IACS2 and the target groups enhances trust between them. Therefore, trust is moved and exchanged between actors involved throughout the project implementation, and it is crucial to keep it controllable to avoid any risk.

The high level of trust generates strong collaboration, and, in most cases, there is no conflict arises based on the friendly relationship and compatibility between both actors. Collaboration among the parties involved helps to smooth the matters about the project implementation. CS1_4 voices this opinion:

> Collaboration with other agencies involved in this project is crucial so that we do not walk alone. Thus, trust and satisfaction are there. Most importantly, the community thinks they are qualified because of the structure and collaboration between agencies involved.

4.2.3 Effective Leadership

Throughout the project implementation, there is evidence in CS1, CS2 and CS3 that leaders at the organisational level are able to motivate and inspire their subordinates in meeting work challenges and demands, which assist them mostly in problem-solving. At the same time, leaders at the ground level can motivate and inspire the target groups to fully utilise the facilities that have been given to them, and the evidence is found in CS1 and CS2. Despite the contextual differences, the leaders' understanding of delivering benefits to the target groups signifies their ability to motivate and inspire. On top of that, leaders must know subject matters, particularly in project management and social welfare. At the top organisational level, leaders are found to understand how the whole programme is executed and the impact on the target groups, while leaders at the management level are found to have basic knowledge in project management and social welfare. These characteristics are found in all cases, despite variations in leadership style according to the project type and complexity.

4.2.4 Decision Making

Generally, decision making can be segregated at two levels, namely, organisational level and project level. Decision making at the organisational level involved the ministry's endorsement of the project before the budget approval at the central agency's level. The main organisation is also responsible for ensuring the budget distribution to the implementing agencies while holding the authority regarding any additional project allocation requested by the implementing agencies. On the other hand, implementing agencies are responsible for decision making on the project activities and the selection of participants. In CS1 and CS2, the authority to manage the project management activities is held by the implementing agency and, they also have to decide on participants' selection but with the assistance of the ministry's officials in the selection committee.

In CS3, the IACS3 has the authority of every project management aspect, and the participants' selection is under the purview of the National Housing Department (JPN). Although the decision-making process is segregated among the main project actors, it is essential to note that the decision-making is generally conducted collectively. Hence, from the evidence found in the analysis, ethical decision making is practised, and most of the time, it is practised collectively. CS3_1 enlightens:

> Here, my bosses are very supportive and encourage us to make a collective decision, think together on how to solve the problem, and they do not say this is my decision or you have to follow my decision... All is done in consensus; we discuss to reach the same decision, and from there, we move forward. We do not make a decision individually because we believe that collective decision yields a better outcome.

4.2.5 Shared Vision

In the context of shared vision, most public officials share the same objectives with the government with regards to help poor people to get a better life through effective project delivery. The sharing of sentiment related to altruistic empathy further motivates the public officials to work on the common ground and collectively share the responsibility in planning and managing projects. However, the means to achieve the objectives could be different, and evidence from the analysis confirms the differences. Nevertheless, the facts that the projects' objectives are aligned with the government's strategic objectives and vision reflect the roles of public officials in setting the right direction in the early stage of project planning and initiation.

5 Discussion

The findings of this study suggest that there is a common interplay among the project governance elements connected in a specific way and contributed to the project governance practices and, subsequently, the project performance. Hence the discussion will highlight these interplays in the context of poverty alleviation projects in Malaysia.

5.1 Altruistic Empathy

This study discovered that altruistic empathy is adopted by officials at every level of the organisation, alike with the suggestion by Kuppelwieser (2011), where altruistic empathy, which is part of stewardship-style behaviour is not limited to the executive levels; instead, it is applied at every level of the organisation for the good of the organisation and stakeholders. Hernandez (2012) explained that the stewardship theory provides the moral commitment needed by the team members where they feel obliged to act in specific ways through the cognitive and affective element. However, the evidence from the case studies suggests that moral commitment is an altruistic empathy element that enables public officials to subjugate their interests for the long-term welfare of poor people.

Essentially, altruistic empathy is part of the motivational factors where officials involved in the project hold a form of altruism based on feeling to others. Evidence from the case studies suggests that empathy induced altruistic motive to increase the welfare of the person for whom empathy is felt, which is consistent with Batson et al. (1995) findings which discovered the importance of three different motives, namely self-interest, collective interest, and other-interest (altruism) in which may operate for the collective good in social dilemmas. Apart from that, altruistic empathy is interplayed with other project governance elements found in the current study. Altruistic empathy drives intrinsic motivation, particularly trust, where the altruistic sense towards poor people forms the backbone of trust, which eliminates any negative feelings among the project actors. This is in tandem with Kluvers and Tippett (2011), who linked the interplay through the lens of stewardship theory, mainly altruism and trust, where employees are actively involved with the operation of the nonprofit organisation despite there is no specific incentive to attract them.

5.2 Intrinsic Motivation

The intrinsic motivation, which comprises trust and collaboration, complements altruistic empathy. Trust, in particular, is intertwined with every project governance element found in this study. Trust encourages mutual respect and avoids negative stigma among the target group and further could accelerate any collaboration efforts needed to complete the project (Edelenbos and Eshuis 2012). Evidence in the case studies suggests that trust is a driven factor that enables various project actors to foster strong collaboration throughout project planning and implementation. While mutual trust drives collaboration among project actors and further improves project performance (Meng 2015), the evidence from case studies are varied, and it is not necessarily translated into project performance improvement.

On the other hand, the findings are consistent with the approach of stewardship theory, which promotes two main intrinsic values, explicitly trust and collaboration as the main idea which the principal should embrace, stewards, project managers and the whole of the project team (Höglund et al. 2019). Apart from the difficulty of measuring the project outcome, poverty alleviation projects are unique and different from other public projects where the priority is to ensure effective project delivery to improve the poor people's lives. Hence, according to Kadefors (2004), the development of trust and tendency to collaborate is strongly influenced by intuitive and emotional reactions and sensitivity to behavioural aspects such as shown respect and concern, which is an altruistic empathy found in this study. Besides, altruistic empathy and intrinsic motivation move in tandem and interplay and influence other project governance elements.

5.3 *Effective Leadership*

The findings have discovered that the effective leadership element found in this study has three main components: empowerment, ability to motivate and inspire, and knowledgeability. As this study found that altruistic empathy leads the officials to act with feelings towards helping people live in poverty and have a sense of what they have gone through, there is strong evidence that the officials can align those motivations with the organisation's working approach. Subsequently, this is translated into better decision making, robust empowerment, shared vision, and integrity among the officials involved. Hence, the positive interplay between altruistic empathy, intrinsic motivation, and effective leadership improves project delivery.

While leaders at the project level coordinate and monitor project activities conducted by the contractor, leaders at the organisational level monitor at the micro-level and intervene only when a problem could not be solved at the project level. Contrary to a business environment where ethical stewardship creates long term wealth and sustainable competitive advantage (Caldwell et al. 2008), ethical stewardship in the context of poverty alleviation projects create meaning and work on the common ground to achieve the project's and organisation's goals. It is also consistent with Pilkien et al.'s (2018) suggestion about the stewardship theory in horizontal leadership governance where stewards are motivated by intrinsic motivation and determined for project goals achievement rather than self-interest.

5.4 *Decision Making*

Decision making is crucial at every level of organisation and project. However, in this study's context, the decision-making process has considered the main stakeholders of poverty, namely the poor people; thus, the decision making is driven mainly by altruistic empathy towards the poor people. Altruistic empathy influences the decision making approach by the public officials, where the interests of the poor people are the main priority in each decision and action to be made by the project actors.

Although the decision-making process is segregated along with the main project actors, it is essential to note that generally, the decision making is conducted collectively. Therefore, the project structure setup, which includes the ministry and implementing agency, influences decision making in these projects. This coincides with Lappi and Aaltonen (2017) arguments where organisational setup and description of roles connected with decision-making authority can influence project governance internally and externally. The evidence in this study suggests that trust has been accepted as a primary culture, particularly between the ministry and the IA, which, in turn, cultivate ethical decision making throughout project planning and implementation. However, as the authority that takes charge of the public interest, the public officials have some reservation about the decision making process which involved the contractor. While the public officials acknowledged the importance of building a trustful and transparent relationship with the contractor as the external stakeholder as proposed by Derakhshan et al. (2019), they exercised extra caution as both parties do not share the same objectives. Nevertheless, this study echoed with Derakhshan et al. (2019) that trust and ethics influence the project's decision making in tandem with the stewardship theory, which could improve project performance.

5.5 Shared Vision

This study reveals shared vision as a project governance element where projects are delivered in line with strategic organisational objectives and stakeholders' expectations, as emphasised by Levie et al. (2017). As the understanding of project objectives is shared among the project actors, they worked together to pursue project outcomes, which are to uplift the quality of life and enhance the socio-economic development of poor people. The shared vision among the project actors complements the altruistic empathy and intrinsic motivation factors that influenced the project governance practices. This sentiment is better explained through the stewardship theory which acts as a suitable model especially with regards to the poverty reduction and clients stability since element such as trust is adopted, and shared among project actors to foster the organisation's goals as shared by the whole members of the organisation (Van Slyke 2007). Following this view, the current study discloses the ability of shared objectives, which subsequently translated into shared outcomes with meaningful existence of altruistic empathy and intrinsic motivation along the line. These elements interplay with each other and can improve project performance.

6 Conclusion

This chapter set out to explore the practices of project governance elements in poverty alleviation projects. Although there is no specific project governance framework adopted by the public officials involved, they informally practised project governance through several elements discovered in the case studies. The most notable findings to emerge from this chapter are the identification of altruistic empathy and intrinsic motivation as two main elements that drive project governance practices, mainly in the context of poverty alleviation projects. These two project governance elements are significantly practised and positively interplayed with other project governance elements, namely effective leadership, decision making, and shared vision throughout the project planning and implementation. These interplays add value to the whole project and guide public officials to effectively deliver the projects to the target groups and facilitate the project outcome achievement towards its sustainability. This new understanding should help to improve the public organisations involved in poverty alleviation projects by adopting project governance elements and a better understanding of how they are practised.

Poverty alleviation projects which involve a substantial amount of investment are always needed to address the poverty issues. Hence, there are considerable challenges to adopting the practices of project governance and the flexible governance structure among the project actors. The tendency of organisations to work in silo and bureaucracy structure is the barriers to providing efficiency and line of accountability in this type of project. Considering the public officials still lack knowledge and understanding regarding project governance, this chapter offers a better understanding of how public officials practice project governance in conducting poverty alleviation projects through the lens of stewardship theory. Hence, the chapter demonstrates the flexibility of stewardship theory in portraying the insights of project governance in the context of poverty alleviation projects towards its successful outcome.

7 Recommendations and Limitation

The current study is subject to limitations in which could lead to further improvement for future research. First, the case study is limited in terms of its generalisation. Although the study utilised multiple case studies to ensure the literal replication, it is still within the context of constructed poverty alleviation projects conducted by the public organisations, and it is limited to the actors involved and the particular local conditions. A further study could explore other project governance elements in other types of poverty alleviation projects and programmes such as cash aid and entrepreneur equipment assistance where more apparent perspectives could be gained from those projects.

Also, there are other factors that may influence the governance practices in poverty alleviation projects that could not be included in this study. For instance, the political influence towards the project implementation at the organisational and project level is not discussed in this study since it involves another type of research subject, namely political substance. As this study is only limited to the public officials who are officially structured in the government machinery system, it is suggested that the influence of a political substance is investigated in future studies to determine its impact on the said projects' governance practices.

Acknowledgements The researcher acknowledges the Ministry of Rural Development Malaysia and Ministry of Housing and Local Government Malaysia for their permission to conduct this study.

References

Abdalla Y, Elhadary E (2012) Political economy and urban poverty in the developing countries: lessons learned from Sudan and Malaysia. J Geograp Geol 4(1):212–223. https://doi.org/10.5539/jgg.v4n1p212

Asian Development Bank (ADB) (1999) Fighting poverty in Asia and the Pacific: the poverty reduction strategy

Barrett CB (1996) Fairness, stewardship and sustainable development. Ecol Econ 19:11–17. https://doi.org/10.1080/02580136.2014.923689

Batson CD, Batson JG, Todd RM et al (1995) Empathy and the collective good: caring for one of the others in a social dilemma. J Pers Soc Psychol 68(4):619–631. https://doi.org/10.1037/0022-3514.68.4.619

Bond-Barnard TJ, Fletcher L, Steyn H (2018) Linking trust and collaboration in project teams to project management success. Int J Manag Proj Bus 11(2):432–457. https://doi.org/10.1108/IJMPB-06-2017-0068

Caldwell C, Hayes LA, Bernal P et al (2008) Ethical stewardship: implications for leadership and trust. J Bus Ethics 78(1–2):153–164. https://doi.org/10.1007/s10551-006-9320-1

Celikay F, Gumus E (2017) The effect of social spending on reducing poverty. Int J Soc Econ 44(5):620–632. https://doi.org/10.1108/IJSE-10-2015-0274

Creswell JW, Poth CN (2018) Qualitative inquiry and research design: choosing among five approaches, 4th edn. SAGE Publications, California

Civil Service Delivery Unit (CSDU) (2017) NTP annual report 2017. Putrajaya

Davis JH, Schoorman FD, Donaldson L (1997) Toward a stewardship theory of management. Acad Manag Rev 22(1):20–47. https://doi.org/10.5465/AMR.1997.9707180258

Derakhshan R, Turner R, Mancini M (2019) Project governance and stakeholders: a literature review. Int J Project Manage 37(1):98–116. https://doi.org/10.1016/j.ijproman.2018.10.007

Department of Statistics Malaysia (DOSM) (2017) Laporan Penyiasatan Pendapatan Isi Rumah Dan Kemudahan Asas 2016. http://www.statistics.gov.my/portal/index.php?option=com_content&view=article&id=1640&Itemid=111&lang=bm. Accessed 15 March 2021

Department of Statistics Malaysia (DOSM) (2020) Laporan Survei Pendapatan Isi Rumah dan Kemudahan Asas 2019. https://www.dosm.gov.my/v1/index.php?r=column/cthemeByCat&cat=120&bul_id=TU00TmRhQ1N5TUxHVWN0T2VjbXJYZz09&menu_id=amVoWU54UTl0a21NWmdhMjFMMWcyZz09. Accessed 15 March 2021

Edelenbos J, Eshuis J (2012) The interplay between trust and control in governance processes: a conceptual and empirical investigation. Adm Soc 44(6):647–674. https://doi.org/10.1177/0095399711413876

Fan S, Chan-Kang C (2008) Regional road development, rural and urban poverty: evidence from China. Transp Policy 15(5):305–314. https://doi.org/10.1016/j.tranpol.2008.12.012

Grindle MS (2004) Good enough governance: poverty reduction and reform in developing countries. Governance 17(4):525–548. https://doi.org/10.1111/j.0952-1895.2004.00256.x

Hernandez M (2012) Toward an understanding of the psychology of stewardship. Acad Manag Rev 37(2):172–193. https://doi.org/10.5465/amr.2010.0363

Höglund L, Mårtensson M, Safari A (2019) The impact of types of trust in the public sector: a case study approach. Int J Public Sect Manag 32(3):247–263. https://doi.org/10.1108/IJPSM-08-2017-0226

Imam H, Zaheer MK (2021) Shared leadership and project success: the roles of knowledge sharing, cohesion and trust in the team. Int J Project Manage. https://doi.org/10.1016/j.ijproman.2021.02.006

Irfan M, Hassan M (2019) The effect of project governance and sustainability on project success of the public sector organisations in Pakistan. Pertanika J Soc Sci Humanit 27(1):177–198

Jindra C, Vaz A (2019) Good governance and multidimensional poverty: a comparative analysis of 71 countries. Int J Policy Adm. https://doi.org/10.1111/gove.12394

Jomo K, Chowdhury A (2016) Is good governance key to eliminating poverty? Inter Press Service News Agency. http://www.ipsnews.net/2016/06/is-good-governance-key-to-eliminating-poverty Accessed 18 April 2020

Joslin R, Müller R (2016) The relationship between project governance and project success. Int J Project Manage 34(4):613–626. https://doi.org/10.1016/j.ijproman.2016.01.008

Kadefors A (2004) Trust in project relationships-inside the black box. Int J Project Manage 22(3):175–182. https://doi.org/10.1016/S0263-7863(03)00031-0

Khan A, Waris M, Ismail I et al (2019) Deficiencies in project governance: an analysis of infrastructure development program. Adm Sci 9(1):1–15. https://doi.org/10.3390/admsci9010009

Khan A, Waris M, Panigrahi S et al (2020) Improving the performance of public sector infrastructure projects: role of project governance and stakeholder management. J Manag Eng 37(2):1–20. https://doi.org/10.1061/(ASCE)ME.1943-5479.0000886

Khan MH (2000) Rural poverty in developing countries: issues and policies

Klakegg O, Haavaldsen T (2011) Governance of major public investment projects: in pursuit of relevance and sustainability. Int J Manag Proj Bus 4(1):157–167. https://doi.org/10.1108/17538371111096953

Kluvers R, Tippett J (2011) An exploration of stewardship theory in a not-for-profit organisation. Account Forum 35(4):275–284. https://doi.org/10.1016/j.accfor.2011.04.002

Kuppelwieser VG (2011) Stewardship behavior and creativity. Manag Revue 22(3):274–295. https://doi.org/10.1688/1861-9908

Kwon HJ, Kim E (2014) Poverty reduction and good governance examining the rationale of the millennium development goals. Dev Chang 45(2):353–375. https://doi.org/10.1111/dech.12084

Lappi T, Aaltonen K (2017) Project governance in public sector agile software projects. Int J Manag Proj Bus 10(2):263–294. https://doi.org/10.1108/IJMPB-04-2016-0031

Levie F, Burke CM, Lannon J (2017) Filling the gaps: an investigation of project governance in a non-governmental organisation's response to the Haiti earthquake disaster. Int J Project Manage 35(5):875–888. https://doi.org/10.1016/j.ijproman.2016.10.007

Li Y, Su B, Liu Y (2016) Realising targeted poverty alleviation in China: people's voices, implementation challenges and policy implications. China Agricu Econ Rev 8(3):443–454. https://doi.org/10.1108/CAER-11-2015-0157

Meng X (2015) The role of trust in relationship development and performance improvement. J Civ Eng Manag 21(7):845–853. https://doi.org/10.3846/13923730.2014.893923

Merriam SB, Tisdell EJ (2016) Qualitative research: a guide to design and implementation, 4th edn. Wiley, San Francisco

Musawir AU, Abd-Karim SB, Mohd-Danuri MS (2020) Project governance and its role in enabling organisational strategy implementation: a systematic literature review. Int J Project Manage 38(1):1–16. https://doi.org/10.1016/j.ijproman.2019.09.007

National Audit Department (2017) Laporan Ketua Audit Negara 2016 Siri 1/2016

Nawi MNM, Nasir NM, Mohd-Danuri MS et al (2018) Client-related factor contributing to delay in the Malaysia public projects. J Soc Sci Res Special Issue:289–293. https://doi.org/10.32861/jssr.spi6.289.293

Nixon S, Asada H, Koen V (2017) Fostering inclusive growth in Malaysia. https://doi.org/10.1787/24837aed-en

Pilkien M, Alonderien R, Chmieliauskas A et al (2018) The governance of horizontal leadership in projects. Int J Project Manage 36:913–924. https://doi.org/10.1016/j.ijproman.2018.06.002

Poverty Task Force (2002) Localising MDGs for poverty reduction in Viet Nam: ensuring good governance for poverty reduction
Project Management Institute (2016) Governance of portfolios, programs and projects. Project Management Institute, Pennsylvania
Renz PS (2007) Project governance: implementing corporate governance and business ethics in nonprofit organisations. Physica, Heidelberg
Sagaran S, Nair S (2015) Poverty in Malaysia: need for a paradigm shift. Inst Econ 7(3):95–123
Sarker MNI, Hossin MA, Min W et al (2018) Poverty alleviation of rural people through good governance in Bangladesh. J Soc Sci Res 4(12):547–555. https://doi.org/10.32861/jssr.412.547.555
Siwar C (2006) Good governance for poverty alleviation: the case of Malaysia. Chin Public Adm Rev 3(4):1–12. https://doi.org/10.22140/cpar.v3i3/4.57
Van Slyke DM (2007) Agents or stewards: Using theory to understand the government-nonprofit social service contracting relationship. J Public Adm Res Theory 17(2):157–187. https://doi.org/10.1093/jopart/mul012
Swanborn PG (2010) Case Study Research. SAGE Publications, London
Thang HNC, Baharuddin AH (2011) Poverty reduction: a continuous social responsibility in Malaysia. Int J Rural Stud 18(2):2–5
UNHR (2020) Visit to Malaysia: Report of the special rapporteur on extreme poverty and human rights. United Nations. https://doi.org/10.1017/s002081830002244x
World Bank (2000) world development report 2000: attacking poverty. World Bank. https://doi.org/10.1596/0-1952-1129-4
Yin RK (2018) Case study research and application: Design and methods, 6th edn. SAGE Publications, Los Angeles
Young R, Chen W, Quazi A et al (2019) The relationship between project governance mechanisms and project success. Int J Manag Proj Bus 13(7):1496–1521. https://doi.org/10.1108/IJMPB-10-2018-0212
Zhang S, Zhang S, Gao Y et al (2016) Contractual governance: effects of risk allocation on contractors' cooperative behavior in construction projects. J Const Eng Manag 142(6):1–11. https://doi.org/10.1061/(ASCE)CO.1943-7862.0001111

Interception Loss of Tree Canopy as Green Infrastructure

A. B. Azinoor Azida

Abstract This chapter discussed the role of trees in the environment and hence on the hydrological cycle. There are many processes involved in the hydrological cycle, namely precipitation, evaporation, infiltration, canopy interception and transpiration. The trees acted as land cover influences the hydrological cycle, including the trees' capacity to intercept, evapotranspiration, purify, store, and infiltrate the rainfall. The interception process in the hydrological cycle can be up to 30–50% of the total gross precipitation. The structure of the canopy forest reduces the amount of runoff which the trees intercept the precipitation and use the water during the interception process. This process thus decreasing the volume of water draining through a catchment area. In this chapter, the amount of interception loss in tropical forests was determined. The interception loss was determined by quantifying the difference between gross rainfall and net rainfall (throughfall and stemflow). Original Gash model was also applied to computed the interception loss value. Two plots were chosen as study area namely Plot 11 and Plot 12. In the direct measurements, interception loss for Plot 11 is 13.6% of the gross rainfall and for Plot 12 produced 10.8% of the gross rainfall for the 12-month periods. Whereas, from the Gash model computation, interception loss from Plot11 produced 14.7% and while for Plot 12, the original contributed 13.6% of the total gross rainfall during the study period.

Keywords Forest canopy · Gash model · Interception loss · Tree canopy · Tropical forest

1 Introduction

Interception loss can be categorized as the interception loss in the forest due to the canopy structure and the interception loss from the building structure. This study is more focused on the interception loss due to the canopy structure in the tropical rainforest. In forest environments, interception loss plays a vital role as the vegetation

A. B. Azinoor Azida (✉)
School of Civil Engineering, College of Engineering, Universiti Teknologi MARA, Kampus Pasir Gudang Bandar Seri Alam, 81750 Masai, Johor, Malaysia
e-mail: azinoor@uitm.edu.my

R. Hassan et al. (eds.), *Green Infrastructure*,
https://doi.org/10.1007/978-981-16-6383-3_15

cover affects the water budget equation used in water catchment volume. van Djik and Bruijnzeel (2001) stated that interception loss is a significant characteristic in hydrological studies, which is relay effects on site and catchment water balance. Interception amounts can increase up to 15–50% of precipitation, a significant part of the water balance. One can distinguish many interceptions, which can also interplay (Astuti and Suryatmojo 2019).

There are three main elements in the forest evaporation process; transpiration or evaporation from dry canopy cover, precipitation interception or evaporation from wet canopy cover and evaporation from the forest floor (Roberts et al. 2005). Interception loss is one of the essential elements in the hydrological cycle. The canopy interception loss, IL generally expressed as the volume of rainwater evaporated from the wetted canopy cover during a specified duration. On a single plant, the interception loss is also known as canopy storage volume, C. Values of interception loss are influenced mainly by the climatological condition. Thus, the volume of water returned to the atmosphere may vary between the hydrological events. Therefore, the amount of interception loss may be greater or lesser than the amount of capacity storage as the value of I is greatly inconstant in a single event. In the forest, the amount of interception losses is higher than the volume from the dryland area as the extensive plant canopy and high frequency of rain events affecting the interception losses occurrence (Dunkerly 2000).

Interception loss is the most neglected process in calculating the water budget equation. As precipitation occurs, part of it will be intercepted by the vegetation before reaching the ground surface. Part of this intercepted volume is retained and eventually evaporates. In forest environments, interception plays a significant role in the partition of rainfall into evaporation and runoff. Evaporation of intercepted rain, usually referred to as interception loss, consumed a significant part of total precipitation. It ranges almost to 25% depending on the nature of the rainfall and the canopy storage capacity (Eltahir and Bras 1992).

Several studies have shown the relationship between forest elements to interception such effects of forest type to interception (Crockford and Richardson 1990; Azinoor Azida and Khairudin 2017), canopy cover relationship to throughfall (Pérez-Suárez et al. 2008), diameter at breast height (dbh) to stemflow (Manfroi et al. 2004) and effects of leaf area index (LAI) and tree height to throughfall (Dietz et al. 2006). However, there is still no research focusing on the effects of canopy cover of the forest stand on the interception loss. Crockford and Richardson (2000) reported that larger crown size and canopy gaps would produce greater interception volume. Therefore, crown size and canopy gaps are crucial for measuring the specified area's canopy cover. Hence, as the canopy cover influences rainfall interception, it also affects the losses (Azinoor Azida and Lu 2015). Besides, quantifying the magnitude of rainfall interception at the global scale is essential to understanding land–atmosphere interactions, the dynamics of the global water cycle, hydrological modelling, and the impacts of deforestation (Zeng and Jia 2019).

Models have been developed to predict the interception loss because of the difficulties in obtaining direct measurements of interception loss from forest to forest. For direct measurements, interception loss can be calculated using throughfall and stemflow volume, requiring frequent data collection from the specific forest. Since difficulties might arise, many models are derived by previous researchers. Rutter, Gash, Mulder and WiMo model is the example of these interception loss models.

Gash model used in this study is the simplification of the Rutter model, which incorporates some simple features of linear regression models within the conceptual background of the Rutter model. Gash model provides equations after considering rainfall as a succession of discrete rainfall events, each having three distinct phases: (1) a wetting phase, from the start of rainfall until the canopy is saturated; (2) a saturation phase; (3) a drying phase, from the end of the rainfall to the completely dried canopy and trunks (Gash 1979). Gash then rearranges the original equations to estimate the interception loss in sparse forests to make it consistent with the conceptual framework for the sparse Rutter model.

1.1 Hydrological Cycle

The hydrologic cycle describes the continuous water movement on, above and below the earth surface that provides constant volume. It is defined as the journey water takes as it circulates from the land to the sky and back to the atmosphere. The cycle stages are evaporation from water bodies, transport of water vapour through the atmosphere, precipitation, interception and transpiration by canopy cover, infiltration and surface runoff, groundwater flow, and streamflow. The volume of water in any given state or zone continuously varies, but the total volume remains constant (Roberson et al. 1989). All of the physical, chemical and biological processes involving water as it travels its various path in the atmosphere, over and beneath the earth's surface and through growing plants, are of interest to those who study the hydrologic cycle.

1.2 Interception Loss, IL

A process whereby rainfall is intercepted by vegetation or buildings, or other canopy covers before reaching the ground surface is the definition of interception. It plays a vital role in hydrological analysis and modelling. When the rainfall is intense enough and exceeded the canopy storage capacity, the excess rainwater drips off as throughfall and along branches and stem as stemflow. Thus, based on the canopy characteristics, a significant amount of rainfall could be intercepted and immediately evaporated back to the atmosphere once the rain has ceased. This fraction of water is called the interception loss.

Interception happens mostly in vegetation because of the canopy storage capacity that is bigger than other capacities. Depending on the density of the vegetation cover, a part of rainfall is intercepted by the leaves and stems and may be able to store on its surface area temporarily. Interception is essential for the calculation of water balances in hydrology because of two reasons. First, the net rainfall or water which reaches the ground is generally less than the total rainfall falling on top of the vegetation cover. The other reason is almost one-third of the total precipitation can be retained as an interception. Therefore, interception significantly reduces precipitation intensity as water is first temporarily sorted and much is lost. The main components of canopy interception for observation in forest areas are throughfall (Tf), stemflow (Sf) and precipitation. When the rainfall is intense enough and exceeded the canopy storage capacity, the excess rainwater drips off as throughfall and along branches and stem as stemflow (Yusop et al. 2003). These parameters are also the main components in estimating interception loss in this study.

1.3 Throughfall, Tf

Throughfall describes the process of precipitation passing through the plant canopy to reach the ground surface (Dunkerly 2000). This process is one of the measured data in calculating interception loss. It can either be from the drips off from canopy cover or the direct precipitation that falls on the ground without touching the canopy. The amount of precipitation passing through the vegetation varies significantly with the vegetation types (Pidwirny 2006). It is controlled by many factors such as plant leaf and stem density, type of precipitation, the intensity of the precipitation, and the duration of the rainfall event. It is also affected by the canopy cover storage of the vegetation. In a previous study, Konishi et al. (2006) stated that the spatial distribution of the throughfall in the tropical rainforest is almost heterogeneous, but the distribution may not 100% random. The fluctuation is most probably due to the canopy structure in the tropical area itself. A wide range of Tf studies under different vegetation canopies has been conducted, as summarized in Table 1.

The values in Table 1 indicate inconsistent rates of Tf obtained by various researchers under different canopies structures. Some of the factors that might contribute to this variation are the numbers of Tf gauges and the arrangements used. In addition, Tf results from natural forests may not apply to other vegetation that grown elsewhere because of the differences in canopy architecture and micrometeorological factors.

1.4 Stemflow, Sf

One of the processes of rainfall that is not intercepted by canopy cover is the stemflow process. This is also one of the main components in determining interception loss using the original Gash model. Stemflow is defined as the process that directs precipitation down plant branches and delivers the water at the edges of the tree in drip form. The redirection of water by this process causes the ground around the plant's stem to receive additional moisture (Pidwirny 2006). The amount of stemflow

Table 1 Throughfall studies under different vegetation canopies

Vegetation structure	Throughfall (%)	References
Tropical regions (Natural forest)		
Tropical rainforest	91.0	Lloyd and de Marques (1988)
Tropical rainforest	82–87	Marin et al. (2000)
Tropical rainforest	80.8	Konishi et al. (2006)
Tropical rainforest	79.0–81.0	Nepstad et al. (2000)
Tropical rainforest	80.7	Sinun et al. (1992)
Lowland tropical rainforest	65.3–94.6	Manokaran (1979)
Lowland tropical rainforest	80.9	Dykes (1997)
Lowland tropical rainforest	82.0	Manfroi et al. (2004)
Lowland tropical rainforest	73.0	Baharuddin (1989)
Unlogged and logged rainforest	87.2; 93.5	Asdak et al. (1998)
Secondary lowland rainforest	72.2	Nik et al. (1979)
Secondary lowland rainforest	85.1	Yusop et al. (2003)

is determined by leaf shape and stem and branch architecture. The stemflow studies under different vegetation canopies is summarized in Table 2.

Although many early studies ignored Sf, it is now known that it can be significant for certain species. In addition, it is significant for water chemistry studies due to its high solute concentrations. Sf is one of the hydrological components responsible for transferring the rainfall from a vegetative canopy to the soil (Levia and Frost 2003). A better understanding of the partitioning of incident gross rainfall into Sf will improve models of its influence on soil solution chemistry, groundwater recharge, leaching of fertilizer applications, and biogeochemical cycles. That could ultimately lead to greater efficiency in managing forests and agricultural lands and optimum forest harvesting.

Table 2 Stemflow studies under different vegetation canopies

Vegetation structure	Stemflow (%)	References
Tropical regions (Natural forest)		
Tropical rainforest	1.8	Lloyd and de Marques (1988)
Tropical rainforest	0.9–1.5	Marin et al. (2000)
Tropical rainforest	1.1	Konishi et al. (2006)
Tropical rainforest	1.9	Sinun et al. (1992)
Tropical rainforest	1.2	Nepstad et al. (2000)
Lowland tropical rainforest	0.32–0.92	Manokaran (1979)
Lowland tropical rainforest	< 1.0	Dykes (1997)
Lowland tropical rainforest	0.4	Baharuddin (1989)
Lowland tropical rainforest	3.5	Manfroi et al. (2004)
Secondary lowland rainforest	1.3	Yusop et al. (2003)
Secondary lowland rainforest	0.89	Nik et al. (1979)
Unlogged and logged rainforest	1.4–0.3	Asdak et al. (1998)

1.5 Original Gash Model

In the original version, Gash considers rainfall to occur as a series of discrete events, each comprising a period of wetting up when the rainfall (P_g) is less than the threshold value necessary to saturate the canopy (P_g') a period of saturation and a period of drying out after rainfall ceases. It is also assumed that the canopy has sufficient time to dry out between two storms. The canopy capacity (S), which describes the forest structure, is defined as the amount of water left on the canopy in zero evaporation conditions when rainfall and throughfall have ceased (Gash and Morton 1978). Meanwhile, a free throughfall coefficient (p) determines the amount of rain that falls directly to the ground surface without touching the canopy and usually assumed to be equal to one minus the canopy cover (c). Evaporation from the trunks is described in terms of a trunk storage capacity (S_t) and the proportion of the rainfall diverted to stemflow (p_t). The mean evaporation rate during rainfall ($\bar{E}$) and the mean rainfall rate ($\underline{R}$) for saturated canopy conditions are also required. The original Gash model had made few assumptions regarding the component of interception loss and the equation used (Table 3).

(a) Rainfall may be represented by series of discrete storms, separated by periods in which the canopy dries completely. Therefore, measurement of incident

Table 3 The original equation of the Gash model

Component of interception loss	
For m storms insufficient to saturate the canopy ($P_g \leq P'_g$)	$(1 - p - p_t)\sum_{j=1}^{m} P_{gj}$
For n storms sufficient to saturate the canopy ($P_g > P'_g$)	
Wetting up canopy	$n\{(1 - p - p_t)P'_g - S\}$
Wet canopy evaporation during the storm	$\frac{\bar{E}}{\bar{R}}\sum_{j=1}^{n}(P_{gj} - P'_g)$
Evaporation after rainfall ceases	nS
Evaporation from stems for q storms $> \frac{S_t}{p_t}$ which saturate the stem	$qS_t + p_t\sum_{j=1}^{m+n-q} P_{gj}$
Parameters	
Rainfall necessary to saturate the canopy	$P'_g = -\frac{RS}{\underline{E}} In\left[1 - \frac{\underline{E}}{(1-p-p_t)\underline{R}}\right]$
Mean wet canopy evaporation rate	$\underline{E} = E_W$
Canopy capacity	S
Canopy cover fraction	$1 - p$

rainfall (P_g), throughfall (Tf) and stemflow (Sf) total on an event basis can be used for modelling rainfall interception.

(b) The average evaporation rate ($\underline{E}$) adequately represents evaporation from the canopy during the storms. Similarly, the implicit assumption is that the ratio of average evaporation rate over average rainfall intensity ($\underline{E}/\underline{R}$) is equal for all storms.

(c) No water drips from the canopy before the canopy capacity (S) is filled.

(d) Evaporation from stems occurs after rainfall has ceased.

2 Methodology

2.1 *Study Area*

This study was conducted at Bukit Lagong Forest Reserve, Selangor (Fig. 1), located at 3° 15′ N latitude and 101° 37′ E longitude. The 485 hectares of reserved forest is governed by FRIM and is covered with primary lowland mixed dipterocarp forest. About 78% of this reserved forest is planted forest, where the forest is subdivided into 53 plots mostly occupied with 60 types of indigenous and exotic species (Ibrahim et al. 2008). From an ecological standpoint, Bukit Lagong Forest Reserve is an important water catchment area with rich forest biodiversity. Therefore, Bukit Lagong Forest Reserve, the study area, has been gazette as a forest reserve (Er et al. 2011). The forest characteristic in class V (vegetation) in Malaysia is suitable for this study and is situated at 176 m above sea level.

Besides being gazette as a forest reserve, Bukit Lagong Forest Reserve is also gazette as a Highly Environmentally Sensitive Area. It is also classified under the International Management Unit for the Conservation of Nature and Natural Resources (IUCN). Bukit Lagong ranges from 290 m to approximately 575 m at the

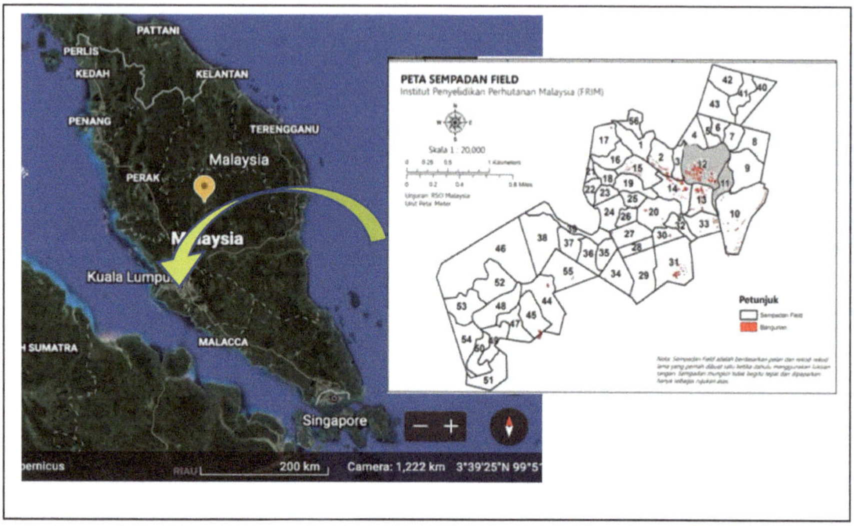

Fig. 1 Site of the study area

peak (Er et al. 2011). The river sources for Buloh River, Jinjang River, Kuang River and Keroh River are located within Bukit Lagong Forest Reserve, which also acts as a water catchment area. However, within this forest reserve, economic activities abound and together with human settlements.

2.2 *Canopy Characteristics*

Plot 11 and Plot 12 have been chosen as study areas to conduct the study on interception loss by the canopy cover, where each plot has an area of $40m^2$ (20 m × 20 m). All trees above 10 cm dbh (diameter at breast height) within the plots were identified, tagged and numbered. For Plot 11, 21 trees with dbh more than 10 cm have been identified, and another 20 trees for Plot 12. Plot 11 is dominated by Kulim species, while Plot 12 is mainly occupied *by Kledan, Keruing, Simpoh* and *Mempisang* species. The canopy height is ranging from 17 m to 29.5 m for both plots (Azinoor Azida et al. 2012). Stand Visualization System (SVS) has been used to visualize the forest standing study area in 3D view based on the selected trees over the interest area.

Fig. 2 The gauge for rainfall collection

2.3 *Rainfall (P) Measurement*

Rainfall, Pg is defined as the precipitation that drops into a catchment and measured above forest canopy or in an open area. Gross rainfall is collected by placing a gauge in the upper place so that no water from other processes will get in the gauge that provides error data. The gauge is placed in the open area for an accurate data collection process. There is an existing rainfall station owned by Malaysia Meteorological Department (MMD) in this study area. The gauge is located on an open area about as shown in Fig. 2. The gauge is placed about 30 m from the plots.

2.4 *Throughfall (Tf) Measurement*

Throughfall, Tf describes the process of precipitation passing through the plant canopy in the rainfall event. Throughfall may be the water that drips after intercepting the tree leaves and branches or the falling rainfall that directly reaches the ground surface through the canopy gap without intercepting the canopy. In this study, throughfall is measured using 25 collectors about 225 mm diameter and 200 mm deep for each plot, as shown in Fig. 3. It is placed at each plot grid with a 5 m interval. Throughfall is measured by several throughfall collectors below the canopy cover, and the data is being averaged out using a measuring cylinder daily. Because the canopy area of the forest is dense, the volume of throughfall was divided with the orifice of the collector to obtain throughfall value in depth (mm).

Fig. 3 Throughfall collectors located at every corner of the grid point

2.5 *Stemflow (Sf) Measurement*

Other than throughfall, another process of the rainfall travels to the ground is called the stemflow process. Stemflow, Sf is defined as the process that directs precipitation down plant branches and stems. There are 15 trees selected for each plot depends on the accessibility in setting up the stemflow collar, as shown in Fig. 4. The collar method is adapted in measuring this process where the selected trees were spirally

Fig. 4 Stemflow collectors at the selected trees

fitted with the spiral rubber collar and draining into a collecting tank. The rubber collar is made of PVC hose was fitted around the tree stem using nails and sealed with silicone glue to seal the space between the stem and the edge of the collar. The 25 L capacities of collecting tanks were emptied daily and measured using a measuring cylinder.

The volume of stemflow is converted into depth (mm) by using the following equation (Bo et al. 1989):

$$\text{Stemflow, } S_f = \frac{1}{2}\left(\frac{(D_1 + D_2}{D_1} + \frac{(B_1 + B_2)}{B_1}\frac{V_c}{A}\right) \tag{1}$$

where D_1 is the total number of trees in the plot, D_2 the number of uncollared trees, B_1 is the total basal area of all trees ($\frac{m^2}{plot}$), B_2 as the basal area of the uncollared trees ($\frac{m^2}{plot}$), V_c is the total volume of stemflow ($\frac{litre}{plot}$), and A is the plot size (m^2).

2.6 Data for Gash Model Procedure

Data analysis is a part of evaluating and analyzing all the parameters where all the parameters are gathered. In this study, the original Gash model was applied to estimate the interception loss and the on-site data measurement and computation of interception loss estimation value. In this study, linear regression techniques are being used to determine the water storage capacity (S), direct throughfall fraction (p) and the ratio of evaporation to rainfall intensity ($\frac{\overline{E}}{\overline{R}}$). Based on the study by Leyton et al. (1967), the storage capacity (S) is usually estimated with straight-line regression. The slope $(1 - p_t)$ was made between throughfall versus gross rainfall graph for storms $\geq$ 3.0 mm as threshold value which is assumed that these points represented situation with minimal evaporation losses. The negative intercept with the throughfall axis gives the value of S. However, according to Asdak et al. (1998), this method seems subjective both in recognition of the inflexion relating to the point of canopy saturation in fitting the upper envelope to the scattered points. Thus, this study adopted the method by using separate linear regressions of gross rainfall versus throughfall for individual small storms (Lloyd et al. 1988), which more appropriate for determining S in the tropical forest. The value of S is given by the slope of the linear regression for zero throughfall.

Besides that, stemflow parameters (p_t) and (S_t) are estimated based on the linear regression between stemflow and gross rainfall using the methods employed by Gash and Morton (1978), where p_t and S_t are assumed to be the slope and intercept, respectively. Throughfall coefficient (p) is estimated by the method of Jackson (1975), where the slope of the regression between gross rainfall and throughfall for storms is too small to fill the canopy storage capacity (S). It is assumed that evaporation loses negligible during these storms. In determining mean evaporation and mean rainfall intensity ($\frac{\overline{E}}{\overline{R}}$), the direct interception loss (IL) and gross rainfall are regressing to

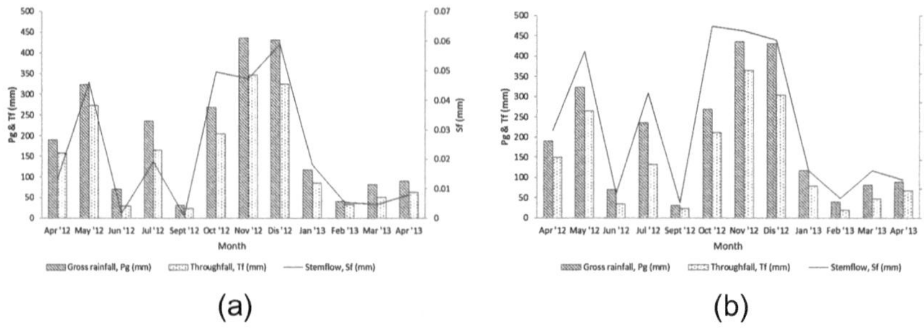

Fig. 5 Monthly gross rainfall, throughfall and stemflow for **a** Plot 11 and **b** Plot 12

obtain the parameter. The gradient of the linear regression between site interception loss and gross rainfall will be the value of the mean evaporation and mean rainfall intensity ($\frac{E}{\underline{R}}$) (Gash 1979).

3 Research Findings and Discussion

3.1 Data from Study Area

The difference between gross rainfall and net rainfall (throughfall and stemflow) is measured as interception loss. The data were collected from 11 April 2013 until 24 April 2013, 12 months of data collection with the excluded August 2012 data due to technical problem. As a result, 94 rainfall events are recorded during the 12 months from April 2012 until April 2013. The minimum gross rainfall recorded is 1.4 mm on 28 September and 12 October 2012, whereas the extreme rainfall events recorded was 109.7 mm on 18 April 2012.

The wettest seasons recorded were from October until December 2012, contributing to the total rainfall of 2095.8 mm within the study periods. For illustration, Fig. 5a, b show that the highest monthly rainfall recorded was 405 mm in November 2012, whereas the lowest was 21 m m in February 2013.

3.2 Interception Loss by Gash Model

Due to the lack of information on the evaporation rate of the study area, another method has been applied to determine the ratio of mean evaporation rate per unit ground area during rainfall over the mean rainfall rate, $\bar{E}/\underline{R}$. The data for the regressions is extracted when the value is over the threshold value of 3 mm. The mean slope of the regressions gave the value of $\bar{E}/\underline{R}$.

To apply the original Gash model, this study adopted the following procedure:

(a) Values of $\bar{E}/\underline{R}$ obtained were used to calculate P'_g for the study period using the value of S = 0.8091 for Plot 11 and S = 0.7378 for Plot 12 obtained in the regression of gross rainfall against throughfall.
(b) Further assumptions made is there is only one storm per rain day. Then, the raindays were divided into those with $P_g \geq P'_g$ and those with $P_g < P'_g$. These two rainfalls set of daily rainfall amounts were then summed to give $\sum_{j=1}^{n} P_{Gj}$ and $\sum_{j=1}^{m} P_{Gj}$.
(c) The number of rain days, n with $P_g \geq P'_g$ was noted.
(d) The numbers of rain days, q with $P_g \geq S_t / p_t$ was noted, and the rainfalls of rain days with $P_g \geq S_t / p_t$ were summed.

The regression of gross rainfall against throughfall, regressions of stemflow against gross rainfall and regressions of interception loss against gross rainfall for (a) Plot 11 and (b) Plot 12 is shown in Fig. 7, Fig. 8 and Fig. 9, respectively.

For Plot 11, from the estimation, the obtained value of interception loss is 308.5 mm from the original Gash model, and the interception loss from forest observation is 284.0 mm. The calculated values from the model overestimated the interception loss by 24.5 mm or 8.6%. On the other hand, for Plot 12, the obtained value of

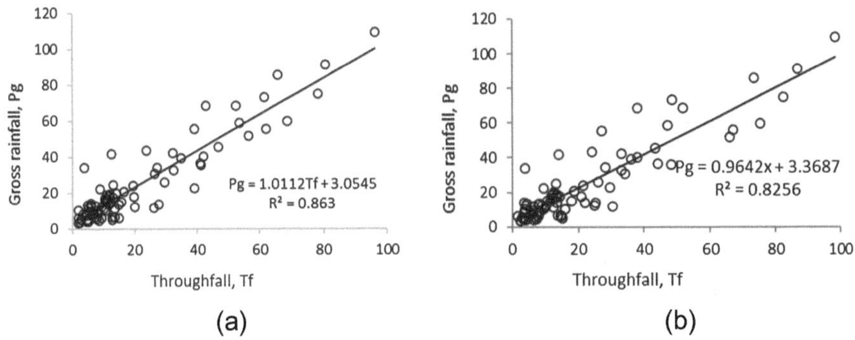

Fig. 7 Linear regression of gross rainfall against throughfall for **a** Plot 11 and **b** Plot 12

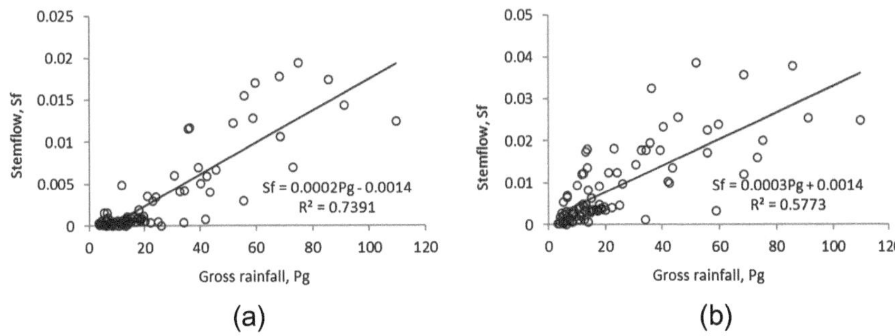

Fig. 8 Linear regressions of stemflow against gross rainfall for **a** Plot 11 and **b** Plot 12

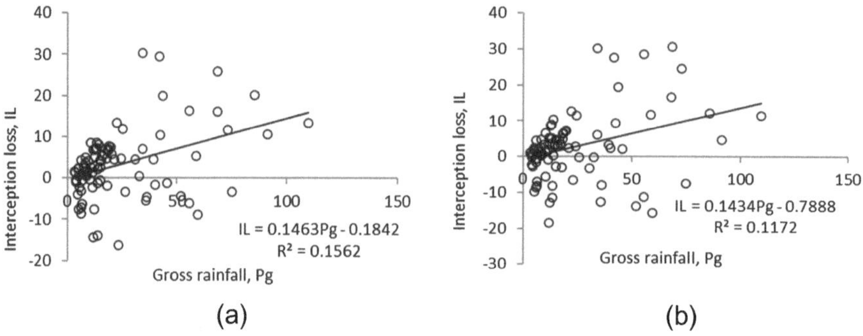

Fig. 9 Linear regressions of interception loss against gross rainfall for **a** Plot 11 and **b** Plot 12

interception loss is 285.3 mm from the original Gash, and the interception loss from forest observation is 226.8 mm, respectively. The calculated values from the model overestimated the interception loss by 58.6 mm or 25.8% of observed interception loss.

The results show that in the direct measurements, interception loss for Plot 11 contributed 13.6% of the gross rainfall and for Plot 12 produced 10.8% of the gross rainfall for the 12-month periods. Whereas, from the Gash model computation, interception loss from Plot 11 produced 14.7% and while for Plot 12, the original contributed 13.6% of the total gross rainfall during the study period. The interception loss values obtained for both plot either from measured and calculated gives a slight difference might due to the forest stand types in both plots. The different stand has typically different sizes of leaves and branches that affected the values of canopy storage capacity, S used in the modelled values.

For throughfall volume, the percentage over gross rainfall for Plot 11 is 78.7% and for Plot 12 is 80.9%. The throughfall volume consumes significant parts of gross rainfall since the precipitation drips from the canopy cover when it's over the storage for the canopy. Other than that, the location of the throughfall measurement tank is placed at each corner of the grid point in the plot area. Thus, some might receive precipitation without touching the canopy. For stemflow data, it gives the value of 0.013% and 0.034% of gross rainfall for Plot 11 and Plot 12, respectively.

Interception loss was measured based on the difference of gross rainfall and the sum of throughfall and stemflow. It can be concluded that the calculated interception loss value overestimates the direct measurement data by 9–26%, with an average of 17.2% for Plot 11 and Plot 12, calculated values overestimate by 17–45% with an average of 30.9%. Thus, the range of interception loss for this study is 10% to 16% and is closer to the interception loss study conducted by Asdak et al. (1998), Carlyle-Moses and Price (1999) and Nik et al. (1979). Asdak et al. (1998) implemented the study at the Wanariset Sangai on the upper reaches of the Mentaya river, Central Kalimantan. A study by Nik et al. (1979) was conducted at Air Hitam Forest Reserve in Puchong, Selangor that shows the site condition almost similar to this study area at Bukit Lagong Forest Reserve Kepong Selangor and the forests are classified as tropical rainforest. Carlyle-Moses and Price (1999) observed within a temperate hardwood forest plot within the Erindale Ecological Research area, University of Toronto at Mississauga, Mississauga, Ontario, Canada.

4 Conclusion

Interception loss was measured based on the difference of gross rainfall and the sum of throughfall and stemflow. During the study period, 94 rainfall events are recorded and assessable in quantifying the interception loss. The interception loss obtained from this study is 13.6% for Plot 11 and 10.8% for Plot 12, respectively. Meanwhile, the Gash model computed the interception loss in Plot 11 as 14.7% and for Plot 12 is 13.6% from the total gross rainfall. The correlation between interception loss and gross rainfall indicates that the interception loss decreases as the rainfall depth increases. In brief, this study has shown great significance to interception loss study. It can provide ground data that can be used to validate developed interception models such as Gash models. On the other hand, for further research purposes, other factors such as LAI and dbh of trees should be considered when correlating the effects of canopy cover to the interception loss in the rainforest.

References

Asdak C, Jarvis PG, Van Gardingen P, Fraser A (1998) Rainfall interception loss in unlogged and logged forest areas of Central Kalimantan, Indonesia. J Hydrol 206:237–244

Astuti HP, Suryatmojo H (2019) Water in the forest: rain-vegetation interaction to estimate canopy interception in a tropical borneo rainforest. IOP Conf Ser Earth Environ Sci 361:012035

Azinoor Azida AB, Lu M (2015) Annual canopy interception at artificial lowland tropical forest. Hydrol Earth Syst Sci Discuss 12:4879–4907

Azinoor Azida AB, Khairudin KM (2017) Preliminary study on tropical forest canopy interception. J Eng Appl Sci 12:5572–5577

Azinoor Azida AB, Baki A, Atan I, Ali F, Yusop Z, Muhammad Khairudin K (2012) Annual interception loss estimation with a simple equation. In: SHUSER 2012–2012 IEEE symposium on humanities, science and engineering research. IEEE, New York, USA, pp 1477–1482. ISBN 978-146731310-0

Baharuddin K (1989) Rainfall interception in dipterocarp forest in Peninsular Malaysia. IHP-UNESCO-FRIM regional seminar on tropical forest hydrology, Kuala Lumpur

Bo MW, Tae HK, Kyong HK (1989) Rainfall interception loss from canopy in forest catchment. IHP-UNESCO-FRIM regional seminar on tropical forest hydrology. Kuala Lumpur

Carlyle-Moses DE, Price AG (1999) An evaluation of the gash interception model in a northern hardwood stand. J Hydrol 214:103–110

Crockford RH, Richardson DP (1990) Partitioning of rainfall in a eucalypt forest and pine plantation in southeastern Australia: IV the relationship of interception and canopy storage capacity, the interception of these forests, and the effect on interception of thinning the pine plantation. Hydrol Process 4(2):169–188

Crockford RH, Richardson DP (2000) Partitioning of rainfall into throughfall, stemflow and interception effect of forest type, ground cover and climate. Hydrol Process 14:2903–2920

Dietz J, Holscher D, Leuschner C (2006) Rainfall partitioning in relation to forest structure in differently managed montane forest stands in central Sulawesi, Indonesia. For Ecol Manag 237:170–178

Dunkerly D (2000) Measuring interception loss and canopy storage in dryland vegetation: a brief review and evaluation of available research strategies. Hydrol Process 14:669–678

Dykes AP (1997) Rainfall interception from a lowland tropical rainforest in Brunei. J Hydrol 200:260–279

Eltahir EAB, Bras RL (1992) A description of rainfall interception over large areas. J Clim 6:1002–1008

Er AC, Raja Datuk Zaharaton Raja Zainal Abidin, Pereira JJ (2011) Sustainable development of Lagong Hill Forest Reserve, Malaysia. Aust J Basic Appl Sci 5(7):364–370. ISSN 1991-8178

Gash GHC (1979) An analytical model of rainfall interception by forests. Quart J R Met Soc 105:43–55

Gash JHC, Morton AJ (1978) An application of Rutter model to the estimation of the interception loss from Thetford Forest. J Hydrol 38:49–58

Ibrahim S, Parlan I, Hamzah KA, Mahmood NNZ, Harun I, Yahya AZ, Omar H, Mohammad A (2008) FRIM Tapak warisan semulajadi. Gemilang Press Sdn. Bhd, Selangor

Jackson IJ (1975) Relationships between rainfall parameters and interception by tropical forest. J Hydrol 24:215–238

Konishi S, Tani M, Sahat MM (2006) Characteristics of spatial distribution of throughfall in a lowland tropical rainforest, Peninsular Malaysia. For Ecol Manage 224:19–25

Levia DF Jr, Frost EE (2003) A review and evaluation of stemflow literature in the hydrologic and biogeochemical cycles of forested and agricultural ecosystems. J Hydrol 274:1–29

Leyton L, Reynolds ERC, Thompson FB (1967) Rainfall interception in forest and moorland. In: Sopper WE, Lull HW (eds) International symposium on forest hydrology. Pergamon Press, Toronto, pp 163–178

Lloyd CR, de Marques OFA (1988) Spatial variability of throughfall and stemflow measurements in Amazonian rainforest. Agric for Meteorol 42:63–73

Lloyd CR, Gash JHC, Shuttleworth WJ, Marques AO (1988) The measurement and modelling of rainfall interception by Amazonian rainforest. Agric Meteorol 43:141–162

Manfroi OJ, Koichiro K, Nobuaki T, Masakazu S, Nakagawa M, Nakashizuka T, Chong L (2004) The stemflow of trees in a Bornean lowland tropical forest. Hydrol Process 18(13):2455–2474

Manokaran N (1979) Stemflow, throughfall and rainfall interception in a lowland tropical rain forest in Peninsular Malaysia. Malays For 42(3):174–200

Marin CT, Bouten W, Sevink J (2000) Gross rainfall and its partitioning into throughfall, stemflow and evaporation of intercepted water in four forest ecosystems in western Amazonian. J Hydrol 237:40–57

Nepstad DC, Moutinho P, Dias-Filho MB, Davidson E, Cardinot G, Markewitz R, Figueiredo R, Vianna N, Chambers J, Ray D, Guerreiros JB, Lefebvre P, Sternberg L, Moriera M, Barros L, Ishida FY, Tohlver I, Belk E, Kalif K, Schwalbe, K (2000) The effects of partial throughfall exclusion on canopy processes, above ground production, and biogeochemistry of an Amazon forest. J Geophys Res: 1–43

Nik MM, Hamzah MB, Ahmad S (1979) Rainfall interception, throughfall and stemflow in a secondary forest. Pertanika 2(2):152–154

Pérez-Suárez M, Fenn ME, Cetina-Alcalá VM, A. Aldrete A (2008) The effects of canopy cover on throughfall and soil chemistry in two forest sites in the Mexico City air basin. Atmósfera 21:83–100

Pidwirny M (2006). Interception, stemflow, canopy drip, and throughfall. In: Fundamentals of physical geography, 2nd edn. May 2021. http://www.physicalgeography.net/fundamentals/8k.html

Roberson JA, Cassidy JJ, Chaudry MH (1989) Hydraulic engineering. Houghton Mifflin Co., Massachusetts

Roberts JM, Tani M, Bruijnzeel LA (2005) Controls on evaporation in lowland tropical rainforest. In: Bonell M, Bruijnzeel LA (eds) Forests, water and people in the humid tropics—book I. Cambridge University Press, Cambridge, pp 287–313

Sinun W, Wong WM, Douglas I, Spencer T (1992) Throughfall, stemflow, overland flow and throughfall in the Ulu Segama rain forest, Sabah, Malaysia. Phil Trans R Soc Lond B 335:389–395

van Djik AIJM, Brujinzeel LA (2001) Modelling rainfall interception by vegetation of variable density using an adapted model analytical model. Part 1. Model description. J Hydrol: 230–238

Yusop Z, Cham SY, Chong JH (2003) Throughfall, stemflow and interception loss of old rubber trees. Jurnal Kejuruteraan Awam 15:24–33

Zheng C, Jia L (2019) Global canopy rainfall interception loss derived from satellite earth observations. Ecohydrology 13(2):e2186

Evaluation of Parameters for Sustainability Assessment of Green Infrastructure in the Urban Water System

A. H. Jemat and N. A. Kamal

Abstract There are essentials on the sustainability assessment to be conducted specifically for the green infrastructure of urban water infrastructure system. The green infrastructure for an urban water system is the evolving concept that linked to the natural and designed ecosystems for stormwater control, expansion of water efficiency, dispersed or on-location treatment of wastewater source. In this study, the method used to conduct the sustainability assessment of the green infrastructure for an urban water system is Analytic Hierarchy Process (AHP). In this case, a parameter from each dimension will be computed as a co-parameter to achieve the goal of sustainable urban water management. Based on the analysis from Kuching city, the co-parameters that be used to achieve sustainable urban water management are *quality of water, water quality control, public participation,* and *public* awareness. It can be concluded that the AHP method can be used as the decision-making for the related stakeholders such as public authority or state government to strategize the planning for achieving sustainable development. As for recommendations, the parties involved should implement more green infrastructure applications such as rainwater harvesting to sustain heavy rainfall during the monsoon season or bio-retention filter to increase the water quality control.

Keywords Sustainability · Green Infrastructure · Urban Water Management

1 Background of the Study

The water system is one of the most important components in any city. When there is a growing population in a city, it is inevitable for water consumption to increase. Also, the tendency for floods to occur will happen when the water infrastructure is not able to sustain a high volume of water. Smart urban water infrastructure can be designed to gather meaningful and actionable data about the flow, pressure, and distribution of a city's water. It is crucial that the consumption and forecasting of

---

A. H. Jemat · N. A. Kamal (✉)
School of Civil Engineering, College of Engineering, Universiti Teknologi MARA, Selangor, Malaysia
e-mail: norashikin7349@uitm.edu.my

R. Hassan et al. (eds.), *Green Infrastructure*,
https://doi.org/10.1007/978-981-16-6383-3_16

water usage are accurate. Be that as it may, infrastructure that is rapidly aging, growth in population, and urbanization growth raise doubt about the momentum of urban water management strategies, particularly in the fast-growing urban regions in Asia (Larsen et al. 2016). Malaysia has one of the most composed water frameworks, with probably the most elevated rate in population access to clean water and sanitation in Southeast Asia.

Malaysia has one of the lowest water tariffs in Asia which is unlikely to be reduced in the short term. Over the last 30 years, the growing population and GDP have increased which led to heavy demand for water. Subjected to rural–urban areas migration and growing urbanization, the problem of increase in population is especially felt within the urban areas. The exponential growth in urban population has stretched the government's ability to answer infrastructure and repair needs along with providing environmental conditions required for a better living. Often the supporting infrastructure for the collection of treatment, disposal of sewage and solid wastes is insufficient to address the amounts generated. This situation raises problems of water and air pollution, public health, and urban environmental degradation. Due to the growth of population, the increased demand for the limited and diminishing supply of clean water available has been a competition among the various water users to contest the continued economic expansion that has increasingly worsened. Besides, because the readily available portion of water resources has already been developed to be used in practically all regions of major water demand, future water resources development will require the construction of more storage dams. These constructions are not only costly to build but will also affect the environment where land must be clear to make way for construction. Furthermore, the acceptable limit of surface water resource development practices has been reached in some regions of high demand, and it is necessary to think about the green infrastructure in the urban water. Thus, for achieving the status of a smart city, this study aims to determine the co-parameters in the urban water infrastructure and to evaluate the parameters for achieving the sustainability status of the urban water infrastructure at Kuching, Sarawak.

2 Related Literature Studies

Smart Cities are cities that incorporate city infrastructure with sustainable technology such as 5G connectivity, cashless economy, automated public transportation, drone distribution, energy-efficient houses, smart water and waste management treatment, and others that will improve public safety and the quality of life of our people. Mohanty et al. (2016) stated that a smart city consists of the connection of physical infrastructures, information-technology infrastructure, the social infrastructure, and also the business infrastructure to leverage the city's collective intelligence. This is a formal definition of a smart city which has been considered as a city with a greener, safer, faster, and user-friendly atmosphere. Table 1 presented the strategy taken for cities in Malaysia towards achieving smart urban water infrastructure at the ASEAN Smart Cities Network Smart City Action Plans Meeting in Singapore in 2018.

Table 1 Strategic taken based on ASEAN (2018) for selected cities in Malaysia (ASEAN 2018)

Cities in Malaysia	Dimension	Dimension description	Current or planned initiative	Based on ASEAN Singapore 2018
• Kuching, Sarawak (Smart City: Physical Infrastructure and Utilities)	• Smart City: Physical Infrastructure and Utilities	• Action 11: Provide efficient water supply services leveraging smart technologies • Action 14: Develop a flood management and response system	None	• To integrate smart technologies into a flood management system
• Kulim, Johor Bahru	• Smart Environment	• Green Development • Green Infrastructure • Clean Environment • Environmental Protection	• Shared responsibility by the private sector • An incentive for green technology and infrastructure • Introduction of green economy and carbon credits	• Upgrade existing water plants with advanced technology and smart water management systems
• Kuala Lumpur	• EPP5	• Achieve green energy by reducing waste generation via various approaches such as installing wastewater treatment plants in a wet market	• Revitalizing the Klang River	• Sustainable Water Management targets: increase the gross grey water recycling rate to 20% by 2030

Other than cities that had been proposed as Smart City which summaries in Table 1, the current city in Malaysia that have smart features are Federal Territory of Putrajaya and Cyberjaya which is located in the state of Selangor (MHLG 2018). The plan of smart city for Cyberjaya was part of Smart Selangor Blueprint. Both cities are recognized as the first cities in Malaysia to be presented with 5G technology, while Melaka city is presenting the smart metering for electricity monitoring. Thus, the state government had set up a Smart City Advisory Council for Smart City policies for all those states (International Trade Administration, 2018). Penang also initiated the plans for Smart City Blueprint specifically for Penang state by the year 2022. As for Kuching, the plans were properly made, however, the initiatives have not been implemented yet.

Concerning the 3 main pillars of sustainability (social, economic, and environmental) perspectives, the parameters for conducting the sustainability assessment had to be conducted due to the various activities that may contribute to the damages to the environment. In the Asian region, the economic growth that is rapidly developing is imposing environmental negative threats to the environmental water security and its natural resources (Chamguri and Begum 2014). This is due to the prioritization of economic development over environmental objectives by governments across the regions. As a result, environmental water security is one of the greatest concerns in the region. In terms of environmental water security, the assessment of the parameters to be used would be assessing the health of rivers and measuring the progress on the restoration of rivers ecosystems on par with the national and regional scale (Ma et al. 2016). The sustainability of economic development and improved lives depends on natural resources (Chamguri and Begum 2014).

Another dimension that needs to be considered for the sustainability assessment is the social dimension. It is one of the dimensions in urban water security particularly in Asia and the Pacific. There are around 43% of the accumulative population in Asia and the Pacific that lives in urban areas. Also, the proportion of urban locality has increased by 29% over the past 20 years (UNESCAP 2016). This social dimension is important to be considered due to social development indicators which may contribute to the sustainability issues, such as urban shrinkage. There are also few studies conducted by Slach et al. (2019) and UNEESCAP (2015) in presenting the several indicators related with the social dimension.

Urban water infrastructure is surface water diversions, wells, pumps, transmission pipes and canals, treatment and storage facilities, and distribution network elements. Sources of water are rivers, reservoirs, seawater, and groundwater. The urban water administration must guarantee appropriate administration of water system and distribution, water and wastewater treatment, and other city-related administrations. Through diversifying or permitting model establishments, the urban water industry can give water and wastewater administrations to urban areas. The urban water utilities are continually expanding the water administration chain, including (however not restricted to) the accompanying zones. For instance, Genma et al. (2014) had concluded on what should be done to improve the water sectors in various services such as wastewater treatment, reclaimed water, water supply services, and raw water administration.

3 Study Area

The study area that has been chosen for this research is the city of Kuching located in Sarawak, Malaysia. Kuching is selected for this study because the city was one of four cities that is targeted to be a smart city. As of 2020, Malaysia has an estimated population of 32.7 million and Sarawak has a population of 2.81 million in 2019 (Department of Statistics Malaysia 2020). This state was chosen due to the annual floods during the monsoon season and occasional flash floods. These floods occurred due to the current urban water infrastructure could not sustain the water flow by heavy rainfall. Based on reports done by the Department of Irrigation and Drainage (DID), Sarawak (2019), most areas that were flooded are due to the urban water infrastructure which has reached its max capacity to convey fast flowing water which led to the water flowing to lower areas causing floods. This study comprises the flow of works started with the data and information that collected from previous research studies and related official documents; data screening for the purpose to select the most suitable and relevant parameters and data analysis by using Analysis Hierarchy Process (AHP) method.

4 Parameters Identification For Assessment Of Urban Water Infrastructure

The first step into identifying parameters for assessment is through the establishment of a hierarchy system. The hierarchy consists of three levels namely goal, dimension, and parameters, respectively as summarized in Table 2. There are few studies conducted by Tomar and Borad (2012); Zhou et al. (2018); Md Yassin et al. (2011); Dong et al. (2017) and Hiessl et al. (2001) which had used the feature of Analytical Hierarchy Process (AHP) method for evaluating the urban water sustainability. With the literature data that is available for screening, it can be said that the dimensions most mentioned are environment, economy, and social with parameters related to it. The parameters that are used for measuring the sustainability of urban water infrastructure should be adequate to cover the whole dimensions. Table 2 shows common parameters according to the 3 main pillars of sustainability that were used in this study as the reference parameters based on the previous studies for assessing the sustainability of urban water infrastructure.

4.1 Analytic Hierarchy Process Method

Urban water infrastructure and sustainability assessment is a multi-parameter decision-making problem that involves multiple evaluations. In this study, the co-parameters of the sustainability assessment were evaluated by using AHP. The Analytical Hierarchy Process (AHP) is one of the most common multi-criteria methods as a decision-making tool for prioritizing parameters of different units of measurement and allows for a minor number of inconsistencies in judgments which was developed by Saaty (2008). AHP is used widely in environmental studies namely,

Table 2 Identified parameters from various literature

Goal (Level 1)	Assessment of Urban Water Infrastructure		
Dimension s (Level 2)	Environment	Economy	Social
Parameters (Level 3)	Quality of Water (Dong et al. 2017; Tomar and Borad 2012; Md Yassin et al. 2011) Ecosystem Conservation (Zhou et al. 2018; Tomar and Borad 2012; Hiessl et al. 2001) Surface Water Quality (Zhou et al. 2018; Hiessl et al. 2001) Groundwater Quality (Zhou et al. 2018; Hiessl et al. 2001) Waste Minimization (Tomar and Borad 2012)	Water Quality Control (Tomar and Borad 2012; Zhou et al. 2018) Water Resources (Zhou et al. 2018) Chemical Cost Reduction (Tomar and Borad 2012) Operation Cost (Dong et al. 2017; Hiessl et al. 2001) Maintenance Cost (Tomar and Borad 2012; Hiessl et al. 2001) Production Rate (Tomar and Borad 2012) Water-Saving (Zhou et al. 2018) Water Intake (Zhou et al. 2018)	Public Participation (Md Yaasin et al. 2011; Hiessl et al. 2001; Tomar and Borad 2012) Public Awareness (Hiessl et al. 2001) Water Coverage (Zhou et al. 2018; Hiessl et al. 2001) Water Use (Zhou et al. 2018; Tomar and Borad 2012)

urban sustainability, environmental impact assessment, energy resource allocation, and water sources security (Aboelnga et al. 2020). However, this study is based on random studies that have been conducted with the AHP method in which the parameters are accumulative of knowledge from related literature findings. The researchers screened all parameters based on the thorough literature study and categorized the parameters into the respective dimensions. Then, the parameters will be evaluated in detail through the implementation of strategic initiatives specifically for Kuching city.

This study used the AHP to assess the sustainability of urban water infrastructure in Kuching, Sarawak. An AHP Excel template that was used for this study was formulated by Goepel (2013) with multiple inputs are used. The template consists of 20 input worksheets for pair-wise comparisons that will show results in Fig. 1 along with the indication of the overall dissonance. Figure 2 shows the example of parameters input and Fig. 3 shows the level of intensity applied on the parameters. The AHP template also provides a graph analysis that does not allow alterations to be made, with the X-axis as the parameter of the dimension and the Y-axis as those shows the calculated weights with error indication (grey bar) as shown in Figs. 4 and 5 is the pairwise comparison matrix with Normalized Principal Eigenvector.

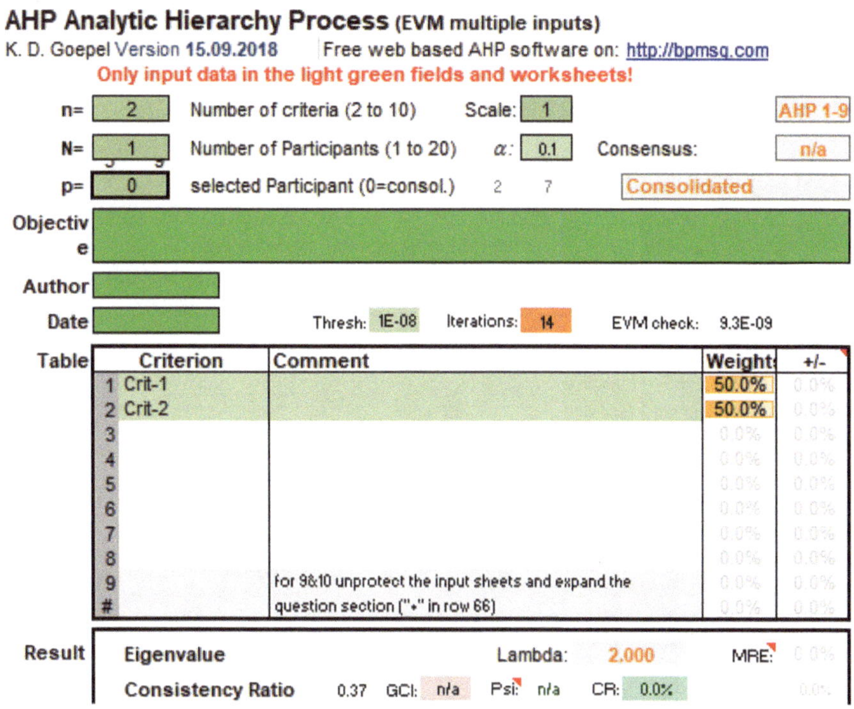

Fig. 1 Sample of input and summary sheet by using the AHP method

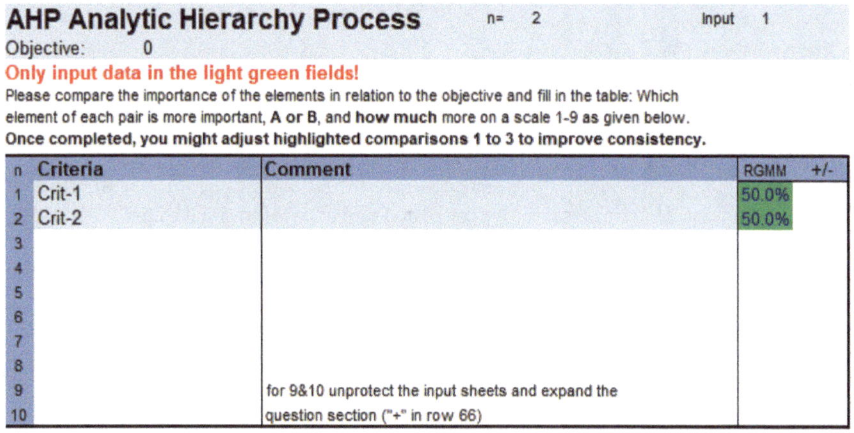

Fig. 2 Empty input sheet

Intensity	Definition	Explanation
1	Equal importance	Two elements contribute equally to the objective
3	Moderate importance	Experience and judgment slightly favor one element over another
5	Strong Importance	Experience and judgment strongly favor one element over another
7	Very strong importance	One element is favored very strongly over another, it dominance is demonstrated in practice
9	Extreme importance	The evidence favoring one element over another is of the highest possible order of affirmation
2,4,6,8 can be used to express intermediate values		

Fig. 3 Table of intensity level

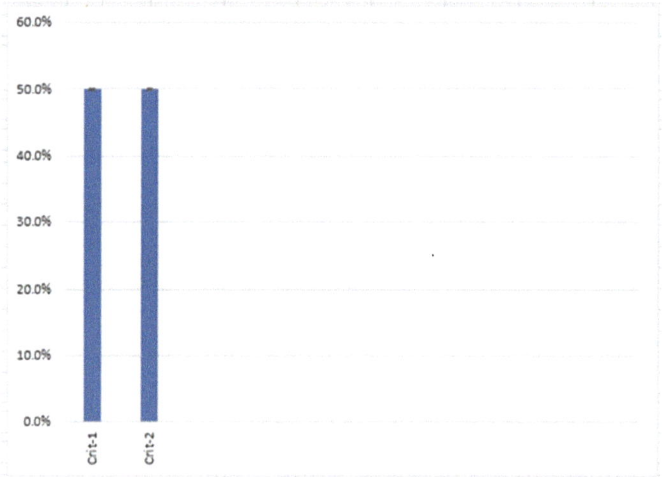

Fig. 4 Empty graph analysis presented the calculated weights with error indication

The AHP Excel Template worksheet is limited to ten (10) parameters and twenty (20) numbers of participants (Goepel 2013). In this study, there were three (3) participants namely the environment, economy, and social dimensions, which comprises the five (5) parameters for the environmental dimension, eight (8) parameters for economy dimension, and four (4) for the social dimension as shown in Fig. 6.

The importance of weightage for criteria and parameters consists of different methods to be measured. A traditional method would be the pair-wise comparison. This method is done by comparing criteria and parameters with each other and the degree of importance for each parameter specified to their respective dimension. In the AHP Excel Template, there are three (3) consistency indices that are calculated in this template namely, the Consistency Ratio (CR), the Geometric Consistency Index

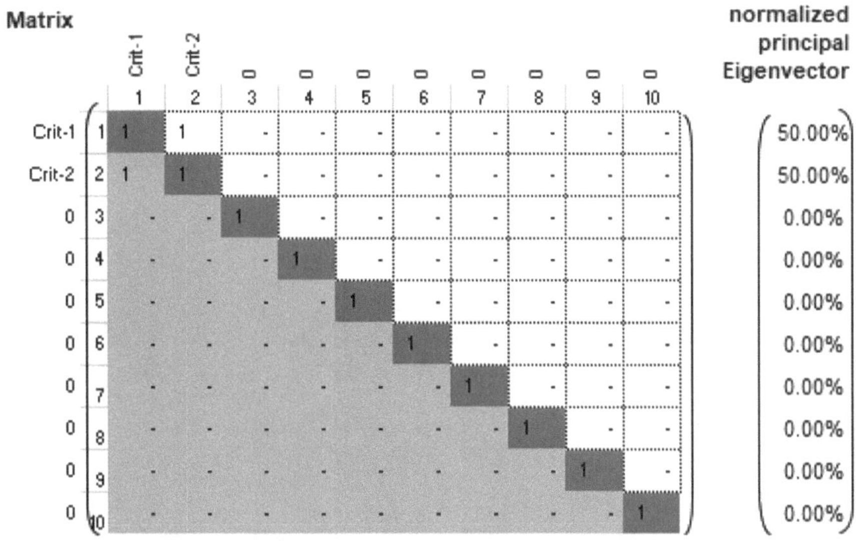

Fig. 5 Pairwise comparison matrix

(GCI), and the Overall Dissonance (Psi). The level of consistency was $\alpha = 0.1$ is implemented as a variable. If the CR exceeds the α, it will be considered unreliable and it will show the top three (3) inconsistent pair-wise comparisons on the excel sheet highlighted in red which allows alterations to scale for a consistent result.

5 Parameters Identification For Assessment Of Urban Water Infrastructure

The hierarchy structure represents the assessment of parameters for sustainable urban water infrastructure. Level 1 is the goal of achieving the sustainability of urban water infrastructure, while level 2 is the dimensions indicating the pillars of sustainability in terms of environment, economy, and social dimensions, and level 3 being its parameters categorized to their respective dimension based on Table 2. Thus, the important parameters were determined to achieve sustainability in urban water infrastructure. Details analysis is described in the next subsection.

In the environmental dimension, it is divided into five (5) parameters which represent the level of important parameters that would be best for urban water infrastructure to become sustainable. The economy dimension has eight (8) parameters that represent the effects of the urban water infrastructure when stakeholders implement the sustainability initiatives. The social dimension consists of four (4) parameters that have their role as society plays an integral part in achieving sustainability. The final priorities were presented based on the calculation of the eigenvector method (EVM)

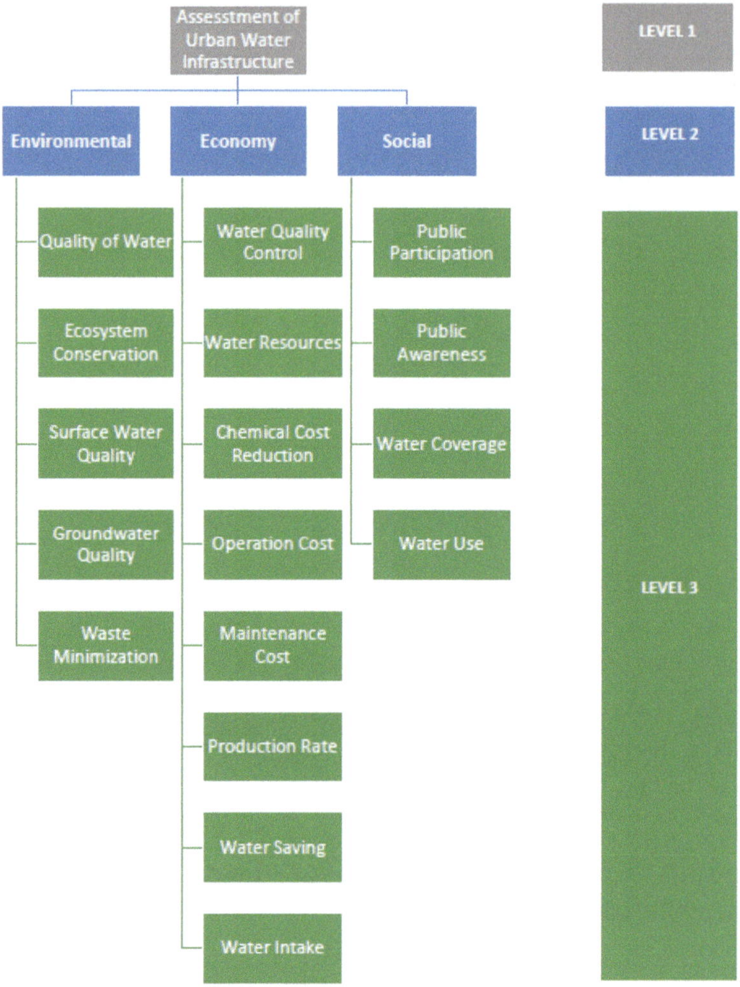

Fig. 6 AHP Hierarchy of dimensions and parameters in their category

and the power method algorithm is the solution for the eigenvalue problem with a fixed number of 20 iterations.

Different weights were implemented based on the level of intensity of smart city initiatives according to the parameters presented in Fig. 6. Results show that all dimensions that have been assessed were reliable as the consistency ratio (CR) falls below 0.1. All results were presented in a 10×10 matrix because it could not be editable to fit the parameters per dimension. The Normalized Principal Eigenvector is known as the priority vector. It is a summation of all parameters in the priority vector and is equal to 1. It is obtained when the summation between dimensions and parameters of its respective dimension.

5.1 AHP Analysis for the Environmental Dimension

Summary analysis that had been conducted in terms of environment perspective is presented in Fig. 7. Based on analysis of all parameters, the parameter with the highest weight is *quality of water* with the weight achieved is 42.1% and the lowest weight is *waste minimization* with the weight achieved is 5.8%. The assessment that had been conducted in terms of environment perspective was reliable as the consistency ratio (CR) is 0.009. The overall dissonance (Psi) is 0.0% which indicates an ordinal consistency in the pairwise comparison matrix.

The *Quality of Water* has the highest weights due to the components covered in this parameter such as initiatives that are implemented to increase the quality of water, the availability of process control of the treatment plant system in the city, and data management of water quality in terms of laboratory capability. All of this information on the components were obtained from the previous literature studies. For instance, based on the Sarawak Urban Stormwater Management (SUStoM) (DID Sarawak, 2018), the Erosion and Sediment Control Plan (ESCP) has been implemented to control erosions and trap sediments that are generated from land clearing activities, earthworks, and construction work. A detailed Guideline for Erosion and Sediment Control in Malaysia was produced by the Department of Irrigation and Drainage (DID), Sarawak, and one of the Best Management Practices (BMP) to provide standards for controlling the erosions and sedimentation of new area to protect the quality

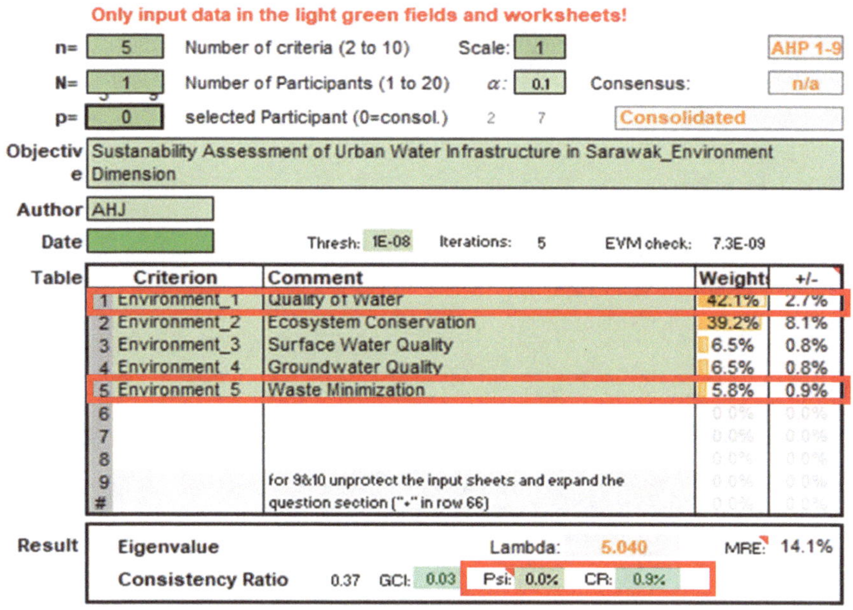

Fig. 7 Summary analysis for the environmental dimension

of water in within the urban area (Al-Ani et al. 2016). There are also study from Al-Ani et al. (2016) which took the hybrid approach to further improve on minimizing erosion and sedimentation caused by urban stormwater in Malaysian development sites. Thus, this is one of the good initiatives implemented by Sarawak State to maintain the good quality of water and reduce environmental degradation in its capital city.

A pairwise comparison matrix for the environmental dimension is shown in Fig. 8. It is the computation in a matrix form of weights that is normalized to obtain the percentage of importance for *the Quality of Water*. The importance of the *Quality of Water* is high which indicates the major contribution performance is needed to achieve sustainability in the urban water infrastructure. Figure 9 shows the weightage calculation using the AHP on the importance of parameters for assessing the sustainability in urban water infrastructure. The pairwise comparison matrix by using the Power Method presented the parameter which is more preferable than the other methods as shown in Fig. 10. The analysis had been programmed to conduct around 20 iterations for ensuring more consistent and reliable results.

Waste Minimization has the lowest weights as the scope covered for the waste minimization in urban water infrastructure is not major as compared to the *Quality of Water*. Moreover, the current system that is used to transport water is constructed along the line that is a one-time use for water, which later would be drained into the sewer and back into the Sarawak River which acts as a wastewater sink (Lai et al. 2008). This shows that little attention is paid to *waste minimization* as the system

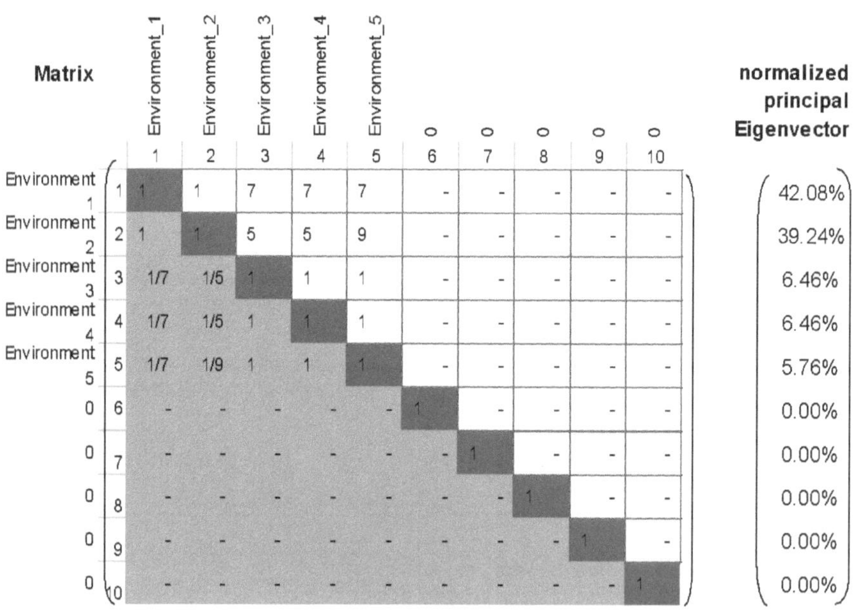

Fig. 8 Pairwise comparison matrix for environmental dimension with normalized principal eigenvector

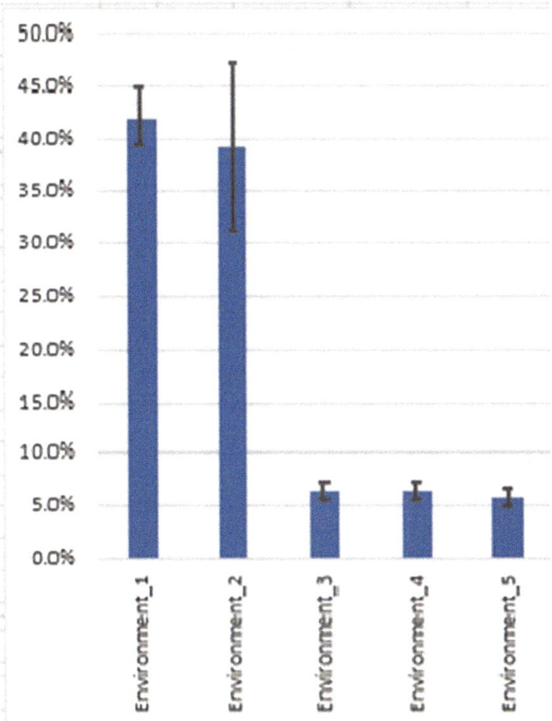

Fig. 9 Graph of calculated weights with error indication for environmental dimension

	1	2	3	4	5	6	7	8	9	10
1	1.00	1.00	7.00	7.00	7.00	-	-	-	-	-
2	1.00	1.00	5.00	5.00	9.00	-	-	-	-	-
3	0.14	0.20	1.00	1.00	1.00	-	-	-	-	-
4	0.14	0.20	1.00	1.00	1.00	-	-	-	-	-
5	0.14	0.11	1.00	1.00	1.00	-	-	-	-	-
6	-	-	-	-	-	1.00	-	-	-	-
7	-	-	-	-	-	-	1.00	-	-	-
8	-	-	-	-	-	-	-	1.00	-	-
9	-	-	-	-	-	-	-	-	1.00	-
10	-	-	-	-	-	-	-	-	-	1.00
Sum (col)	2.4286	2.5111	15	15	19	0	0	0	0	0

Iterations	
0	20
2.30	5.04
2.10	4.70
0.33	0.77
0.33	0.77
0.33	0.69
0.10	0.00
0.10	0.00
0.10	0.00
0.10	0.00
0.10	0.00

Fig. 10 Power method (dominant eigenvalue) matrix with 20 iterations made to each pair of parameters for the environmental dimension

could have been left with debris that was not discarded. In another study from India that was conducted by Madhu and Borad (2012) *Waste Minimization* was not taken into consideration to achieve sustainability of urban water infrastructure due to its less prioritization.

5.2 *AHP Analysis for the Economic Dimension*

A summary analysis of parameters for the economic dimension is shown in Fig. 11. The parameters selection was made based on the economical values of the operation system, local economic development, and reduction effects due to the pollution control activities in the urban water infrastructure at Kuching City. All related information was gathered and the important parameters were sorted based on prioritization. Parameter with the highest weights is *Water Quality Control* with 21.1% and the lowest weightage is *Water Saving* with 9.7%. Parameter of *water quality control* consists of the highest weightage due to its important effects in terms of the economic dimension in referred to few perspectives such as the sustainable cash flow on the operation process, local economic development, and cost for reduction of pollution discharge. Result for the economy dimension is reliable as the Consistency Ratio (CR) is 0.047, which is less than 0.1. The overall dissonance (Psi) is 0.0% indicates an ordinal consistency in the pairwise comparison matrix.

Pairwise comparison matrix shown in Fig. 12, represented the computation in a matrix form of weights which is normalized to obtain the percentage of importance for *Water Quality Control*. This indicated that Water*Quality Control* needs to be continuously improved to achieve urban water infrastructure sustainability. The weightage calculation by the AHP is shown in Fig. 13 which highlighted the impor-

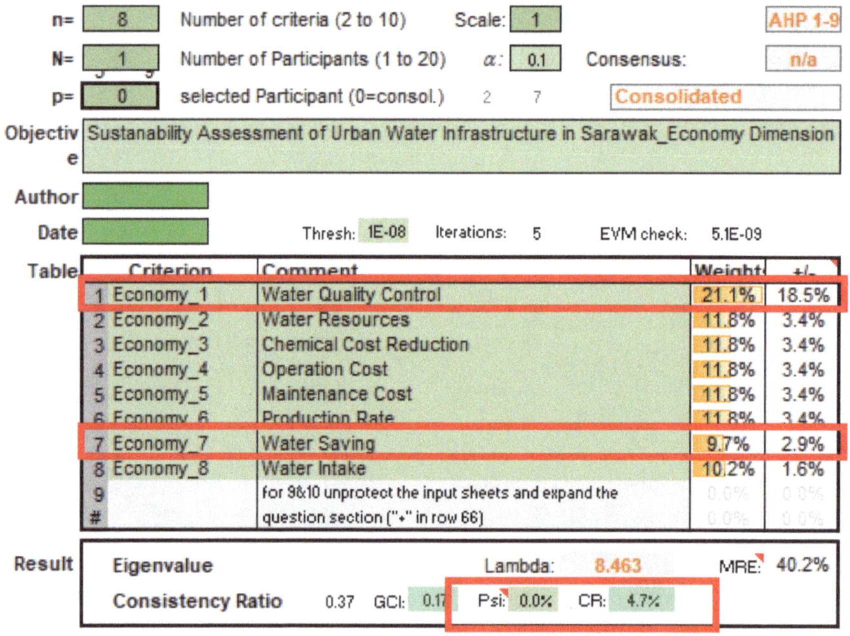

Fig. 11 Summary sheet for the economy dimension

Matrix		Economy_1 1	Economy_2 2	Economy_3 3	Economy_4 4	Economy_5 5	Economy_6 6	Economy_7 7	Economy_8 8	0 9	0 10	normalized principal Eigenvector
Economy_1	1	1	1	1	1	1	1	7	3	-	-	21.08%
Economy_2	2		1	1	1	1	1	1	1	-	-	11.82%
Economy_3	3	1	1	1	1	1	1	1	1	-	-	11.82%
Economy_4	4	1	1	1	1	1	1	1	1	-	-	11.82%
Economy_5	5	1	1	1	1	1	1	1	1	-	-	11.82%
Economy_6	6	1	1	1	1	1	1	1	1	-	-	11.82%
Economy_7	7	1/7	1	1	1	1	1	1	1	-	-	9.68%
Economy_8	8	1/3	1	1	1	1	1	1	1	-	-	10.16%
0	9	-	-	-	-	-	-	-	-	1	-	0.00%
0	10	-	-	-	-	-	-	-	-	-	1	0.00%

Fig. 12 Pairwise comparison matrix for economy dimension with normalized principal eigenvector

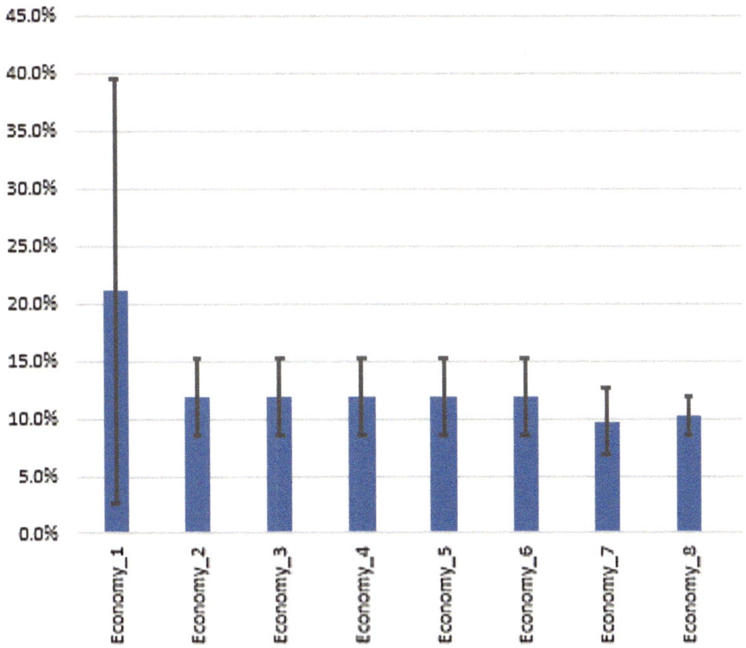

Fig. 13 Graph of calculated weights with error indication for economy dimension

tance of parameters for assessing the urban water infrastructure in terms of economic dimension.

Water Quality Control has the highest weight for the Economy Dimension as illustrated in Fig. 13. It is important to ensure the quality of water is maximized before it enters the urban water infrastructure system. This parameter works closely with *the Water Quality* parameter which analyzes in the Environment Dimension. According to the Department of Irrigation and Drainage Sarawak (2021) one of the Best Management Practices (BMP) that has been implemented is stormwater management. In relation to this, a structural and non-structural combination is under development and implementation to address the conveying and storage function of stormwater systems in the spaces provided and expansion for the required urban infrastructure.

Study conducted by Nguk et al. (2018) concluded that the water and wastewater treatment plants in Sarawak are still inadequate. The treatment plants use a traditional method to control the quality of water and the drinking water standard must comply with the National Drinking Water Standard. While the National Water Quality Standards for Malaysia is used to maintain the quality of water in Class II. Most reports from National Resources and Environment and Kuching Water Board in Sarawak that had been referred to in this study are focusing more on the water quality control for all rivers in Sarawak. With that, an allocation of RM4.5 billion will be allocated to Sarawak which will be used to upgrade various utilities including water infrastructure (Bernama 2021).

Water-Saving has the lowest weights because the parameter is more focusing on the initiative to save the water from pollution to avoid it from affecting the future generation that will use these resources. The Department of Irrigation and Drainage of Sarawak collects data for the development and management of water resources to maximize the utilization of water resources to encourage economic growth and to improve the livelihood of citizens (Department of Irrigation and Drainage Sarawak 2021).

Water Quality Control has the highest percentage of errors compared to the other parameters as shown in Fig. 13. The high discrepancy also might be contributed due to a lot of parameters in the economic dimension that needed to be considered. The analysis is programmed to conduct the 20 iterations for ensuring the result is more consistent and reliable as shown in Fig. 14.

5.3 AHP Analysis for the Social Dimension

Based on Fig. 15, parameters with the highest weights were tied between Public Participation and Public Awareness with 43.7% and the lowest weightage is between Water Coverage and Water Use with 6.2%. The result for the Social Dimension is reliable as the CR labelled with a red box is 0 which is less than 0.1. This is also due to the small parameters that have been evaluated in this analysis. The overall

	1	2	3	4	5	6	7	8	9	10	Iterations	
1	1.00	1.00	1.00	1.00	1.00	1.00	7.00	3.00	-	-	0	20
2	1.00	1.00	1.00	1.00	1.00	1.00	1.00	1.00	-	-	1.60	8.46
3	1.00	1.00	1.00	1.00	1.00	1.00	1.00	1.00	-	-	0.80	4.74
4	1.00	1.00	1.00	1.00	1.00	1.00	1.00	1.00	-	-	0.80	4.74
5	1.00	1.00	1.00	1.00	1.00	1.00	1.00	1.00	-	-	0.80	4.74
6	1.00	1.00	1.00	1.00	1.00	1.00	1.00	1.00	-	-	0.80	4.74
7	0.14	1.00	1.00	1.00	1.00	1.00	1.00	1.00	-	-	0.80	4.74
8	0.33	1.00	1.00	1.00	1.00	1.00	1.00	1.00	-	-	0.71	3.89
9	-	-	-	-	-	-	-	-	1.00	-	0.73	4.08
10	-	-	-	-	-	-	-	-	-	1.00	0.10	0.00
Sum (col)	6.4762	8	8	8	8	8	14	10	0	0	0.10	0.00

Fig. 14 Power method (dominant eigenvalue) matrix with 20 iterations made to each pair of parameters for the economy dimension

n= 4 Number of criteria (2 to 10) Scale: 1 AHP 1-9

N= 1 Number of Participants (1 to 20) α: 0.1 Consensus: n/a

p= 0 selected Participant (0=consol.) 2 7 Consolidated

Objective Sustainability Assessment of Urban Water Infrastructure in Sarawak_ Social Dimension

Author AHJ

Date 20-Nov-20 Thresh: 1E-08 Iterations: 6 EVM check: 8.0E-09

Table

	Criterion	Comment	Weights	+/-
1	Social_1	Public Participation	43.7%	0.0%
2	Social_2	Public Awareness	43.7%	0.0%
3	Social_3	Water Coverage	6.2%	0.0%
4	Social_4	Water Use	6.2%	0.0%
5			0.0%	0.0%
6			0.0%	0.0%
7			0.0%	0.0%
8			0.0%	0.0%
9		for 9&10 unprotect the input sheets and expand the	0.0%	0.0%
#		question section ("+" in row 66)	0.0%	0.0%

Result

Eigenvalue Lambda: 4.000 MRE: 0.0%

Consistency Ratio 0.37 GCI: 0.00 Psi: 0.0% CR: 0.0% 0.0%

Fig. 15 Summary sheet for social dimension

dissonance (Psi) is 0.0% which indicates an ordinal consistency in the pairwise comparison matrix.

Public participation and *public awareness* are the important parameters for achieving sustainability in urban water infrastructure in terms of the social dimension. This finding was based on the related information based on the previous report or studies related to programs and activities of the urban water system which involve public participation and awareness. According to Md Yassin et al. (2011), stakeholders or parties have to be involved and need to up the stakes and cooperate when it comes to sustainability. A study by Lai et al. (2008) stated that ecological sanitation was one of the initiatives that get public involvement and was supported by the State Government of Sarawak. More of these initiatives need to be implemented for achieving sustainability. Campaign, forum, or webinar need to be conducted more frequently for the public to be aware and participate in ensuring the sustainability of urban water infrastructure. For instance, study conducted by Amado et al., (2010) had suggested on the implementation of public participation process could allow the population's intervention which occurring from the first stage of urban planning, which in this study it reflects the urban water infrastructure system that may help in contributing to the definition, determination, and validation of its terms of reference and SWOT analysis.

The pairwise comparison matrix and calculated weight of social dimension shown in Figs. 16 and 17, respectively, indicated that *Public Participation* and *Public Awareness* are equally important based on the information from previous literature studies to achieve urban water infrastructure sustainability. Water Coverage and Water Use have the same weightage which both of these co-exist with each other in terms of the availability and consumption of water. For example, when there is a lower amount of water use, there will be higher water coverage for others. When Sarawak was experiencing the El Nino Phenomenon in 2014, the water demand from consumers became high. Consumers that come from the higher ground area had experienced issues with low water pressure. To overcome the several issues on water management in Sarawak, a single entity had been established in 2020 to improve the efficiency and effectiveness to operate and manage water supply throughout the state planned by Sarawak under the 12th Malaysia Plan (12MP) (Borneo Today 2020). Same with the analysis in environmental and economic dimensions, the analysis was programmed using AHP to conduct the 20 iterations for getting the more consistent and reliable result as shown in Fig. 18.

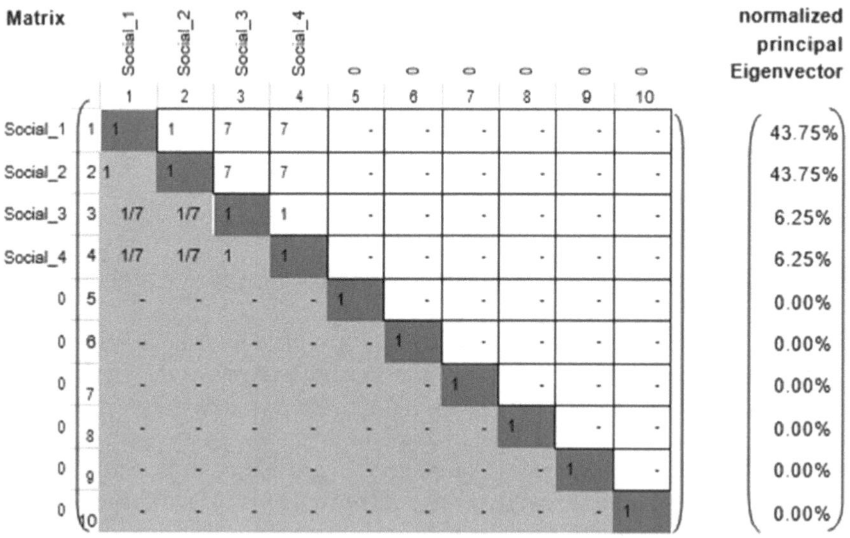

Fig. 16 Pairwise comparison matrix for social dimension with normalized principal eigenvector

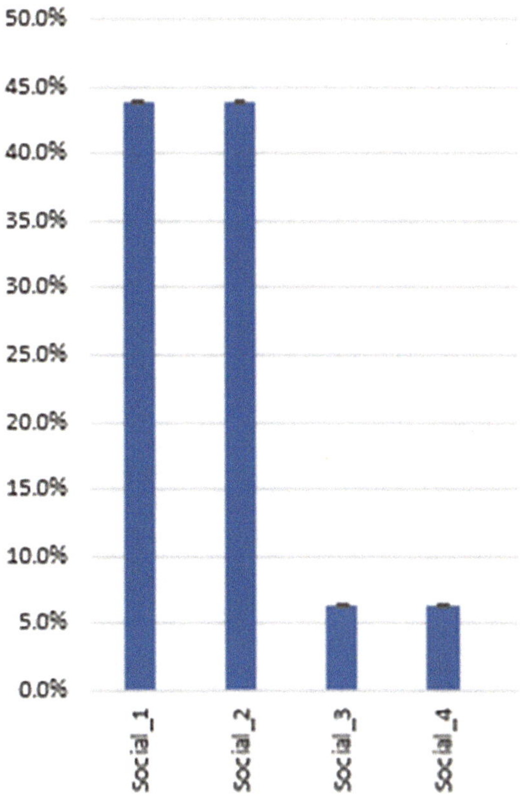

Fig. 17 Graph of calculated weights with error indication for Social Dimension

	1	2	3	4	5	6	7	8	9	10	Iterations	
1	1.00	1.00	7.00	7.00	-	-	-	-	-	-	0	20
2	1.00	1.00	7.00	7.00	-	-	-	-	-	-	1.60	4.00
3	0.14	0.14	1.00	1.00	-	-	-	-	-	-	1.60	4.00
4	0.14	0.14	1.00	1.00	-	-	-	-	-	-	0.23	0.57
5	-	-	-	-	1.00	-	-	-	-	-	0.23	0.57
6	-	-	-	-	-	1.00	-	-	-	-	0.10	0.00
7	-	-	-	-	-	-	1.00	-	-	-	0.10	0.00
8	-	-	-	-	-	-	-	1.00	-	-	0.10	0.00
9	-	-	-	-	-	-	-	-	1.00	-	0.10	0.00
10	-	-	-	-	-	-	-	-	-	1.00	0.10	0.00
Sum (col)	2.2857	2.2857	16	16	0	0	0	0	0	0	0.10	0.00

Fig. 18 Power method (Dominant Eigenvalue) Matrix with 20 iterations made to each pair of parameters for the Social Dimension

6 Conclusions

The purpose of this study was to determine the co-parameters in the urban water infrastructure for evaluating the sustainability status in Sarawak. Based on the results, the co-parameters that can be concluded as the most important parameters for conducting the sustainability assessment specifically for urban water infrastructure systems are *water quality* for the environmental dimension, *water quality control* for the economic dimension, and *public participation & public awareness* for the social dimension.

This study evaluated the sustainability of urban water infrastructure in Kuching based on the identified co-parameters. Based on the evaluation that has been done, it can be said Sarawak does have plans towards a sustainable urban water infrastructure as stated in the strategic actions taken based on ASEAN 2018 for Kuching to integrate smart technologies into the flood management system. However, it has not yet been under current and planned initiatives (Sarawak Multimedia Authority 2020).

The AHP was very useful in the decision-making study to single out the co-parameter needed for assessing the sustainability of urban water infrastructure in Kuching, Sarawak. As for recommendations, the related stakeholders can use this study as the reference to implement Best Management Practices (BMP) or green technology such as rainwater harvesting to sustain a heavy rainfall during the monsoon season or bio-retention filter to increase the water quality control, besides focusing on wastewater management system (Borneo Post 2018) and Sarawak's water supply system (Borneo Today 2020). Society should be more aware of the importance of sustainable urban water infrastructure by raising their awareness, so they can do their part in making the goal a reality.

References

Aboelnga HT, El-Naser H et al (2020) Assessing Water security in water-scarce cities: applying the Integrated Urban Water Security Index (IUWSI) in Madaba, Jordan. Sustainability J

Al-Ani IA, Mohtar WM et al (2016) Development of a hybrid expert system-multi criteria analysis on erosion and sediment control in the construction industry. https://doi.org/10.1007/s12205-016-0589-z. Accessed 1 Mar 2020

Amado M, Vitorino C et al (2010) Public participation in sustainable urban planning

ASEAN (2018) ASEAN smart cities network. https://asean.org/asean/asean-smart-cities-network/. Accessed 20 Jan 2021

Bernama (2021) Budget 2021: RM5.1 bln development allocation for Sabah, RM4.5bln for Sarawak. https://bernama.com/en/business/news_budget.php?id=1898183. Accessed Jan 2021

Borneo Post (2018) Sarawak to improve the wastewater management. https://www.theborneopost.com/2018/10/11/sarawak-to-improve-wastewater-management/. Accessed Nov 2020

Borneo Today (2020) Sarawak to have single body to manage water supply. https://www.borneotoday.net/sarawak-to-have-single-body-to-manage-water-supply/. Accessed Nov 2020

Chamguri S, Begum RA (2014) Water resources in Malaysia: issues and challenges. https://www.researchgate.net/publication/287929596_Water_resources_in_Malaysia_Issues_and_challenges. Accessed November 2020

Department of Statistic Malaysia (2020) Sarawak. https://www.dosm.gov.my/v1/index.php?r=column/cone&menu_id=clJnWTlTbWFHdmUwbmtSTE1EQStFZz09. Accessed Nov 2020

Department of Irrigation and Drainage (DID) Sarawak (2020) Sarawak Urban Stormwater Management (SUStoM). https://did.sarawak.gov.my/page-0-0-1332-SUStoM.html#tag1. Accessed Nov 2020

Department of Irrigation and Drainage Sarawak (DID) (2021) https://did.sarawak.gov.my/page-0-119-156-Water-Resources-Management.html. Accessed Jan 2021

Dong X, Du X et al (2017) Benchmarking sustainability of urban water infrastructure systems in China. https://www.sciencedirect.com/science/article/pii/S095965261732036X. Accessed Jan 2021

Goepel KD (2013) Implementing the Analytic Hierarchy Process as a Standard Method for Multi-Criteria Decision Making In Corporate Enterprises – A New AHP Excel Template with Multiple Inputs, Proceedings of the International Symposium on the Analytic Hierarchy Process, Kuala Lumpur 2013. https://doi.org/10.13033/isahp.y2013.047. https://bpmsg.com/new-ahp-excel-template-with-multiple-inputs/

Genma P, Sang Z et al (2014) Smart Water Management in Cities (Rep.). International Telecommunication Union website

Hiessl H, Walz R et al (2001) Design and sustainability assessment of scenarios of urban water infrastructure systems. http://citeseerx.ist.psu.edu/viewdoc/summary?doi=10.1.1.487.7839

Lai S, Mah DY et al (2008) Ecological sanitation, sustainable strategy as an alternative urban water source. http://redac.eng.usm.my/html/publish/2008_20.pdf. Accessed Apr 2021

Larsen TA, Hoffmann S et al (2016) Emerging solutions to the water challenges of an urbanizing world. https://science.sciencemag.org/content/352/6288/928.abstract. Accessed Jan 2021

Ma X, Xue X et al (2016) Sustainable water systems for the city of tomorrow—a conceptual framework. Sustainability 7:12071–12105; https://doi.org/10.3390/su70912071

Md Yassin AB, Bond S et al (2011) Waterfront development in Malaysia: Do we have sustainable governance? https://www.tandfonline.com/doi/abs/10.1080/14445921.2011.11104338. Accessed Jan 2021

Ministry of Housing and Local Government (MHLG) (2018) Final report: Malaysia city framework. Ministry of Housing and Local Government (MHLG)

Mohanty SP, Choppali U et al (2016) Everything you wanted to know about smart cities. https://www.researchgate.net/publication/306046857_Everything_You_Wanted_to_Know_About_Smart_Cities. Accessed Jan 2021

Nguk T, Abul Bashar Sarker M et al (2018) Inadequate water treatment quality as assessed by protozoa removal in Sarawak, Malaysia. https://www.ncbi.nlm.nih.gov/pmc/articles/PMC5995738/. Accessed Jan 2021

Privacy Shield Framework (n.d.) Malaysia-Smart Cities Malaysia-Smart Cities. Malaysia-Smart Cities | Privacy Shield. https://www.privacyshield.gov/article?id=Malaysia-Smart-Cities

Tomar MM, Borad NN (2012) Use of AHP method in efficiency analysis of existing water treatment plants. https://www.idc-online.com/technical_references/pdfs/civil_engineering/Use%20of%20AHP.pdf. Accessed Jan 2021

Sarawak Multimedia Authority (2020) Digital economy sectors: smart city (infrastructure and utilities). https://www.sma.gov.my/page-0-0-16-Smart-City.html

Saaty TL (2008) Decision making with the Analytic Hierarchy Process (AHP). Int J Serv Sci 1(1)

Slach O, Bosák V et al (2019) Urban shrinkage and sustainability: assessing the nexus between population density, urban structures and urban sustainability. MDPI. https://doi.org/10.3390/su11154142

UNESCAP (2015) The state of asian and pacific cities 2015: urban transformations Shifting from quantity to quality. ESCAP

UNESCAP (2016) Economic and social survey of asia and the pacific. https://www.unescap.org/sites/default/files/Economic%20and%20Social%20Survey%20of%20Asia%20and%20the%20Pacific%202016_0.pdf. Accessed 25 Jan 2021

Zhou J, Xu Q et al (2018) Water resources and sustainability assessment based on group AHP-PCA Method; a case study in Jinsha River Basin. https://www.researchgate.net/publication/329790744_Water_Resources_and_Sustainability_Assessment_Based_on_Group_AHP-PCA_Method_A_Case_Study_in_the_Jinsha_River_Basin. Accessed 25 Jan 2021

Freshwater Lobster Production-Green Aquaponics Perennial System (FLP-GAPS): A Green Community-Based Social Enterprise (CBSE) for B40 in Sabah

Rozita [a] Uji Mohammed, Cyril Supain [a] Christopher, Hendry Joseph, and Surail Abdul Kahar [a] Eting

Abstract Freshwater Lobster Production-Green Aquaponics Perennial System (FLP-GAPS) project combines three essential components, i.e., green, aquaponics, and perennial in its framework. FLP-GAPS adopts an eco-system that promoting a green-aquaponics-perennial system among B40 in the Sabah region. It also implements a Community-Based Social Enterprise (CBSE) as the key to its existence and *Koperasi Pembangunan Komuniti Usahawan Lobster Sabah* [KoPKULS] Berhad becomes the platform in executing the enterprise. Green CBSE FLP-GAPS, an integrated project can be defined as a subset of a green social enterprise, an independent, and a non-profit-oriented entity in freshwater lobster farming managed by the community members. This paper aims to develop a better understanding of the formation of the integrated project. Six villages in a West Coast Division of Sabah namely Kampung Kayu Madang, Telipok; Kampung Shabandar, Tuaran; Kampung Bantayan, Tamparuli; Kampung Ratau, Menggatal; Kampung Duvanson, Putatan; and Kampung Muhibbah, Petagas were selected based on the purposive homogenous sampling technique, pooling 30 participants registered under eKasih and six landlords that willing to let their lands to be utilized with zero rental to materialize the project. Based on this tripartite component, it is seen that FLP-GAPS could be an impactful and significant project to uplift the involvement of B40 towards the green CBSE.

Keywords FLP-GAPS · Social enterprise · Green CBSE · B40 · Local community

Rozita [a] Uji Mohammed (✉) · Cyril Supain [a] Christopher · H. Joseph · Surail Abdul Kahar [a] Eting
Universiti Teknologi MARA Sabah Branch, Sabah, Malaysia
e-mail: rozlim97@uitm.edu.my

Cyril Supain [a] Christopher
e-mail: cyril489@uitm.edu.my

H. Joseph
e-mail: hendry@uitm.edu.my

Surail Abdul Kahar [a] Eting
e-mail: surailab@uitm.edu.my

R. Hassan et al. (eds.), *Green Infrastructure*,
https://doi.org/10.1007/978-981-16-6383-3_17

1 Introduction

> Freshwater lobsters are considered a high-end "white tablecloth" product and global demand for red claw crayfish—native to northern Australia—is growing at a rapid rate.
>
> Nichols and Major (2019)

Malaysia's food import amount is a whopping RM50 billion every year. For a country that has vast reserved land, it is a wasted opportunity that is not to be used for food production. This condition reflects the phenomenon of the food crisis as the demand for food from outside is getting higher (The Sun Daily 2020). The demand for food in Malaysia is increasing continuously year after year and the supply of food is unable to meet the demand by its citizens. Malaysia produced 70% of its rice, vegetables (78%), beef (325), mutton (23%), and fish (86%) (DOA 2020). Malaysia imported US9.7 billion of food and beverage products in 2017 and the import of food products will continue to grow for the next ten years (GAIN 2018). One way to recover from the food crisis is by optimizing the natural resources that the state, in particular, Sabah has at the first hand which is its land. Many can be done including agriculture and inland fisheries, or a combination of these two (aquaculture) operated commercially to lessen the food crisis of Malaysia and Sabah per se. Overexploitation of marine fish causing a decrease in marine biomass resources up to 80% because of the use of trawlers in 1960.

Aquaculture is defined as the process of breeding, rearing, and harvesting fish, shellfish, algae, and other organisms in all types of water environments (NOAA 2021). As the demand for seafood has increased, technology has made it possible to grow food in coastal marine waters and the open ocean. Aquaculture is a method used to produce food and other commercial products, restore habitat, and replenish wild stocks, and rebuild populations of threatened and endangered species. Aquaculture is a nascent commercial venture in Sabah. There are two main types of aquacultures, i.e., marine and freshwater (Li and Liu 2019). For this study, freshwater aquaculture will be chosen. It is estimated that the total production from both brackish water and freshwater aquaculture is 35,000 metric tons per year. The single most important enterprise is the culture of tiger prawns in brackishwater ponds. This aquaculture produces about 60 million Ringgits worth of prawns annually. Prawn farms produce on an average of four to five tonnes per hectare of ponds. Major producers are Tawau, Semporna, Lahad Datu, and Sandakan (East Coast of Sabah). The Malaysian government had identified aquaculture to fulfil the domestic and global demand. Based on Fig. 1, Malaysia is expected to produce 1.443 metric tons of aquaculture production under the initiative of *Program Transformasi Agromakanan* in order to ensure enough food supply for the nation (Department of Fisheries Sabah 2021). This is followed by the same trend made for Sabah's fish aquaculture production from 2012 to 2019 (refer to Fig. 2). The current government of Sabah is committed to enhance the agricultural, subsector aquaculture and seek out its potential for Sabah to focus on as stated in the newly launched blueprint of Sabah, i.e., *Pelan Pembangunan Sabah Maju Jaya* (2021–2025).

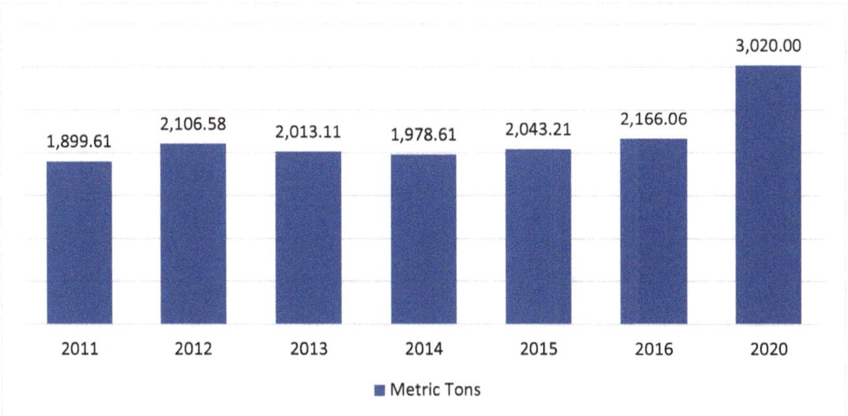

Year ('000 Metric Tons)	2011	2012	2013	2014	2015	2016	2020	RMK-10	RMK-11
CAPTURE	**1,373.10**	**1,472.20**	**1,482.90**	**1,458.10**	**1,415.00**	**1,455.00**	**1,577.00**	**-0.20%**	**2.20%**
Laut Pantai	1,086.00	1,136.20	1,156.70	1,193.00	1,100	1,100	1,100	-0.20%	0
Laut Dalam	287.1	336.1	326.2	265	315	355	477	-0.30%	8.70%
AQUACULTURE	**526.51**	**634.38**	**530.21**	**520.51**	**628.21**	**711.06**	**1,443**	**1.60%**	**18.10%**
AIR TAWAR	122.22	163.76	132.89	106.73	169.75	189.85	313	1.80%	13.00%
Ikan	121.99	163.34	132.44	106.33	169	189	311.65	1.80%	13.00%
Udang Galah	0.33	0.41	0.46	0.398	0.75	0.85	1.35	3.90%	12.50%
AIR PAYAU	164.84	139.13	127.88	168.45	168.41	191.21	400	-5.00%	18.90%
Udang Laut	67.47	55.57	49.96	61.34	75.87	85.15	189.04	-2.70%	20.00%
Ikan Laut	36.55	38.38	35.97	64.34	47.54	61.04	165.86	2.90%	28.40%
Kerang-Kerangan	60.81	45.18	41.96	42.68	45	45.02	45.1	-12.80%	0.00%
RUMPAI LAUT	239.45	331.49	269.43	245.33	290.05	330	730	6.90%	20.30%
TOTAL FISHERIES	**1,899.61**	**2,106.58**	**2,013.11**	**1,978.61**	**2,043.21**	**2,166.06**	**3,020.00**	**0.30%**	**8.10%**

Fig. 1 The performance of fishing production (2011–2020) in Malaysia

Freshwater aquaculture has been underrepresented in the mushrooming literature on global environment and food system interactions since 2000 despite its dominant contribution to aquatic food supplies and nutrition security. Freshwater aquaculture refers to raising and breeding aquatic animals such as fish, shrimp, crab, shellfish, lobster, and plants for economic purposes using ponds, reservoirs, lakes, rivers, and other inland waterways, which play an important role in the aquaculture industry. Global aquaculture production more than tripled in live-weight volume from 34 million tons in 1997 to 112 million tons in 2017 (see Fig. 3). Its production in the past few years was worth more than 40 million tons and 95% of freshwater aquaculture production is from Asia, and China is the leading country. Carp species are most cultured, but tilapias, catfishes, and a few other species are produced in large quantities. Ponds are the most common systems for aquaculture, but raceways and other flowing water systems, cages, and net pens are also important (Naylor et al. 2021).

The species under the Freshwater Lobster Production-Green Aquaponics Perennial System (FLP-GAPS) project is called Redclaw Cherax Quadricarinatus, a species of freshwater crayfish. In Malaysia, the redclaw is locally known as freshwater

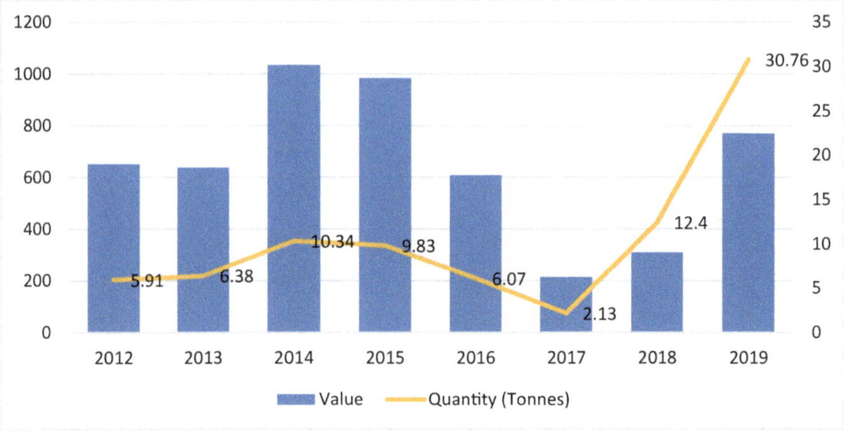

Year	Wholesale Value (RM'000)	Quantity (Tonnes)
2012	650.1	5.91
2013	638	6.38
2014	1,034	10.34
2015	983	9.83
2016	607.2	6.07
2017	212.6	2.13
2018	310	12.4
2019	769	30.76

Fig. 2 Estimated gross fish aquaculture production Sabah (2019) *Source* Naylor et al. (2021) *Note* **a** the species composition is shown for 1997 and 2017. Green, plants, and algae; blue, freshwater fish; pink, shellfish; orange, diadromous fish. **b**, **c**, Growth is shown from 1997 to 2017 for the following product categories (**b**): total, freshwater fish, algae, molluscs, and CDMM, which comprises crustaceans, diadromous fish, marine fish, and miscellaneous species and is expanded in **c**. Algae comprised more than 99% of the production weight of 'algae and aquatic plants' production in 2017. NEI is not elsewhere included for species identification in question

lobster due to its lobster-like appearance and habitat. The exact year of the redclaw introduction into Malaysia is unknown but commercial-scale culturing activity of this species was recorded since 2003 in the southern part of the Malaysian peninsula (Alimon et al. 2003). The texture and flavour of the flesh compare very favourably with other commonly eaten marine crustaceans, and having the appearance of a lobster, is positioned at the premium end of the crustacean market spectrum. It is a new and very promising aquaculture species. According to Gosbell in Nichols (2017), the industry has had its ups and downs. For instance, in Australia, an estimated 30 Queensland redclaw crayfish operators cannot keep up with the growing Australian and overwhelming international demand for freshwater aquaculture species that are raised in netted ponds to keep out predators. He claimed that freshwater lobster is now becoming popular. As seafood gets harder to get, they have become a consistent live animal that people know is fresh and clean. Many operators cannot fulfil the high

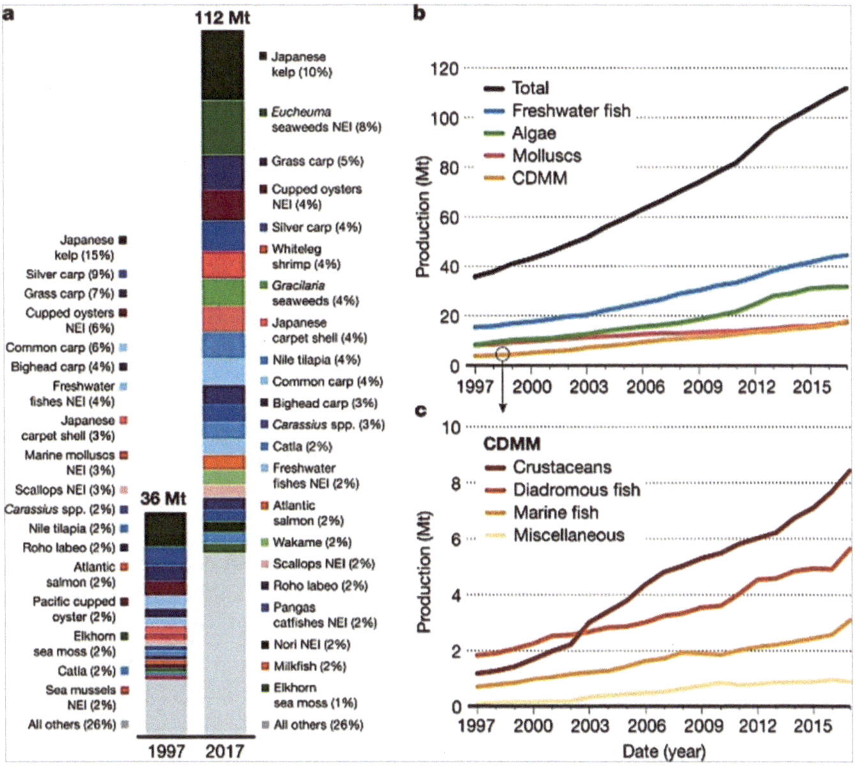

Fig. 3 Composition and growth of global live-weight aquaculture

demand for this species. This is a species with considerable potential for commercial culture done with a minimal set of tools and applied by the community to create a social entrepreneurship model. The features that made it suitable for aquaculture and the aquarium are its ability to withstand low oxygen levels and warm water associated with the tropics.

Australian Red Claw crayfish are blue to a blue, brown, or green colour (refer to Fig. 4). Because of their selective breeding in the hobby, they will not change to their original coloration, except in times of stress or when feed or housed incorrectly. The blue colour seems to intensify when the water is clean and be greener or brown when left in brackish water. The Blue Lobsters' colour intensifies as the crayfish matures. Red claw lobsters are scavengers. They are a bottom dweller and will eat anything they come across, as they are opportunistic eaters (Jones 1998).

Fig. 4 Redclaw Cherax quadricarinatus species. (*Source* Amazon.com (n.d.))

2 Freshwater Lobster Production—Green Aquaponics Perennial System Project

The need for green-based aquaculture is crucial nowadays to boost a healthier and more secure way of freshwater production. In this study, the FLP-GAPS project was introduced to fill the gap in high-end demand and a more practical way of production during a situation like a pandemic. It is officially practiced in February 2020 by a group of lecturers from Universiti Teknologi MARA (UiTM) Sabah, Malaysia named SIG-ToCoDeS (Special Interest Group-Team of Community Development Studies). The fund called *Dana Pembangunan Usahawan Bumiputera* (DPUB) Sabah was awarded by *Unit Peneraju Agenda Bumiputera* (TERAJU) to UiTM in February 2020 worth RM3.5 million. The group members of four are known as the *Pelaksana* who are responsible for running the project with three essential components, i.e., green, aquaponics, and perennial (refer to Fig. 5).

Many aquaculture systems, however, still lack the motivation to meet sustainability criteria. The FLP-GAPS promotes biosecurity in lobster hatcheries by minimizing risk factors and the presence of pathogens and their potential dispersion is key to good egg yield and post-larvae production of eggs. Generally, lobsters will go through 11 larval stages before metamorphosing to become post-larvae. During this period, balanced nutrition, optimum water temperature, suitable water quality, and ventilation must be strictly maintained (Abdullah 2020). Any sudden temperature changes can cause mortalities when lobsters are stocked, and the FLP-GAPS exists due to this reason. The three main components are:

a. Green

Green refers to any technology intended to reduce the impact of humans on the environment, economy, society, and politics. This can include technologies that reduce resource usage as well as incorporate renewable resources. Technically for this study, the usage of solar energy is crucial due to the logistic limitation. Solar energy begins with the sun. Solar panels or PV panels are used to convert light from the sun, which is composed of particles of energy called ‘photons’,

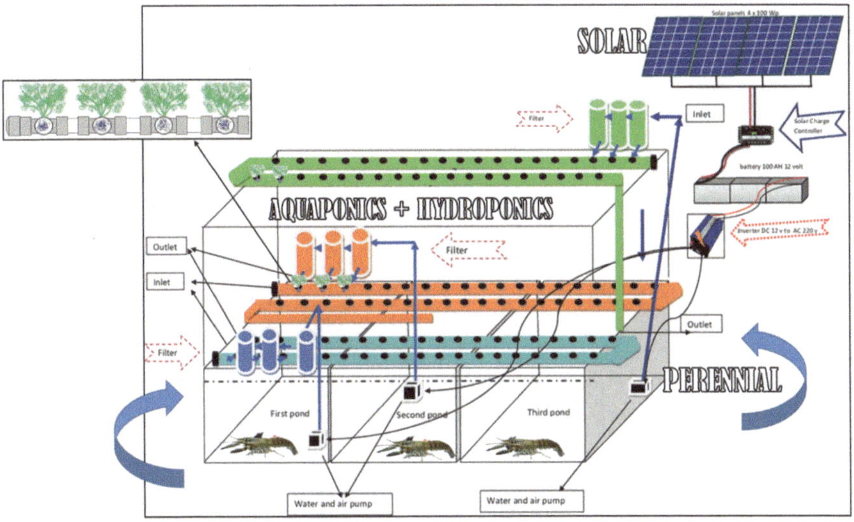

Fig. 5 FLP-GAPS model. *Source* SIG-ToCoDeS (2021)

into electricity that can be used to power electrical loads. Besides, the impact of this project is beyond expectation. The implementation of the green agenda is not only about solar power. What is more important is how this study could frame an integrated green value consisting of four major elements, i.e., green technology, green economy, green society, and green politics. According to the OECD, the World Bank and the United Nations (2012), the element of green is related to the term 'sustainable development'—*non-declining human welfare over time*. Any decision made by the government, for instance, will take into consideration the impact of its decision on humans, the environment, and other living creatures. Green practices in every aspect of life will preserve a high quality of living standard.

b. Aquaponics

Aquaponics consists of two main parts, with the aquaculture part for raising aquatic animals and the hydroponics part for growing plants. It refers to a system of aquaculture in which the waste produced by farmed lobsters or other aquatic creatures supplies the nutrients for plants grown hydroponically, which in turn purifies the water. Aquatic effluents, resulting from uneaten feed or raising animals, accumulate in water due to the closed-system recirculation of most aquaculture systems. The effluent-rich water becomes toxic to the aquatic animal in high concentrations, but this contains nutrients essential for plant growth.

c. Perennial

This means the FLP-GAPS project is lasting for a very long time or happening repeatedly or all the time. The lobster itself is a type of animal with a relatively long life span. Lobsters have a longevity of about 40 to 50 years in the wild if not caught or adversely affected by the environment. Lobsters can live long because they are organisms that can grow without aging. The shell of the lobster is very sturdy, it takes a few years to grow up and develop for a very long time. It is an animal that is very difficult to be affected by pollution sources. The system built also reflects perennial or continuous process from breeding to rearing and harvesting (internal process) up to commercialization and joint venture with the third parties (external process).

In Malaysia, the culture of redclaw crayfish or freshwater lobster has become an interest for the local population and viewed as a way to gain extra income. Culture is usually for food with target size of 6 to 8 inches of length but high demand on the juveniles prompts more focus on breeding rather than grow out activities or go for commercial. Commercial freshwater lobster farming is a very profitable business, and its production is not a very old type of business. It is relatively a new business idea, and modern commercial farming method has been used since the early twentieth century. Commercial freshwater lobster farming business yet to be practised widely, particularly in Sabah. This is mainly because of lobster's tendency towards cannibalism and the slow growth of the species. For those who have started the modern freshwater lobster farming business, many advantages gained as stated by Roy's Farm (2021) including:

a. Commercial lobster farming is relatively a new business idea. And whoever around the world can take this opportunity for making a profit.
b. Everyone can start this business and a bit distance from the coastal areas.
c. Commercial lobster production is a profitable business. So, it can be a good employment source, especially for unemployed people.
d. Both demand and price of lobsters are very good, a great advantage to go commercial.
e. Marketing lobsters is very easy. Lobsters, especially freshwater lobsters are classified as premium food and already very popular throughout the world. Hence, for commercial purposes, not so much worry on the marketing part.
f. Freshwater lobster is very healthy, tasty, and delicious. Everyone can enjoy some fresh lobsters if started with ones' very own lobster farming business.
g. The price of the crayfish turns up 9 oz going for $57 before shipping. Most of the time the fresh crayfishes are out of season due to their cycle of harvesting causing short of supplies. Still, it could be very lucrative in the right market Clayton (2013).

According to Naqiuddin et al. (2016), hatcheries and small-scale culture facilities of the redclaw were recorded in Kelantan, Terengganu, Perak, Pulau Pinang, and Negeri Sembilan. According to the owners, the broodstocks were obtained from Australia and Indonesia (international translocation) or Johor (domestic). While most of the owners suffered from high maintenance and overhead resulting inactive or closed hatcheries, the ones in Perak and Terengganu remain active. Not many studies done on the same background of Naqiuddin et al. which is based in Malaysia and researchers could not find a formal study has been done on the freshwater lobster project that is based on a commercial purpose, the benefits or advantages out of it, and the impact made from such activities. This gap has made the FLP-GAP project relevant in accommodating the government aspiration to eradicate poverty and uplift the standard of living of B40 through agricultural-based activities including freshwater lobster farming. Gunnarsson and Wingborg (2018) ensured that agriculture-based activities will build strong communities which mean social and economic sustainability. Basically, agriculture has a key role to play in food security throughout the world and lifting people out of poverty, but at the same time, there are a number of challenges to be faced. One of the key challenges is to create a solid project that integrates the local communities with the relevant stakeholders. The big hurdle to pursue on the long-term community project is to find a group of landlords allowing their lands to be used for such projects without any monthly rental paid. Therefore, a brilliant formula should be introduced to make everyone benefited from the community project. For the land incentive, the researchers have formulated Islamic *Wakalah-Mudharabah* Partnership Contract as a basis to administer the said project and a win–win concept through Islamic governance.

3 Community-Based Social Enterprise under FLP-GAPS

The implementation of the FLP-GAPS is based on the social enterprise (SE) concept. SE has recently attracted great attention due to its ability to combine social and economic goals through entrepreneurial and innovative strategies. Social entrepreneurship in Malaysia is a growing sector that has the potential to contribute to the socio-economy of the nation. Many social enterprises have been actively delivering social values and addressing social and environmental issues in the community. While they have delivered significant impact to the community and the environment, there are still many challenges and barriers in their journey to scale and increase their impact (British Council 2019). Due to those issues, this study has narrowed down its focus on a subtype of SEs that predominates in rural areas, i.e., Community-Based SEs (CBSEs). The FLP-GAPS implements a green CBSE as the key to its existence. Green CBSE FLP-GAPS project then can be defined as a subset of a green social enterprise, an independent, and a non-profit-oriented entity in freshwater lobster farming that is managed by the community members (participants), *Shariah* compliance, and committed to delivering long-term benefits to the local community.

CBSE responds to austerity and policy reforms by providing services, jobs, and other amenities for residents in underprivileged communities, thus contributing to neighbourhood redevelopment. Here, CBSE is considered as a Shariah-compliant joint venture between all parties involved, and they are called partners. This paper aims to develop a better understanding of CBSE in the rural area and sub-rural area, and how CBSE is applied in the FLP-GAPS project to achieve all its Key Performance Indicators (KPIs). FLP-GAPS project through its cooperative *Koperasi Pembangunan Komuniti Usahawan Lobster Sabah* (KoPKULS) Berhad, acts as a Special

Vehicle Purpose (SPV), a business entity that occupies an intermediate space between the ministry, the state, the market, the university, the agencies, and the civil society in ensuring the project sustainability (SIG-ToCoDeS 2021; Nyssens 2006). Moreover, the FLP-GAPS project is characterized by combining multiple goals/KPIs (social, economic, training, entrepreneurship, political, Shariah, and/or environmental) and producing 'quasi-collective' lobster-based goods and related services through the combination of a wide range of local resources in such innovative ways (Nyssens and Petrella 2015; Steyaert and Hjort 2006).

The implementation of the Green CBSE FLP-GAPS project needs support from the local people and the right area for freshwater lobster farming. The six selected areas have undergone the Logistic Suitability Study and SWOT Analysis to confirm their feasibility, amenities, land, and weather appropriateness in breeding freshwater lobsters. The land endorsement was made by the appointed consultant of the FLP-GAPS Project, i.e., Fastharvest Aquaculture, a well-experienced Bumiputera operating freshwater lobster together with the technical experts from the *Bahagian Pengurusan Fasiliti*, UiTM Sabah Branch. Figure 6 shows the locations of each area or village which are arranged from workstation 1 (WS1) until workstation 6 (WS6). In addition, the elaboration on the business concept and key activities of each WS are shown in Fig. 7. All these workstations are managed by *Pusat Pentadbiran*, *Unit Ladang* UiTM Sabah Branch, and the replica for all workstations following what is demonstrated at the *Mini Inkubator LAT* (MITLAT). As mentioned before, besides the suitability of the selected lands, the willingness of landlords to allow their lands

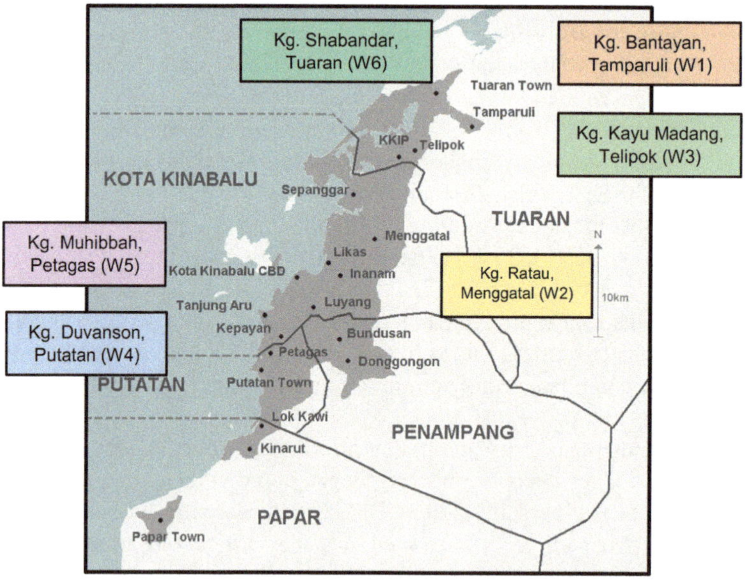

Fig. 6 The locations of All Workstations. *Source* https://commons.wikimedia.org/wiki/File:GKK_map.PNG

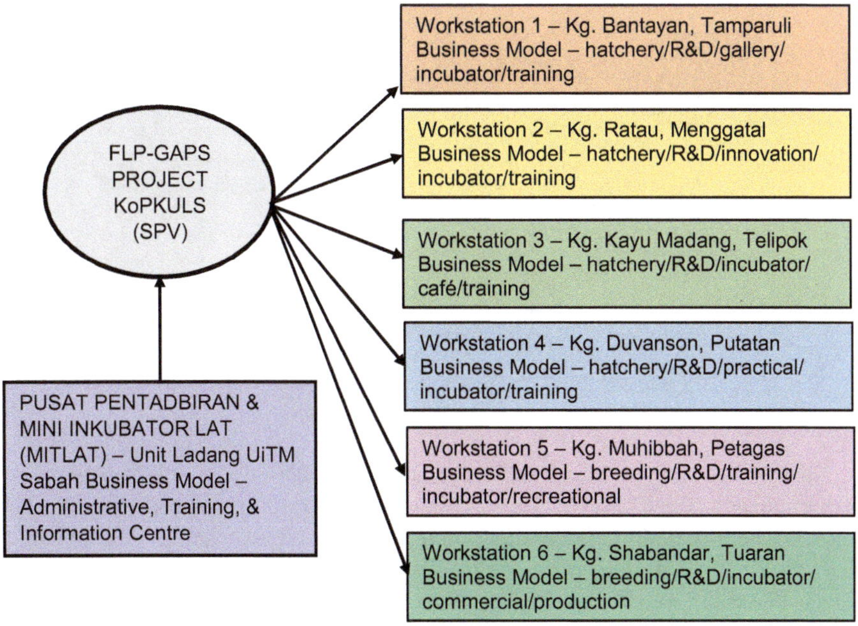

Fig. 7 Six Workstations of the Green CBSE FLP-GAPS Project. *Source* SIG-ToCoDeS (2021). *Note* The participants' profile: **a** 73% (Male), 27% (Female). **b** 13% (Pensioner), 87% (Self-Work/Unemployed). **c** 47% (Youth), 53% (Adult)

to be used for such projects without the monthly rental paid to them is part of the key success of the social enterprise project. Even though at the very first place many locations were listed and proposed by the District Officers in charge in the particular districts and also suggested by the local leaders, but at end of the day, the willingness of the landlord of that particular area became the vital factor of selection.

After the selection of land, the next step is to seek the relevant participants. The determination of six locations with five participants for each area is based on the budget given and endorsed by the Project Steering Committee (PSC) chaired by the CEO of TERAJU. Purposive homogenous is used as a sampling technique. A homogeneous purposive sample is about someone who is selected for having a shared characteristic or set of characteristics (Crossman 2020). Based on the KPI set by the fund provider, the researchers need to identify five candidates of B40 registered under eKasih program and with Bumiputera status for all six areas. Hence, this sampling technique using the homogenous samples that are created based on eKasih system and Bumiputera status. Everyone has two criteria in common. eKasih is Malaysia's national poverty database system developed by the Implementation Coordination Unit (ICU) under the Prime Minister's Department in 2008 to collect the relevant data of households living in poverty and also the vulnerable at the bottom income percentile.

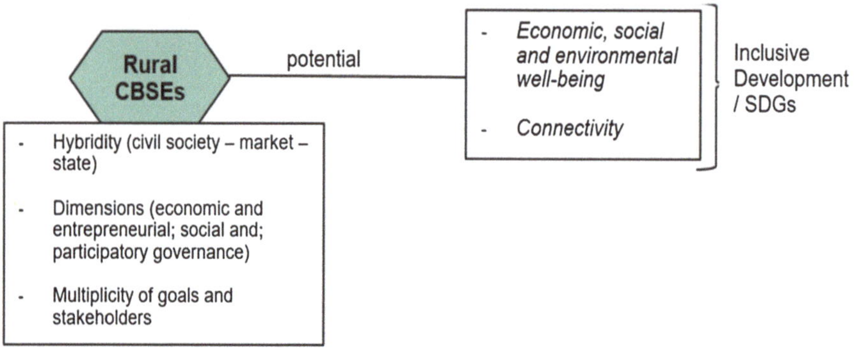

Fig. 8 Existing CBSE model in rural areas. *Source* Olmedo et al. (2019)

All areas are known as workstations located on the West Coast of Sabah involving 30 participants. The FLP-GAPS was introduced to change the socio-economic landscape of the B40 and the government-university involvement as the game changer to transformation. As illustrated in Fig. 8 which is based on the empirical research across Europe, it was stated that the SEs comprised three different dimensions—economic and entrepreneurial, social, and participatory governance (Defourny and Nyssens 2016). The indicators within the economic and entrepreneurial dimension are a continuous activity producing goods and/or selling services as the productive activity, a significant level of economic risk where the financial viability depends on the resources secured by its members and workers, and a minimum amount of paid work which there are some employed workers although normally also volunteers. The social dimension indicators are an explicit aim to benefit the community where the main mission is to serve the community or a specific group within it and promote social responsibility at the local level, an initiative launched by a group of citizens or civil society organizations as the results from collective dynamics that must be maintained, and a limited profit distribution, i.e., there is non-distribution or constrained distribution of profits.

Finally, the participatory governance is composed of a high degree of autonomy governed by the people involved in it, not by public authorities or other organizations, a decision-making power not based on capital ownership, i.e., normally one member, one vote, and a participatory nature which involve various parties affected by the activity, i.e., representation of various stakeholders in decision-making (Borzaga and Defourny 2001; Defourny and Nyssens 2012).

Looking at the practice under the FLP-GAPS project, the above model which was proposed by Olmedo et al. has been enhanced to achieve broad KPIs or objectives that might turn FLP-GAPS into the holistic CBSE model in achieving the long-term and short-term aims:

a. To increase the standard of living of the B40 by increasing job opportunity, income level, and business opportunity through FLP-GAPS. Indirectly, the FLP-GAPS project meets the need under the United Nations Sustainable

Fig. 9 United Nations Sustainable Development Goals. *Source* United Nations (n.d.)

Development Goals (SDGs). Based on Fig. 9, The FLP-GAPS project could tackle five SDGs such as SDG 1—No poverty, SDG 4—Quality education, SDG 8—Decent work and economic growth, SDG 9—Industry, innovation and infrastructure, SDG 13—Partnerships for the goals.

b. To enhance the feasibility of the areas chosen as the training centre, incubator, research centre, industrial training students, and sales centre.
c. To uplift the participants' ability to be self-funded and independent entrepreneurs.
d. To improve the knowledge and skill of participants in freshwater lobster production.
e. To promote the Islamic joint venture called *Wakalah-Mudharabah* Partnership between tripartite parties, i.e., the *Pelaksana (wakil)*, the participants *(mudharib)*, and the landlord *(mudharib)*—a win–win situation with all the stakeholders of the FLP-GAPS project through profit-sharing principle.
f. To establish a cooperative entity named *Koperasi Pembangunan Komuniti Usahawan Lobster Sabah* [KoPKULS] Berhad as a Special Purpose Vehicle (SPV) to continue the FLP-GAPS project for the benefits of the participants and the local community.
g. To penetrate the freshwater lobster market in Malaysia and create a new global market in order to strengthen up a reputation as the centre for the Freshwater Lobster hub.
h. To build engagement with the government agencies and state government to provide support in many forms to ensure the achievement of the Shared Prosperity Vision 2030 (*Dasar Wawasan Kemakmuran Bersama 2030*) and *Sabah Maju Jaya (2021 – 2025)* blueprint, particularly in the agriculture sector.

i. To boost the Bumiputera's ownership/equity in business entity based on the freshwater lobster production—fresh lobster, frozen lobster, lobster cake, lobster ball, etc.
j. To obtain accreditation and certification, particularly MyGAP (Malaysian Good Agricultural Practices) and halal to gain the public trust in the products produced for the mass market.
k. To create local experts in the aquaculture field which promote the aquaponics and hydroponics systems parallelly, in turn promote more cost-effective production of freshwater lobster.
l. To foster a green system with solar energy as the power resources—reduce the global warming effect.
m. To improve graduate employability rate at the tertiary level among the local universities.
n. To shift the teaching and learning (T&L) approach nowadays, i.e., from physical to online, from online to fieldwork platform to gain real experience, especially when dealing with the aquaculture production—promotes the life-long learning education.
o. To transform universities into the entrepreneurial universities in order to generate more fund to finance universities' expenses and expansions.
p. To transform lecturers as the entrepreneurial leaders so that they could explore more opportunities outside the university to gain more experience, exposure, and knowledge.
q. To introduce and implement different business models among the areas involved instead of as the breeding and hatchery centres. For instance, a recreational model for Kg. Muhibbah, a hipster café/restaurant for Kg. Kayu Madang, an agricultural centre for Kg. Duvanson, an innovative and research station for Kg. Ratau, a sales centre for Kg. Shabandar, and Kg. Bantayan as an aquaponics centre for the FLP-GAPS project.
r. To boost the culture of research and publication in aquaculture farming among academicians, researchers, and practitioners worldwide. The reputation of the local universities could be uplift in such a way as to be known by other prominent universities or research institutions.
s. To cultivate the entrepreneurial characteristics among the young generation and the local community to become more agile and sustainable in the situation of a pandemic.
t. To be the main player of supplying/exporting downstream and upstream products worldwide including both the Muslim and non-Muslim countries.

Therefore, the above objectives could only be achieved if it has such holistic interconnectivity among stakeholders from different bodies and the level of engagement can be illustrated as in Fig. 10:

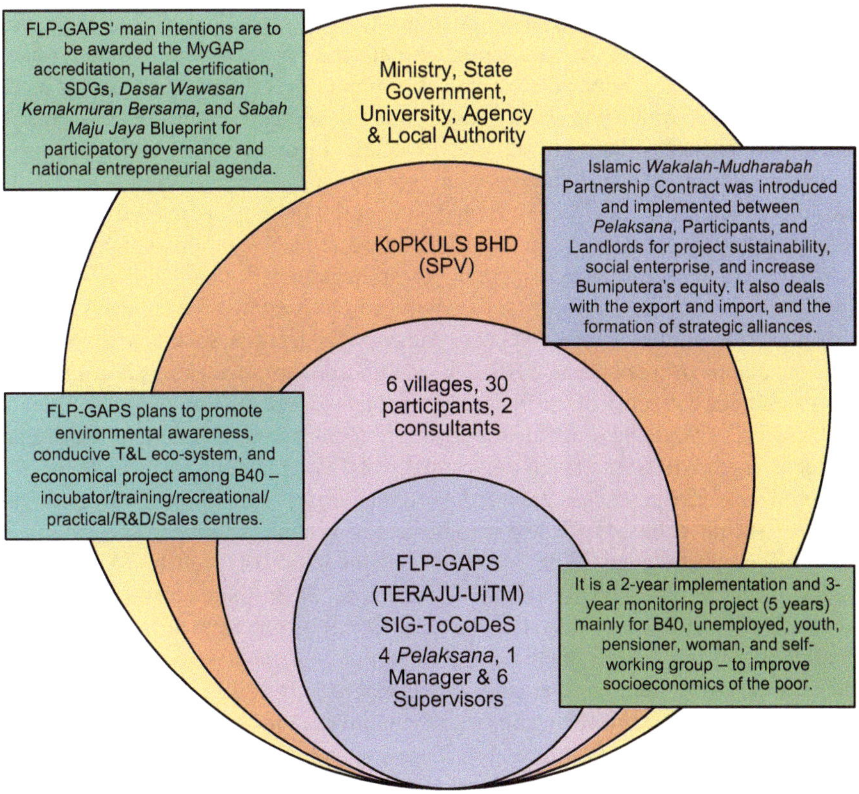

Fig. 10 The Holistic Interconnectivity of Green CBSE under the FLP-GAPS. *Source* SIG-ToCoDeS (2021). *Note* After the 3 years of execution of the project (2020-2023), the FLP-GAPS will be continued completely by the KoPKULS, of which the members are among the *pelaksana* (SIG-ToCoDeS), manager, supervisors, participants, landlords, freshwater lobster Small Medium Enterprise (SME) operators, individuals, and whoever have interest in the freshwater lobster/aquaculture farming

4 Suggestion and Conclusion

Green CBSE through the execution of the FLP-GAPS project is practically to solve the major problems related to aquaculture farm besides other socio-economic problems. Aquaculture farm's disposal contributed to the contamination of natural water bodies. The main contaminants of the wastewater effluent are suspended solids, ammonium, organic nitrogen, and phosphorus (Turcios and Papenbrock 2014) and most of the effluent was not treated (Baruah et al. 2006; Boyd 2003; Bunting 2004). The untreated effluent causing oxygen depletion and increased algal blooms among others (Boyd 2003). The removal of contaminants in soil and groundwater happens naturally by various plants and this process is called phytoremediation (Hegazy et al. 2011; Mojiri 2012). By implementing the FLP-GAPS project, the concept of it using the aquaponics system, i.e., the integration of aquaculture and hydroponics whereby the aquaculture wastes are used as a nutrient source of hydroponic production beds for growing vegetables. This system reduces fertilizer costs and is more environmentally sustainable. Furthermore, one of the major recurring problems in aquaculture is energy, i.e., fuel and electricity that incurred in the production cost. As mentioned before, the issue regarding energy can be reduced by using renewable energy sources. Renewable energy is an energy that is generated from natural processes that are continuously replenished for example by sunlight which is known as solar energy and considered as 'green power' because it does not pollute the environment. In aquaculture farming, solar energy can be used in different ways such as solar power generators, solar water heat systems, solar aerators, solar feed dispensers, and solar pumps (Bharathi et al. 2019). The multipurpose of solar energy is going to be implemented in five areas of the FLP-GAPS project except for Kampung Shabandar, the last workstation, due to its function is more focusing on producing downstream products which will be marketed domestically and internationally.

In conclusion, as discussed before, the original model of CBSE has been modified in order to achieve the 20 KPIs and with the intention to be recognized as the holistic interconnectivity of green CBSE under the implementation of the FLP-GAPS project (Green CBSE FLP-GAPS) in the six identified areas. In the execution stage, the green CBSE will be represented by the KoPKULS as the SPV to materialize the whole KPIs in turn achieving the local community, state, and national's aspirations. The adaptation of the three elements as mentioned before, i.e., green-aquaponics-perennial system are seen as parallel with the motives of the green CBSE formation. It is considered as a mega project for freshwater lobster production in the Sabah region, many have seen this as a catalyst to the existing industry players who majority produce in a small volume and cannot fulfil or meet the demand of the market not only in Sabah but also for the whole Malaysia.

> ...two key elements, economic growth and equity, and it will seek to foster income growth among the bottom 40 percent of a country's population. Without sustained economic growth, poor people are unlikely to increase their living standards. But growth is not enough by itself. Improvement in the Shared Prosperity Indicator requires growth to be inclusive...
>
> The World Bank

The FLP-GAPS exists to liaise with the policy of the federal government named Shared Prosperity Vision 2030. The statement made by the World Bank above was inserted in the vision booklet, particularly to address the 40 percent of the poor group. Based on Fig. 11, it can be seen that the vision is so much inclusive as compared to the previous Vision 2020 with three objectives, 15 guiding principles, and eight enablers to achieve the said vision.

FLP-GAPS fulfils the first layer of the above framework of the Shared Prosperity Vision 2030 restructuring the microeconomic towards more progressive, knowledge-based, and high-valued with the participation from the local community of B40. The second objective is addressing wealth and income disparities – FLP-GAPS initiates a project to close the gap of income disparities particularly among the Bumiputera coming from the underprivileged group in six rural/sub-urban areas regardless of

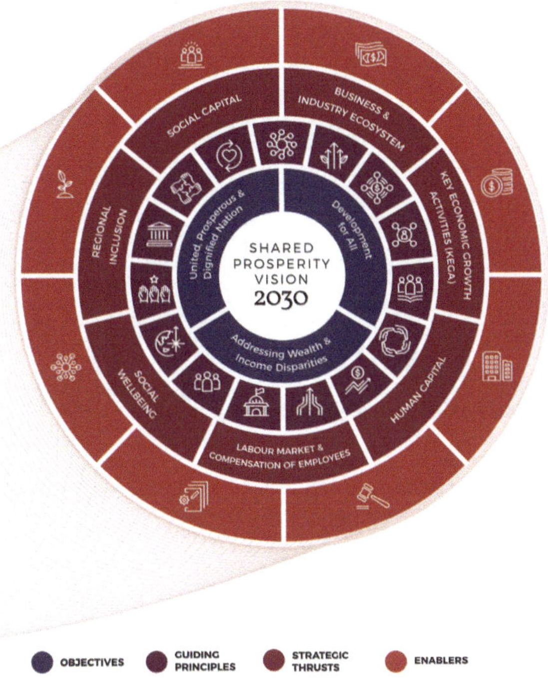

Fig. 11 Framework of Shared Prosperity Vision 2030. *Source* Ministry of Economic Affairs (2019)

their believes and ethnicities. The last objective related to 'building Malaysia as a united, prosperous and dignified nation and subsequently becoming an economic centre of Asia' is in line with the long-term objective of FLP-GAPS to become the centre of a freshwater lobster in the Sabah/Malaysia region and the main player of supplying downstream and upstream products worldwide.

Finally, as in Fig. 12, FLP-GAPS meets the newly launched blueprint of Sabah state named *Pelan Pembangunan Sabah Maju Jaya (2021-2025).* Three pillars in addressing a way forwards for Sabah are the focused prime sectors including agriculture, industrial, and tourism. It is clearly stated in the blueprint that shrimp farming is one of the potential industries for Sabah to increase its Gross Domestic Product (GDP) rate. The second pillar is developing the human capital to uplift the standard of living and in the end achieving prosperity in life. As far as the knowledge transfer is concerned, the focal points as the training and practical centre, FLP-GAPS supports lifelong learning among school leavers, graduates, and adults. Learning a skill in aquaculture farming up until the arts of JV and commercialization is so much crucial to reserve sustainability in entrepreneurship.

Fig. 12 *Pelan Pembangunan Sabah Maju Jaya* (2021-2025). *Source* Jabatan Perkhidmatan Awam Negeri Sabah (2021)

After all, the last pillar is important to penetrate new ways of doing a 'healthy' product and 'environment-friendly' kind of production system which support the green, aquaponics, and perennial system, and at the same time providing basic needs to the underprivileged society for them to self-sustain. The basic need is not only the breeding/hatchery centre as their workplace but also the certifications required for the project to be endorsed halal and MyGAP.

Acknowledgements The completion of this preliminary study pertaining to the Freshwater Lobster-Green Aquaponics Perennial System (FLP-GAPS) project together with the concept of Green Community-Based Social Enterprise (CBSE) which has lifted the national agenda of Social Entrepreneurship. Social enterprise plays a noteworthy role in society as it endlessly delivers social value very much needed by the community. This project was inspired by the Malaysian Social Enterprise Blueprint in 2015, a three-year blueprint that entailed a pivotal strategy to promote and accelerate the development of the local community to be self-sustaining, equitable, and people-centric in order to empower impact-driven entrepreneurs. A zillion of thanks to TERAJU, a strategic unit to spearhead progress through meaningful, resilient, and sustainable Bumiputera economic participation for awarding us with the *Dana Pembangunan Usahawan Bumiputera* (DPUB) Sabah in 2020, which will be ended up in 2024. We are also very much grateful for all the support from the management/Chancellery, legal advisor office, *Bahagian Pengurusan Fasiliti, Bendahari, Unit Ladang*, and *Bahagian Hal Ehwal Akademik* of Universiti Teknologi MARA in ensuring the execution of the FLP-GAPS project to be run smoothly. A special thank goes to YAB Datuk Seri Panglima Haji Hajiji bin Haji Noor, Chief Minister of Sabah for supporting our FLP-GAPS project located in six areas of the West Coast Division of Sabah, Malaysia. To our beloved families, nothing much we can say, and hope Allah will always bless them all for their undivided love, understanding, and support. May our journey to achieve a social entrepreneurial community based on freshwater lobster production will be a reality, increasing the standard of living among the B40 and make them independent throughout the rest of their life.

References

Abdullah D (2020) Freshwater Aquaculture: overcoming challenges of prawn farming in Rajshahi. Dhaka Tribune. https://www.dhakatribune.com/bangladesh/nation/2020/10/15/freshwater-aquaculture-overcoming-challenges-of-prawn-farming-in-rajshahi. Accessed 2 Apr 2021

Alimon AR, Roustaian P, Saad CR, Kamarudin MS (2003) Lipid content and fatty acid composition during early and late embryonic development of Redclaw Crayfish, Cherax quadricarinatus (Crustacea, decapoda). J Appl Ichthyol 19:397–398

Amazon.com (n.d.) Live aquaponics 10 live Australian Red Claw Crayfish 2+ Inch, Freshwater Crayfish Species. Amazon. https://www.amazon.com/Live-Australian-Crayfish-Freshwater-Species/dp/B0179ZDT1K. Accessed 22 Apr 2021

Bailey N, Kleinhans R, Lindbergh J (2018) An assessment of community-based social enterprises in three European countries: case study appendix. Power to Change Research Institute, England

Baruah K, Norouzitallab P, Sorgeloos P (2006) Seaweeds: an ideal component for wastewater treatment for use in aquaculture. Aquac Eur 31:3–6

Bharathi S CA, Uma A AB, Aanand S, Somu Sunder Lingam R (2019) Application of renewable energy in aquaculture. Aqua International 48–54

Borzaga C, Defourny J (2001) The emergence of social enterprise. Routledge, London

Boyd CE (2003) Guidelines for aquaculture effluent management at the farm-level. Aquaculture 226:101–112

British Council (2019) The state of social enterprise in Malaysia 2018. British Council, Ministry of Entrepreneur Development, United Nations Economic and Social Commission for Asia and the Pacific, and Yayasan Hasanah. https://www.britishcouncil.org/sites/default/files/the_state_of_social_enterprise_in_malaysia_british_council_low_res.pdf

Bunting SW (2004) Wastewater aquaculture: perpetuating vulnerability or opportunity to enhance poor livelihoods? Aqua Resourc Cult Dev 1(1):51–75

Clayton T (2013) Raising aquaponic shrimp in freshwater. https://gardenculturemagazine.com/raising-aquaponic-shrimp-in-freshwater/ Accessed 5 April 2021

Crossman A (2020) Understanding purposive sampling: an overview of the method and its applications. https://www.thoughtco.com/purposive-sampling-3026727. Accessed 24 Apr 2021

Defourny J, Nyssens M (2016) Fundamentals for an international typology of social enterprise models, ICSEM Working Papers No. 33. Liege: The International Comparative Social Enterprise Models (ICSEM) Project

Defourny J, Nyssens M (2012) El enfoque emes de empresa social desde una perspectiva comparada, ciriec-españa. Revista De Economía Pública 75:7–34

Department of Fisheries Sabah (2021) The fisheries industry in Sabah - a short profile. https://fishdept.sabah.gov.my/?q=en/content/fisheries-profile. Accessed 2 April 2021

DOA (2020) Statistics of food crops in Malaysia 2020. Department of Agriculture, Kuala Lumpur

Farming Crawfish (2019) Live freshwater lobster. https://www.farmingcrawfish.com/crawfish_redclaw_crayfish/lobster.html. Accessed 1 Apr 2021

GAIN (2018) Global agricultural information network: Malaysia retail foods. Report Number My8004

Gunnarsson A, Wingborg M (2018) Reducing poverty through agriculture. WE Effect Stockholm, Sweden

Hegazy AK, Abdel-Ghani NT, El-Chaghaby GA (2011) Phytoremediation of industrial wastewater potentiality by Typha Domingensis. Int J Environ Sci Technol 8(3):639–648

Jabatan Perkhidmatan Awam Negeri Sabah (2021) Ringkasan eksekutif hala tuju Sabah Maju jaya: pelan pembangunan Sabah maju jaya 1.0 (2021–2025). Jabatan Cetak Kerajaan Sabah

Jones C (1998) Redclaw crayfish in Hyde KW (ed.) The new rural industries: a handbook for farmers and investors. Rural Industries Research and Development Corporation, Canberra

Li D, Liu S (2019) Water quality monitoring and management: basis, technology and case studies, 1st edn. Elsevier Inc., China

Ministry of Economic Affairs (2019) Shared prosperity vision 2030. Attin Press Sdn. Bhd, Selangor

Mojiri A (2012) Phytoremediation of heavy metals from municipal wastewater by typhadomingensis. Afr J Microbiol Res 6:643–647

Naqiuddin AS, Rahim A, Mohd KA, Long S, Nicholas FFF (2016) The spread of the Australian redclaw crayfish (cherax quadricarinatus von martens, 1868) in Malaysia. J Sustain Sci Manag 11(2):31–38

Naylor RL, Hardy RW, Buschmann AH, Bush SR, Cao L, Klinger DH, Little DC, Lubchenco J, Shumway SE, Troell M (2021) A 20-year retrospective review of global aquaculture. Nature 591:551–563

Nichols J, Major T (2019) Crayfish could be next boom seafood industry with breakthrough breeding technology. https://www.abc.net.au/news/rural/2019-12-22/cutting-edge-australian-technology-to-power-crayfish-aquaculture/11806194. Accessed 1 May 2021

Nichols J (2017) Redclaw crayfish industry struggles to meet booming demand, but dry conditions make it hard. https://www.abc.net.au/news/rural/2017-09-14/redclaw-crayfish-demand-boom-dry-weather-impact/8895726. Accessed 5 Apr 2021

NOAA (2021) What is aquaculture? https://oceanservice.noaa.gov/facts/aquaculture.html. Accessed 28 Mar 2021

Nyssens M (2006) Social enterprises at the crossroads of market, public policies and civil society. Routledge, London

Nyssens M, Petrella F (2015) The social and solidarity economy and Ostrom's approach to common pool resources: towards a better understanding of institutional diversity? in civil society, the third sector and social enterprise: governance and democracy. Routledge, London

Olmedo L, Twuijver MV, O'Shaughnessy M (2019) Community-based social enterprises fostering inclusive development in peripheral European rural areas. UNTFSSE, Ireland

Roy's Farm (2021) Lobster farming: guide for starting the business for beginners. https://www.roysfarm.com/lobster-farming/. Accessed 20 June 2021

SIG-ToCoDeS (2021) Freshwater lobster production-green aquaponics perennial system (FLP-GAPS): 2020 progress report. Universiti Teknologi MARA, Sabah

Steyaert C, Hjort D (2006) Entrepreneurship as social change. Northhamptom Edwar Elgar, Cheltenham

The Sun Daily (2020) Food crisis may hit Malaysia, says ISI. https://www.thesundaily.my/local/food-crisis-may-hit-malaysia-says-isi-IN2395219. Accessed 26 June 2021

The OECD, The World Bank, The United Nations (2012) Incorporating green growth and sustainable development policies into structural reform agendas. the G20 summit. https://www.oecd.org/g20/topics/energy-environment-green-growth/G20_report_on_GG_and_SD_final.pdf. Accessed 25 June 2021

Turcios AE, Papenbrock J (2014) Sustainable treatment of aquaculture effluents—what can we learn from the past for the future? Sustainability 6:836–856

United Nations (n.d.) Sustainable development goals. https://sdgs.un.org/goal. Accessed 5 Apr 2021

Wikipedia (n.d.) Greater Kota Kinabalu. https://en.wikipedia.org/wiki/Greater_kota_kinabalu. Accessed 25 June 2021

An Edible Cutleries Using Green Materials: Sorghum Flour

Mohd Hafizalrisman Kabir, Nuramidah Hamidon, Mariah Awang, Mohammad Ashraf Abdul Rahman, and Suraya Hani Adnan

Abstract Plastic cutlery may be a convenient alternative and detrimental to the health. The presence of toxins and carcinogens that can quickly be ingested into the body is a petroleum by-product. As toxins and carcinogens can leech into food through the natural ecosystem, the processing of plastic cutlery has become harmful for the environment. Therefore, to minimize the negative impact towards the environment, and to support the green materials product, this study proposes an edible cutlery from sorghum flour. This study aims to calculate the exact quantity of the product used to manufacture edible cutlery, to identify the expense of making edible cutlery, and to compare the ability of edible cutleries and plastic cutleries to resist water absorption test and soil burial test. The edible cutleries are produced from the combination of natural ingredients like green materials, which are sorghum flour, wheat flour and rice. The tests involved in this study are soil burial test and water absorption test. Thus, from this study, it can minimize the negative impact towards the environment and be able to replace the conventional plastic cutleries. For commercialized purposes in future, the edible cutlery should be moulded in perfect shape and size to get the same quality for every single unit.

Keywords Edible cutleries · Plastic cutleries · Biodegradable cutleries · Green materials · Sorghum flour · Water absorption · Soil burial test

M. H. Kabir · N. Hamidon (✉) · M. Awang · M. A. A. Rahman · S. H. Adnan
UTHM Kampus Pagoh, Hab Pendidikan Tinggi Pagoh, KM1, Jalan Panchor, 84,600 Johor, Panchor, Malaysia
e-mail: nuramidah@uthm.edu.my

M. H. Kabir
e-mail: cn170005@siswa.uthm.edu.my

M. Awang
e-mail: mariah@uthm.edu.my

M. A. A. Rahman
e-mail: ashrafr@uthm.edu.my

S. H. Adnan
e-mail: suraya@uthm.edu.my

R. Hassan et al. (eds.), *Green Infrastructure*,
https://doi.org/10.1007/978-981-16-6383-3_18

1 Introduction

Cutlery has been one of the most basic, but very effective, food consumption devices produced and used worldwide. It is claimed that spoons are one of the oldest eating facilities that human beings have used and are made of natural elements such as wood, animal bones, seashells. The first recorded evidence of spoons was in England far back in the year 1259 (Michel et al. 2015).

At that time, spoons were not merely used for eating, but often used to represent wealth and influence in ceremonies. By the turn of the eighteenth century, forks and knives for consuming food were also added. Silver was the most common metal for cutlery since, before the advent of stainless steel, it was non-reactive to most foods. For most of the cutlery, stainless steel was the chosen metal, as it was simple to maintain, non-reactive and durable. It significantly lowered the prices of cutlery with the introduction of plastics into the market and made its availability very simple at the same time. For people to choose from, a lot of variations and sizes were added, such as cups, bowls, spoons, forks, knives, etc. The price of stainless-steel cutlery today is much higher than plastic or edible cutlery (Patil and Sinhal 2018).

The use of these plastics and the issue of disposing of them is a major problem that is currently being seen in our world (Parker 2019). In this study, natural edible cutleries will be produced, and it is fully biodegradable and does not require any special conditions. This study is to support the green materials and sustainability of products that will be used by people in the twentieth century. Moreover, at the end of the meal, the edible cutleries can be eaten. This could theoretically lead to plastic cutlery being substituted, but not traditional metal cutlery.

The idea of edible cutleries was first introduced in India as an advertisement product in 2010 by an organization called Bakey's (Reddy 2016). The merchandise has been further produced and the company is now able to supply 50,000 units a day as of 2016 and has earned worldwide orders of 25 million units. Although bamboo-based, sugar-based, and corn-based spoons are reported in numerous reports, this study will only specialize in cutleries based on sorghum. Sorghum is an ancient African crop that utilizes little water and its super absorbent properties for cultivation (Rajah 2018). According to Mr. Narayana Peesapaty, a former ICRISAT researcher, sorghum needs 60 times less water than rice (TwentieFour 2018). Sorghum-based edible cutleries are also a potential competitor to the single plastic cutlery that plagues the ecosystem of the planet. Single-use plastic cutlery contributes 4.24 percent of the marine litter on European beaches in line with the figures provided by the ecu Commission in 2016 (Welle 2018).

Sorghum is the key ingredient used in this study and compared to rice, it takes 60 times less water to develop sorghum (Erik Loots 2018). In 95% of the world's arable land, the crop has the power to rise (Munir 2017).

Sorghum has a super absorbent capacity that makes it extremely flexible to use as edible cutleries, it would not only be suitable for rice or wheat-based cuisine that is popular in Bangladesh, but it will complement frozen dessert, yogurt, and soup type well, because it does not degrade in hot or cold liquids. Benefit of this sorghum, it is green materials and biodegradable (Carson and Sun 2000).

The specific objective of this study is to produce edible cutlery that is suitable by studying the proportion of the materials quantity of the product research. Then, the study also calculates the total cost of making edible cutlery. Finally, this material product is tested by comparing the ability of edible cutlery and plastic cutlery to resist water absorption test and soil burial test.

The scope of this study is to produce edible cutlery from natural green material organic ingredients, where sorghum flour, wheat flour, and rice are used to produce edible cutlery. Edible cutlery is a food that is environmentally friendly for the climate, humans, livestock, and the atmosphere. In addition, the researcher decided to analyze the time for the edible cutleries to be degradable and to resist water absorption by the substance. This is because natural ingredients are made to produce this product and it is supposed to degrade quickly, ensuring that it consumes less degradation time compared to traditional plastic cutleries. In addition, it is assured that this product is safe and stable for the climate, humans, and other living things.

2 Literature Review

Sorghum is an ancient African crop that utilizes little water and its super absorbent properties for cultivation (Reddy 2016). A former ICRISAT researcher, in conjunction with Mr. Narayana Peesapaty, reports that sorghum requires 60 times less water than rice (TwentieFour 2018). UNESCO's quest on 'The green, blue and gray water footprint of crops and derived crop products' between 1996 and 2005 found that sorghum accounted for less than 2% of the worldwide water footprint of crop production. Compared to rice with quite a lot of crop production ((Mekonnen and Hoekstra 2011), In Bangladesh, farmers waste 800 L of water to form 1 kg of paddy field (Ahmed 2019) and the amount of water wasted can be a mammoth 15.6 trillion liters once a year with an estimated annual production of 19.5 million metric tonnes of rice (Hoque 2018). According to a recent Bangladesh Agricultural Development Corporation study, farmers use 75 percent of groundwater, 25 percent of surface water, and in 1960–70 it was just 20 percent for groundwater, while 80 percent for surface water (Hoque 2018).

Edible cutleries based on sorghum are also a potential competitor to the single-use plastic cutleries that plague the earth's environment. In line with the figures provided by the ecu Commission in 2016, single-use plastic cutleries (Welle 2018) contribute 4.24 percent of the marine litter on European beaches. There is a massive market for one-time cutlery; u. u. s. As an example, 40 billion plastic cutleries are consumed annually (Alexander 1999), while India annually throws away 120 billion pieces of plastic cutleries (Reddy 2016). The market worldwide is 640 billion p.a. According to information published by the Digital Journal, the global plastic cutlery market was worth US$ 2.62 billion in 2017 and is projected to hit about US$ 3 billion by 2025 (Journal 2019).

2.1 Sorghum

As described earlier, edible cutleries are visiting the existing single-use plastic cutleries as an environmentally friendly alternative. It consists of edible ingredients such as rice, wheat, and sorghum (jowar), which are the core elements.

Sorghum is the key ingredient being hired. Compared with rice, it takes 60 times less water to develop sorghum (Erik Loots 2018). In 95% of the arable land in the world, the crop has the power to rise (Munir 2017).

Sorghum has a super absorbent capacity that makes it extremely flexible to use edible cutleries, that is, it would not only be suitable for rice or wheat-based cuisine that is popular in Bangladesh, but it will complement frozen dessert, yogurt, and soup form well because it does not degrade in hot or cold liquids (Munir 2017).

In 2016, Kickstarter, an American crowd-funding organization, conducted a study on the energy consumption and waste generated when producing sustainable sorghum-based spoons and their comparison with maize-based and plastic counterparts, and their result indicated that sorghum was the most effective of the three (Table 1).

In addition, the values indicated for sorghum were taken from water use in the processing and generation of sorghum crops as the crop was transformed into spoons. Kickstarter concluded that 100 units of sorghum-based spoons would be produced with the energy required to supply 1 plastic utensil, while 50 units will be produced as opposed to Corn/PLA. It also found that the low tide use of the spoon also improved the shelf life to 2 years without losing the crispiness of the spoon (Mekonnen and Hoekstra 2011).

Sorghum-based edible cutlery production saves 88% of CO_2 emissions when placed next to plastic cutleries. Many producers of plastic cutlery use vaseline-like substances on the spoons in order not to mix with the mold during manufacturing (Rajah 2018). The oil film inside the cutleries remains unwashed until the ultimate product enters the market and contaminates the food and drinks, thus risking the consumer's health and safety. Sorghum-based cutleries have no preservatives, are fully organic, green materials, vegetarian-friendly, non-toxic and completely environmentally friendly (Munir 2017), making the product highly qualified and ahead of the curve.

Table 1 Energy used, wastage, and emissions per one pound of material produced (*Source* Kickstarter)

Manufacturing 1 lb of material	Energy used (kWh)	Water used (gals)	Solid waste (lbs)	CO_2 Emissions (lbs)
Plastic (Polypropylene)	9.34	5.12	0.029	1.67
Corn/PLA based	5.37	8.29	0.042	1.3
Sorghum based	0.18	1.15	n/a	0.19

3 Methodology

The methodology covers the main aspects of the research process, preparing the materials, conducting tests, and collecting data. The methodology contains reports and explanations to produce evidence that can support a conclusion. Figure 1 shows a methodology chart for this study. The first phase begins with identifying the problems,

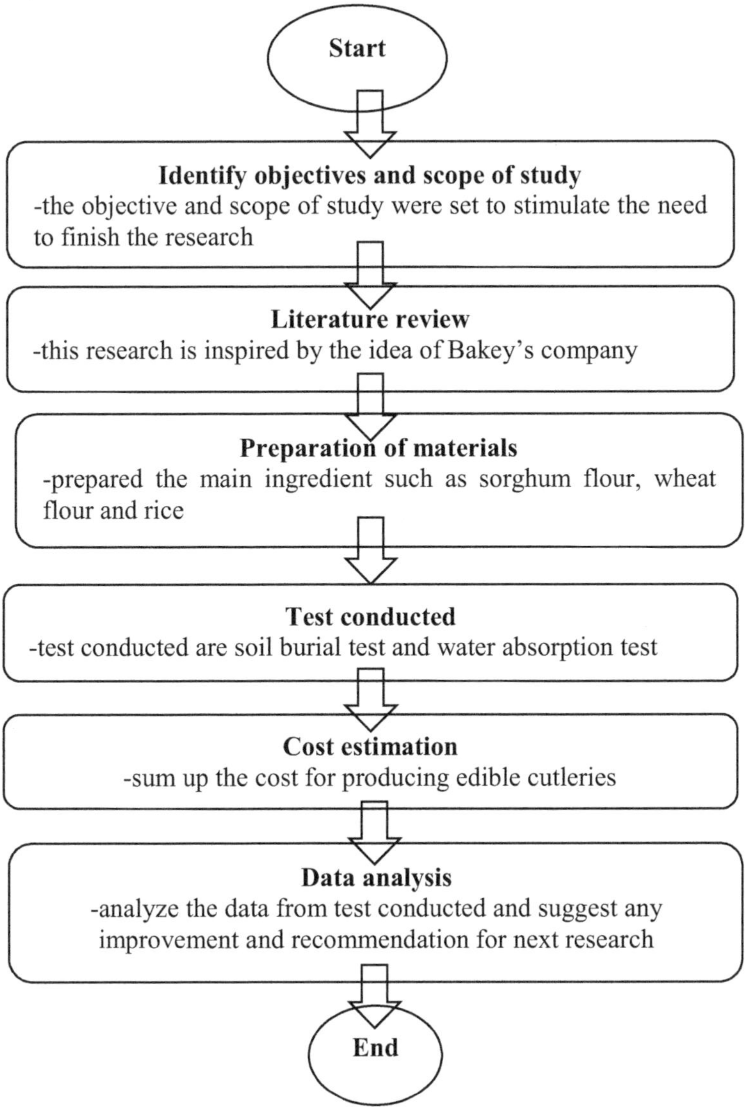

Fig. 1 Methodology chart

the study's objectives, and continuing the literature review. Subsequently, the second phase begins with the material preparation and testing. Besides, at the end of this phase, the result will be identified. Next, the third phase is about the analysis and discussion of this project. Finally, the fourth phase is the conclusion and proposal for any recommendation in future research.

3.1 Materials and Method

Sorghum contains a super absorbent capacity that makes it extremely flexible to use as edible cutleries, it would not only be ideal for rice or wheat-based cuisine, but it will complement frozen dessert, yogurt, and soup form well, since it does not degrade in hot or cold liquids. The process of baking an edible cutlery and reworking it is shown in Fig. 2.

In this study, there are five different amounts of mixture between the ingredients, which are sorghum flour, wheat flour, and rice to determine which amount would give the best result for edible cutleries. The different amounts of mixture obtained from the DIY video. The person in the video is using 3.5 cups of flour, 1 cup of water, and 1 tsp of salt. Meanwhile, in this study, the new mixtures were obtained.

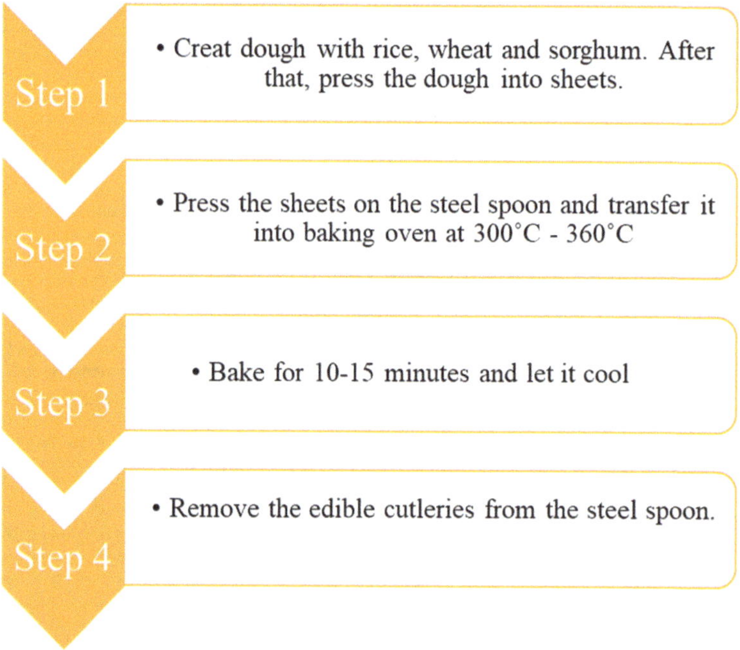

Fig. 2 Baking process of edible cutleries

Table 2 Samples of edible cutleries

Ingredients	Sample				
	1	2	3	4	5
Sorghum flour (g)	100	150	100	100	50
Wheat flour (g)	150	100	100	200	100
Rice (g)	100	100	150	50	200
Water (cups)	1	1	1	1	1
Observations	Sturdy and water resistance	Sturdy and water resistance	Sturdy and water resistance	Dry looking but fragile	Dry looking but too hard

The amounts of edible cutleries able to be produced from the ingredients are around 20–30 pieces. The difference amount between the ingredients is stated in Table 2.

Based on Table 2, samples 1, 2, and 3, were found to be stable and water-resistant relative to the other ratios. In order to obtain the most effective method of manufacturing edible cutlery, these three ratios are compared in analysis of the water absorption test and soil burial test.

4 Result and Discussion

4.1 Water Absorption Test

Three samples of edible cutlery were tested for water absorption test and three of the samples were made from different ratios of sorghum flour, wheat flour and rice. The duration of the test depends on the ability of edible cutleries to resist the water absorption (Fig. 3). From the graph obtained in Figs. 4, 5, and 6, it can be analyzed that sample 1 had the higher percentage of water absorption compared to sample 2 and sample 3. Sample 1 was made from 150 g of wheat flour and 100 g of sorghum flour and rice respectively. The higher amount of wheat flour is probably the cause of the sample absorbing more water. However, sample 2 and sample 3 have approximately the same percentage of water absorption. Sample 2 had a high amount of sorghum flour while sample 3 had a high amount of rice. The lower percentage of water absorption was shown in sample 2 which the last reading of the percentage was only 31.60% and the edible cutleries manage to resist water absorption until almost 3 h compared to samples 1 and 3. Sample 1 had the higher percentage of water absorption with 36.69% and only can resist water absorption until 110 min. Meanwhile, the percentage of water absorption for sample 3 was only 28.97% and resists the water absorption for 130 min. This showed that the sorghum flour had a lack of moisture content or fat and it allowed having a long shelf life without the need for extra

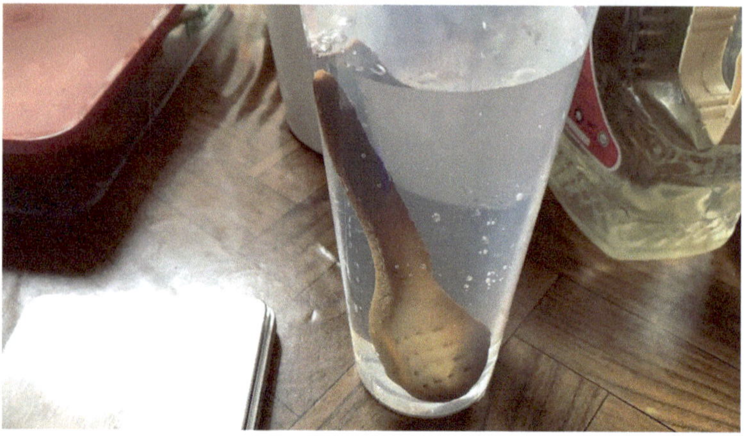

Fig. 3 Water absorption test for edible cutleries

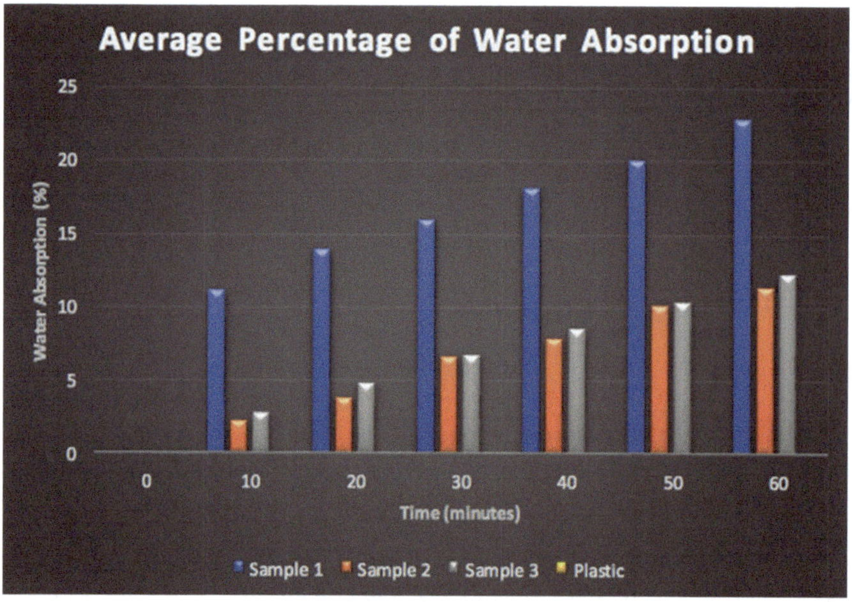

Fig. 4 Average percentage of water absorption from 0 to 60 min

preservatives and not degrading within liquid easily. Gluten proteins have been linked to the difference in water absorption between sorghum flour and wheat flour. In essence, the rise in sorghum flour eventually weakened the protein network, resulting in longer growth time and lower dough strength (Carson and Sun 2000).

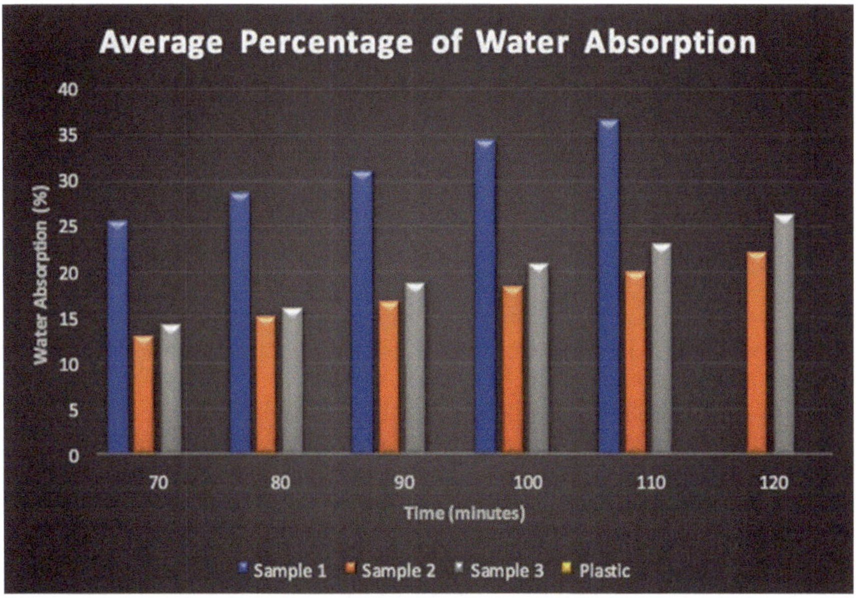

Fig. 5 Average percentage of water absorption from 70 to 120 min

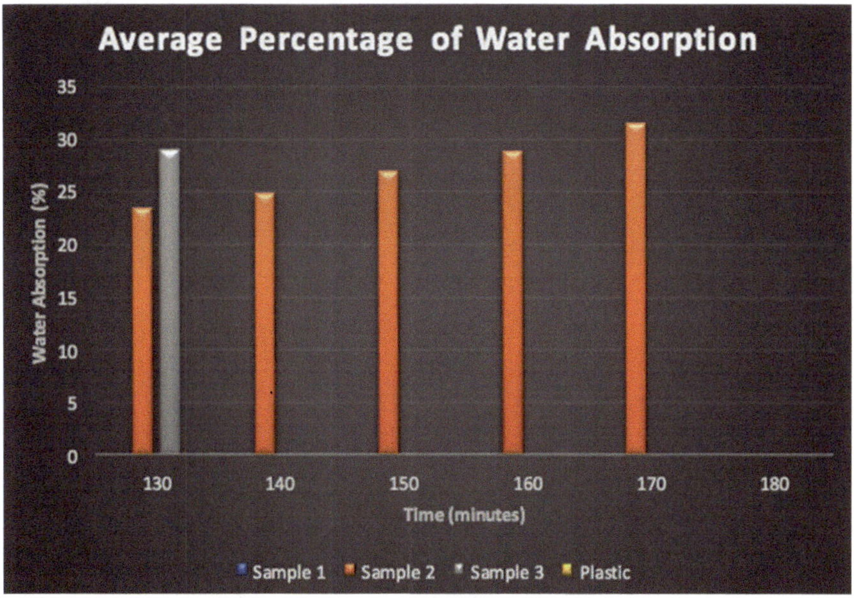

Fig. 6 Average percentage of water absorption from 130 to 180 min

4.2 Soil Burial Test

The test was conducted for the edible cutleries from samples 1, 2, and 3 to get the biodegradability rate. Method for burial test is shown in Figs. 7, 8 and 9. The result was obtained as in Table 3 after being observed for a week. From the graph obtained in Fig. 10, the edible cutleries from these three samples had not shown a huge difference in biodegradability rate. The reason for this is because all the three samples used natural ingredients without any preservative or chemical. It makes the edible cutleries combined with water easier. Thus, the biodegradability rate is also affected by the heavy rain. The soil will absorb the water from the rain as well as the edible cutleries. The more moisture in the soil, the faster degradation happens because water is important for the biodegradation process Yotube video (2016). Due to the moisture absorption during the soil burial test, the weight of the edible cutleries increased for 24 h. Then, the edible cutleries began to break out of their shape after 3 days of observation. This makes the weighing process become challenging because the edible cutleries already broke into pieces. Some of the pieces already mixed well with the soil and only a few pieces left. The edible cutleries were fully gone after 7 days of observation. It is also better to dispose of the edible cutleries in a commercial compost heap or to send them to a recycling plant.

Fig. 7 Edible cutleries after 24 h of soil burial

Fig. 8 Edible cutleries after 3 days of soil burial test

Fig. 9 Edible cutleries after 7 days of soil burial test.

Table 3 Average biodegradable rate of edible cutleries

Sample	Average biodegradability rate (%)			
	Initial	After 24 hours	After 3 days	After a week
Sample 1	0	33.89	−62.30	None
Sample 2	0	26.15	−68.17	None
Sample 3	0	25.37	−39.12	None
Plastic	0	0	0	0

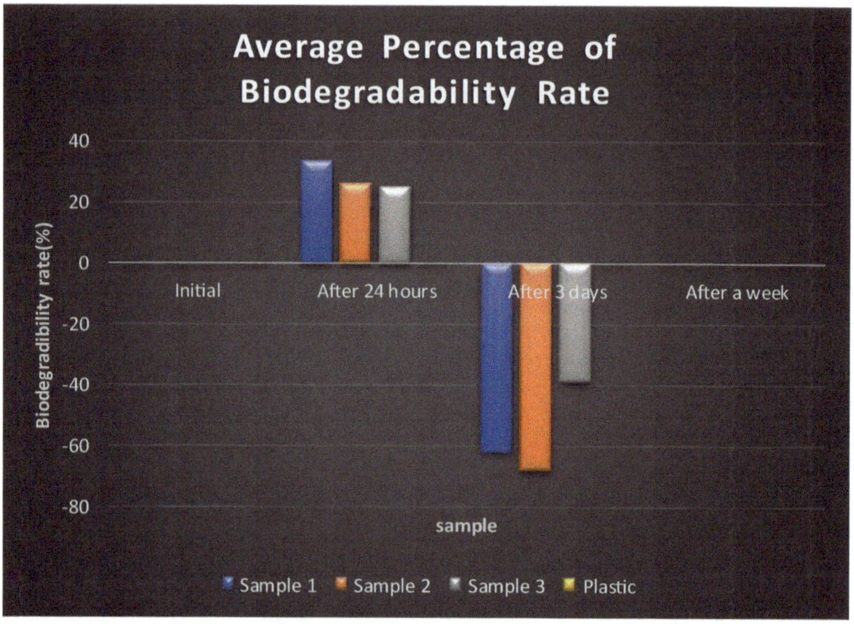

Fig. 10 Average biodegradability rate

5 Conclusion

Study of edible cutleries from sorghum flour shows that the production of edible cutleries is successful. The edible cutleries were produced from different amounts of sorghum flour, wheat flour, and rice. Finally, samples 1, 2, and 3 were tested for soil burial test and water absorption test to determine the best ratio of ingredients among the sample. After obtaining the testing result, it is shown that sample 2 is more stable than the other sample and manages to resist water absorption tests and degrade easily. The second objective is to identify the expense of making edible cutlery. The ingredient was bought from any organic store with the total price RM36.40. Each of the mixture manages to produce around 20–30 pieces of edible cutleries. The third objective is to compare the ability of edible cutlery and plastic cutlery to resist water absorption test and soil burial test. As a result of the test conducted, sample 2 managed to resist water more than sample 1 and 2 due to the higher amount of sorghum flour in sample 2. Meanwhile, the result for the soil burial test is approximately the same among samples 1, 2, and 3. These samples are completely degraded within a week.

After conducting this study, there are several suggestions for future experiments that could be made to further the development of this Project. Follow-up suggestions to suggest are to perform Strength test for edible cutleries and to improve the scale and sizing of edible cutleries by preparing a mold to make sure every edible cutlery will have the same size and shape. Other than that, the composition of the material used can also be altered to make it stronger and look better.

Acknowledgements The authors would like to thank the Faculty of Engineering Technology, Universiti Tun Hussein Onn Malaysia, Grant K222 Fundamental Research Grant FRGS.

References

Ahmed KA (2019) Farmer waste 800 litres of water to grow a kg of paddy. https://www.dhakatribune.com/bangladesh/agriculture/2019/01/13/farmers-waste-800-litres-of-water-to-grow-a-kg-of-paddy. Accessed 24 Dec 2020

Alexander M (1999) Biodegradation and bioremediation. Gulf Professional Publishing

Carson LC, Sun XS (2000) Breads from white grain sorghum: Rheological properties and baking volume with exogenous gluten protein. Appl Eng Agric 16(4):423

Erik Loots LV (Director) (2018) Fix the world and make money [Motion Picture]. https://medium.com/@twentiefour/fix-the-world-and-make-money-f386c9066b41. Accessed 6 Sept 2020

Hoque M (2018) Why Farmers in Bangladesh Waste 800 Liters of Water to Produce 1 Kg of Paddy. http://www.ipsnews.net/2018/12/farmers-bangladesh-waste-800-liters-water-produce-1-kg-paddy/. Accessed 24 Dec 2020

Journal D (2019) Plastic cutlery market is expected to reach at a 2920 Million USD by End of 2025-QYR. https://www.globenewswire.com/news-release/2019/03/07/1749694/0/en/Plastic-Cutlery-Industry-Global-Market-to-Grow-Over-US-2920-Mn-by-2025-end-QY-Research-Inc.html Accessed 20 Dec 2020

Michel C, Velasco C, Spence C (2015) Cutlery matters: heavy cutlery enhances diners' enjoyment of the food served in a realistic dining environment. Flavour 4(1):1–8

Mekonnen MM, Hoekstra AY (2011) The green, blue and grey water footprint of crops and derived crop products. Hydrol Earth Syst Sci 15(5):1577–1600

Munir S (2017) Edible cutlery: the future of eco friendly utensils. https://www.kickstarter.com/projects/1240116767/edible-cutlery-the-future-of-eco-friendly-utensils. Accessed 10 Aug 2020

Patil HN, Sinhal P (2018) A study on edible cutlery: an alternative for conventional ones. Atithya J Hos 4(1):45–51

Parker L (2019) The world's plastic pollution crisis explained. National Geographic. https://www.nationalgeographic.com/environment/article/plastic-pollution. Accessed 7 June 2020

Reddy BD (2016) Bakeys: You can use and eat this innovative cutlery. India. https://www.business-standard.com/article/companies/bakeys-you-can-use-and-eat-this-innovative-cutlery-116062200024_1.html Accessed 22 June 2020

Rajah L (2018) Cutlery you can eat: one company's approach to the plastic pollution problem. https://www.cbc.ca/news/science/bakeys-edible-cutlery-1.4763171. Accessed 29 July 2020

TwentieFour (2018) Fix the world and make money. https://medium.com/@twentiefour/fix-the-world-and-make-money-f386c9066b41. Accessed 6 Sept 2020

Welle D (2018) EU reaches agreement on single-use plastic ban. https://www.dw.com/en/eu-reaches-agreement-on-single-use-plastic-ban/a-46797494. Accessed 19 Dec 2020

Yotube video (2016) How to make: edible utensils you can eat after using. https://www.youtube.com/watch?v=xK1eptaAFzk. Accessed 31 Mar 2020